# APPLIED MATHEMATICAL METHODS for CHEMICAL ENGINEERS

# APPLIED MATHEMATICAL METHODS for CHEMICAL ENGINEERS

Norman W. Loney

**CRC Press**

Boca Raton   London   New York   Washington, D.C.

## Library of Congress Cataloging-in-Publication Data

Loney, Norman W.
   Applied mathematical methods for chemical engineers / by Norman W. Loney.
      p.   cm.
   Includes bibliographical references and index.
   ISBN 0-8493-0890-9
   1. Chemical engineering--Mathematics. I. Title.
   2. Biology—molecular.  I. McLachlan, Alan.  II. Title.

   TP155.2.M36 .L66 2000
   660'.01'51--dc21                                                    00-044454
                                                                       CIP

### Visit the CRC Press Web site at www.crcpress.com

© 2001 by CRC Press LLC

No claim to original U.S. Government works
International Standard Book Number 0-8493-0890-9
Library of Congress Card Number 00-044454
Printed in the United States of America     3  4  5  6  7  8  9  0
Printed on acid-free paper

# *Preface*

The purpose of this book is to introduce students of chemical engineering to several mathematical methods that are often essential to successfully solve real process engineering problems. The book emphasizes analytical methods even though most realistic models will be solved using numerical methods. However, prior to an extensive and expensive numerical analysis of a model, it is very useful to develop some understanding of the gross tendencies of the model. This type of understanding usually comes from the derivation of analytical solutions of a modified version of the problem under consideration.

Typical chemical engineering curriculums consist of the equivalent of three semesters of calculus capped off by a course in elementary ordinary differential equations. This usually occurs within the first two years of a four-year program (five years, if Co-op is an option). The next two or three years are usually dedicated to solving unit operations problems using pre-derived formulae. The point being, with few exceptions, the use of the four semesters' worth of mathematics is not applied until the first year of graduate school.

Those graduates who do go on to industry and later encounter the need to understand and apply the results of computer algebraic systems — indeed, to choose a software package for their own applications — have to rely heavily on the salesperson's judgment.

This book provides worked-out examples using a number of solution techniques while exposing the use of mathematics in chemical engineering to the reader.

The first chapter provides an introduction to the three classes of transport common to chemical engineering.

Chapter 2 deals with select first order ordinary differential equations and provides chemical engineering examples that demonstrate the use of solution techniques. A section addressing the formulation of some physically applicable first order ordinary differential equations (problem setup) is included.

The third chapter addresses linear second order ordinary differential equations. A brief discourse, it reviews elementary differential equations, and the chapter serves as an important basis to the solution techniques of partial differential equations discussed in Chapter 6. An applications section is also included with ten worked-out examples covering heat transfer, fluid flow, and simultaneous diffusion and chemical reaction. In addition, the Residue theorem as an alternative method for Laplace transform inversion is introduced.

Chapters 4 and 5 introduce Sturm-Liouville problems and Fourier Series and Integrals, respectively. These topics contain essential background material for use in solving linear partial differential equations. Applications of these are postponed until partial differential equations are discussed.

The sixth chapter provides instruction in a number of solution techniques for linear partial differential equations. Also included is a section introducing regular perturbation, a common approach to solving some nonlinear differential equations. Since the material in this chapter will be a new experience for a large segment of the readership, a substantial number of drill-type examples is included.

Chapter 7 is dedicated entirely to worked-out examples taken from the chemical engineering research literature. This chapter relies on the mathematics of the previous six chapters to solve problems in heat transfer; mass transfer; simultaneous diffusion and convection; simultaneous diffusion and chemical reaction; simultaneous diffusion, convection, and chemical reaction; and viscous flow.

The eighth chapter briefly discusses dimensional analysis and scaling of boundary value problems. This is an important topic in chemical engineering. The practicing engineer is continually faced with justifying the simplifying assumptions invoked in deriving a solution to some process model of concern to him or her.

Chapter 9 introduces selected numerical methods and available software packages. Because methods that were previously too effort-consuming are now commercially available in many software packages, it is more important to mention those packages and leave the algorithmic details to the numerical analysis literature. Here it is hoped that enough of an introduction to numerical methods is made so that the interested reader can independently pursue the subject.

Since a goal of this text is to remove the mathematics phobia that usually exists among some of our bright young chemical engineers, rigor is sacrificed in favor of exposition. Therefore, the references at the end of each chapter have been carefully selected to aid the reader who wishes to pursue further study in the discussed subject matter. However, I do wish to point out that those references are not in any way exhaustive. Boldface or italic type is used to draw attention to a term or statement that is significant to the concept under discussion.

Others and I have successfully used this book as a text for both undergraduate and first-year graduate courses. Most of the students in the graduate course have been chemical engineers with varying background in elementary differential equations. For an undergraduate one-semester course, the applications in Chapters 2 and 3 are emphasized, and knowledge of those solution techniques is treated as a prerequisite. Chapters 4, 5, and 6 are covered in their entirety with some applications taken from Chapter 7. Also, parts of Chapter 9 are covered based on the audience needs. The graduate class uses the entire book in a one-semester course.

# *Author*

**Norman W. Loney** is an Associate Professor of Chemical Engineering at New Jersey Institute of Technology (NJIT). He has authored or co-authored more than 27 publications and presentations relating to the use of applied mathematics in chemical engineering since joining the department in 1991. Dr. Loney has mentored or conducted research with undergraduate students resulting in more than six journal articles in the last four years. He was awarded certificates of recognition from the National Aeronautics and Space Administration and the American Society for Engineering Education for research contributions.

Prior to joining NJIT, Dr. Loney, a licensed professional engineer, practiced engineering at Foster Wheeler, M. W. Kellogg Company, Oxirane Chemical Company, and Exxon Chemical Company.

# Acknowledgments

I am indebted to many who have encouraged me on this project and to all my students who sat through the developing manuscript. Thanks especially to my student Rose Mogerman, who typed the first three chapters, and to Ritesh Ramraj and Laurent Simon, who each read large parts of the manuscript.

I would like to thank Ronald Gabbard of BASF Corporation for reading the first eight chapters and giving me positive feedback. Mistakes that remain are entirely my own.

Finally, I would like to extend my thanks and gratitude to my department chairman, Gordon Lewadowski, for his helpful suggestions.

# Contents

*This work is dedicated to my sons*

*Alexander David and Michael Oliver*

*and to my parents*

*Ella Esedora and David Alexander*

# 1

## Differential Equations

### 1.1 Introduction

There are several significant problems in chemical engineering that require a fundamental understanding of differential equations in order to fully appreciate the underlying transport issues. In this text, *differential equation* means an equation containing derivatives of the unknown function to be determined (Boyce and DiPrima[1]). For example, Fourier's law (Geankopolis,[2] Bennett and Myers[3]) for molecular transport of heat in a fluid or solid can be written as a first order differential equation

$$\frac{q_z}{A} = -\alpha \frac{d(\rho C_P T)}{dz} \tag{1.1.1}$$

for constant density $\rho$ and heat capacity $C_P$. Here $q_z/A$ is the heat flux $(J/s \cdot m^2)$, $\alpha$ is the thermal diffusivity $(m^2/s)$ and $\rho C_P T$ represents the concentration of heat $(J/m^3)$, with the subscript $z$ indicating that energy is transferred in the $z$ direction. The unknown function is the temperature $T(z)$. A second example that is familiar to chemical engineers is Fick's law (Geankopolis,[2] Bennett and Myers[3]) for molecular transport of mass in a fluid or solid for constant total concentration in the fluid. This fundamental transport process can be written as

$$J_{AZ} = -D_{AB} \frac{dC_A}{dz} \tag{1.1.2}$$

where $J_{AZ}$ is the flux of species A $(kmol /s \cdot m^2)$, $D_{AB}$ is the molecular diffusivity $(m^2/s)$ of species A in B and $C_A$ is the concentration of A $(kmol/m^3)$. In this case the unknown function to be determined is $C_A(z)$. A third example is Newton's law (Geankopolis,[2] Bennett and Myers[3]) of viscosity, written as follows for constant density $\rho$

$$\tau_{zx} = -\gamma \frac{d(v_x \rho)}{dz} \tag{1.1.3}$$

where $\tau_{zx}$ is the flux of x-directed momentum in the z direction [(Kg·m/s)/s.m²], $\gamma$ is the kinematic viscosity ($\mu/\rho$) or momentum diffusivity. Transport or diffusion is in the z direction and $\mu$ is the viscosity (kg/m·s). Here the unknown function to be determined is the x component of velocity $v_x(z)$.

Differential equations are usually divided into two classes. If the unknown function depends on a single independent variable, then the differential equation is classified as an ordinary differential equation (ODE); if there are two or more independent variables, then the equation is called a partial differential equation (PDE).

Equations (1.1.1) to (1.1.3) are examples of ordinary differential equations. A general equation for the conservation of momentum, thermal energy, or mass can be written as

$$\frac{\partial \Gamma}{\partial t} - \delta \frac{\partial^2 \Gamma}{\partial z^2} = R \tag{1.1.4}$$

where $\Gamma$ represents the concentration of the property (momentum, energy, or mass), $\delta$ is a proportionality constant (e.g., diffusivity), t is time, z indicates the distance in the direction of flow, and R is a source term (generation). Here the unknown function to be determined is $\Gamma(z, t)$ which depends on both distance and time. This equation is an example of a PDE. There are several other examples of partial differential equations in Chapter 6.

## 1.2   Ordinary Differential Equations

In discussions involving differential equations, the word *order* is very prevalent. The working definition of order of an ordinary differential equation is the order of the highest derivative that appears in that equation. Hence, the equation

$$f\left[t, \rho(t), \rho'(t), \cdots, \rho^{(n)}(t)\right] = 0 \tag{1.2.1}$$

is an ordinary differential equation of the *n*th order. Here Equation 1.2.1 represents a relation between the variable t (independent) and the values of the dependent variable $\rho$ and its first n derivatives $\rho'$, $\rho''$, ..., $\rho^{(n)}$. An explicit example of Equation 1.2.1 is

$$\rho''' + 2e^t \rho'' + \rho\rho' = t^4 \tag{1.2.2}$$

a third order differential equation for $\rho = \rho(t)$.

In this text we will avoid the common assumption that it is always possible to solve a given ordinary differential equation for the highest derivative. However, most of the discussion will be expedited when the form

$$\rho^{(n)} = f\left(t, \rho, \rho', \rho'', \cdots, \rho^{(n-1)}\right) \tag{1.2.3}$$

can be obtained. It should also be noted that even when the form given by Equation 1.2.3 is achievable, it does not generally mean that there is a function $\rho = \phi(t)$ that satisfies it. Thus a *solution* of the ordinary differential equation (1.2.3) on $\alpha < t < \beta$ is a function $\phi$ such that $\phi'$, $\phi''$, ..., $\phi^{(n-1)}$ exist and satisfy

$$\phi^{(n)}(t) = f\left[t, \phi(t), \phi'(t), \cdots, \phi^{(n-1)}(t)\right] \tag{1.2.4}$$

for all t in $\alpha < t < \beta$. In other words, a solution of a differential equation is a function that satisfies the differential equation and the domain of definition of the differential equation. By way of direct substitution into the first order equation

$$\frac{dQ}{dt} = kQ \tag{1.2.5}$$

it can be shown that

$$Q(t) = ce^{kt}, \qquad -\infty < t < \infty \tag{1.2.6}$$

is a solution, where c is an arbitrary constant and k is a given constant.

There are three very important issues that an engineer will want to have resolved when given a differential equation:

1. Existence of a solution
2. Uniqueness of the solution
3. How to determine a solution

These issues are covered in detail in Ince[4] and other standard differential equation texts. However, if an engineering problem is to be formulated as a differential equation, it is expected to have a solution. This solution provides one way of verifying the correctness of the mathematical formulation.

The remaining chapters of this book will be dedicated to the exposition of some of the common mathematical techniques that are useful in chemical engineering.

## References

1. Boyce, W.E. and DiPrima, R.C., *Elementary Differential Equations and Boundary Value Problems*, 3rd ed., John Wiley & Sons, New York, 1977, chap. 1.
2. Geankopolis, C.J. *Transport Processes and Unit Operations*, 3rd ed., Prentice-Hall, Englewood Cliffs, NJ, 1978.
3. Bennett, C.O. and Myers, J.E., *Momentum, Heat, and Mass Transfer*, McGraw-Hill, New York, 1962.
4. Ince, E.L., *Ordinary Differential Equations*, Dover, New York, 1956.

# 2

## First Order Ordinary Differential Equations

### 2.1 Linear Equations

Examples of linear first order differential equations occur frequently in chemical engineering practice through unsteady state mass balances or first order chemical reaction problems.

Here we will review a few methods for solving first order ordinary differential equations. Following each method are examples demonstrating the application of that method. Also, the notion of translating prose into mathematical symbolism is introduced as "Problem Setup" in Section 2.4.

Presented below is a brief recap of the definition of linear equations in the context of differential equations. Following the recap are examples of unsteady mass balances, which lead to linear first order problems. Also presented are examples involving chemical reactions that can be treated as linear first order problems.

In this chapter, attention will be focused on differential equations of the form

$$\rho' = f(t, \rho) \tag{2.1.1}$$

where f is a given function of t and $\rho$. By *linear equations*, we mean any equation that can be expressed in the polynomial form:

$$a_n(t)\rho^{(n)} + a_{n-1}(t)\rho^{(n-1)} + \cdots + a_2(t)\rho' + a_1(t)\rho^{(0)} + a_0(t) = g(t) \tag{2.1.2}$$

where $\rho^{(\cdot)}$ symbolizes the derivative of $\rho$ with respect to t. Consequently, the equation

$$a_2(t)\rho^{(1)} + a_1(t)\rho^{(0)} + a_0 = g(t) \tag{2.1.3}$$

is a linear first order differential equation and is more familiar in the form

$$a_2(t)\rho' + a_1(t)\rho + a_0(t) = g(t) \tag{2.1.4}$$

5

The general solution of Equation 2.1.4 can be obtained by employing the following steps:

1. **Rewrite Equation 2.1.4 as**

$$\rho' + \frac{a_1(t)}{a_2(t)}\rho = \frac{g(t) - a_0(t)}{a_2(t)}; \qquad a_2(t) \neq 0 \quad \text{for all } t \tag{2.1.5}$$

2. **Determine**

$$\mu(t) = exp\left(\int \frac{a_1(t)}{a_2(t)} dt\right) \tag{2.1.6}$$

where $\mu(t)$ is called an integrating factor.

3. **Multiply both sides of the Equation 2.1.5 by $\mu(t)$**

$$\left[\rho' + \frac{a_1(t)}{a_2(t)}\rho\right]\mu(t) = \frac{g(t) - a_0(t)}{a_2(t)}\mu(t) \tag{2.1.7}$$

**and observe that the left-hand side of Equation 2.1.7 can be written as**

$$\frac{d}{dt}[\rho\mu(t)]$$

or

$$\rho' exp\left(\int \frac{a_1(t)}{a_2(t)} dt\right) + \rho \frac{a_1(t)}{a_2(t)} exp\left(\int \frac{a_1(t)}{a_2(t)} dt\right) = \frac{d}{dt}[\rho\mu(t)] \tag{2.1.7A}$$

Thus, Equation 2.1.7 can be recast as

$$\frac{d}{dt}\left[\rho exp\left(\int \frac{a_1(t)}{a_2(t)} dt\right)\right] = \frac{g(t) - a_0(t)}{a_2(t)} exp\left(\int \frac{a_1(t)}{a_2(t)} dt\right) \tag{2.1.8}$$

4. **Integrate both sides of Equation 2.1.8 with respect to the independent variable to get**

$$\rho exp\left(\int \frac{a_1(t)}{a_2(t)} dt\right) = \int \frac{g(t) - a_0(t)}{a_2(t)} exp\left(\int \frac{a_1(t)}{a_2(t)} dt\right) dt + c$$

or

$$p(t) = \exp\left(-\int \frac{a_1(t)}{a_2(t)}\, dt\right)\int \frac{g(t) - a_0(t)}{a_2(t)} \exp\left(\int \frac{a_1(t)}{a_2(t)}\, dt\right) dt + c \, \exp\left(-\int \frac{a_1(t)}{a_2(t)}\, dt\right) \quad (2.1.9)$$

where c is the constant of integration.

### Example 2.1.1 (Boyce and DiPrima[1])

Water containing 0.5 kg of salt per liter (l) is poured into a tank at a rate of 2 l/min, and the well-stirred mixture leaves at the same rate. After 10 minutes, the process is stopped and fresh water is poured into the tank at a rate of 2 l/min, with the new mixture leaving at 2 l/min. Determine the amount (kg) of salt in the tank at the end of 20 minutes if there were 100 liters of pure water initially in the tank.

*Solution*

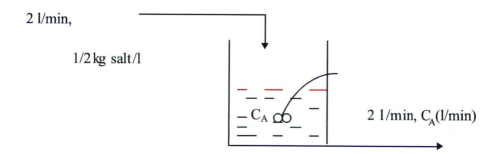

2 l/min,

1/2 kg salt/l

$- C_A \, \phi \phi$

2 l/min, $C_A$(l/min)

Let $C_A$(kg/lit) be the concentration of salt in the tank at any time t. Then, from material balance (Felder and Roussau[5]), a salt balance gives

Rate of accumulation = Rate of input − Rate of output     (2.1.10)

or in symbols,

$$100 \, liters \frac{dC_A}{dt} = \left(\frac{1/2 \, Kg}{liter}\right)\left(\frac{2 \, liter}{min}\right) - C_A\left(\frac{2 \, liter}{min}\right) \qquad (2.1.11)$$

with initial condition

$$C_A(0) = 0 \qquad (2.1.12)$$

Equation 2.1.11 can be rewritten as

$$\frac{dC_A}{dt} + \frac{1}{50}C_A = \frac{1}{100} \tag{2.1.13}$$

as suggested in step 1. Then, following step 2 above,

$$\mu(t) = \exp\left(\int \frac{1}{50}dt\right) = e^{\frac{t}{50}}$$

and Equation 2.1.13 is recasted as

$$\left(e^{\frac{t}{50}}C_A\right)' = \frac{1}{100}e^{\frac{t}{50}} \tag{2.1.14}$$

which is in the form of Equation 2.1.8 as given in step 3. Following step 4 we get

$$C_A(t) = \frac{50}{100} + ae^{-\frac{1}{50}t} \tag{2.1.15}$$

where $a$ is an arbitrary constant. Using the given initial condition, Equation 2.1.12, we get

$$C_A(t) = \frac{50}{100}\left(1 - e^{-\frac{1}{50}t}\right) \tag{2.1.16}$$

Equation 2.1.16 is the salt concentration profile for the first 10 minutes of the process. For the subsequent time during which pure water is added, Equation 2.1.10 reduces to

$$\text{Rate of accumulation} = - \text{ Rate of output} \tag{2.1.17}$$

and the new initial condition is

$$C_A(10) = \frac{50}{100}\left(1 - e^{-\frac{10}{50}}\right)$$

Thus

$$\frac{dC_A}{dt} = -\frac{2}{100}C_A \tag{2.1.18}$$

and

$$C_A(10) = \frac{50}{100}\left(1 - e^{-\frac{10}{50}}\right) \tag{2.1.19}$$

describe the process where no salt (pure water) is being poured in. Solution of Equations 2.1.18 and 2.1.19 gives

$$C_A(t) = be^{-\frac{1}{50}t} \tag{2.1.20}$$

where

$$b = \frac{1}{2}\left(1 - e^{-\frac{10}{50}}\right)e^{\frac{10}{50}}$$

Then at t = 20 min or after the second 10 min period,

$$C_A(20) = 1/2e^{-\frac{10}{50}}\left(1 - e^{-\frac{10}{50}}\right)$$

and the amount (kg) of salt in the tank at the end of this time period is

$$100C_A(20) = 50e^{-0.2}\left(1 - e^{-0.2}\right)kg$$

## Example 2.1.2

Consider a tank with a 500 L capacity that initially contains 200 L of water with 100 kg of salt in solution. Water containing 1 kg of salt/L is entering at a rate of 3 L/min, and the mixture is allowed to flow out of the tank at a rate of 2 L/min. Determine the amount (kg) of salt in the tank at any time prior to the instant when the solution begins to overflow. Determine the concentration (kg/L) of salt in the tank when it is at the point of overflowing. Compare this concentration with the theoretical limiting concentration if the tank had infinite capacity (Boyce and DiPrima[1]).

*Solution*

Let $C_A(t)$ (kg/liter) be the concentration in the tank at any time t, and let V(t) (l) be the volume of the tank contents, with $V_0$ the initial volume. Then Equation 2.1.10 becomes

$$\frac{d(VC_A)}{dt} = \left(\frac{1\,Kg}{l}\right)\left(\frac{3\,l}{min}\right) - C_A\left(\frac{2\,l}{min}\right) \tag{2.1.21}$$

but

$$V(t) = V_0 + \left(\text{Volumetric rate in} - \text{Volumetric rate out}\right)t \tag{2.1.22}$$

and

$$\frac{dV}{dt} = \text{Volumetric rate in} - \text{Volumetric rate out}$$

Therefore, Equation 2.1.21 becomes

$$C_A\frac{dV}{dt} + V\frac{dC_A}{dt} = 3 - 2C_A \tag{2.1.23}$$

or

$$C_A + (200 + t)\frac{dC_A}{dt} = 3 - 2C_A \tag{2.1.24}$$

subject to

$$C_A(0) = 1/2\frac{Kg}{l}$$

Equation 2.1.24 can be solved using the 4 steps previously given as follows:

**Step 1:**
$$\frac{dC_A}{dt} + \frac{3}{200+t}C_A = \frac{3}{200+t}$$

**Step 2:**
$$\mu(t) = exp\left(\int\frac{3}{200+t}dt\right) = (200+t)^3$$

**Step 3:**
$$\left[(200+t)^3 C_A\right]' = 3(200+t)^2$$

$$(200+t)^3 C_A = 3\int (200+t)^2 dt + k$$

or

**Step 4:**
$$C_A(t) = 1 + \frac{k}{(200+t)^3}$$

at t = 0,
$$C_A = 1/2 = 1 + \frac{k}{(200)^3}$$

thus

$$k = -\tfrac{1}{2}(200)^3$$

$$\therefore \quad C_A(t) = 1 - \tfrac{1}{2}\left(\frac{200}{200+t}\right)^3$$

Then the amount of salt in the tank at any time t prior to the instant when the solution begins to overflow is $V(t)C_A$.

That is

$$V(t)C_A(t) = 200 + t - \frac{100(200)}{(200+t)^2} \text{ for } t < \text{instant of overflow}$$

Noting that the tank's capacity is 500 liters, then at the instant of overflow

$$500 = 200 + t \ \therefore \ t = 300 .$$

at t = 300

$$C_A(300) = 1 - \frac{1/2(200)^3}{(500)^3} = \frac{121}{125}$$

in comparison to the theoretical limiting concentration of

$$\lim_{t \to \infty} C_A(t) = 1$$

**Example 2.1.3**

Consider the consecutive second order, irreversible reactions occurring in a batch reactor (Rice and Do[6]):

$$A + S \xrightarrow{\kappa_1} X$$

$$X + S \xrightarrow{k_2} Y$$

If one mole of A and two moles of S are initially added, determine the mole fraction of X remaining after half the A is consumed. Assume that $\frac{k_2}{k_1} = 2$.

*Solution*

$$\frac{dC_X}{dt} = k_1 C_A C_S - k_2 C_X C_S \tag{2.1.25}$$

is the net rate of formation of X in terms of the appropriate concentrations.

$$\frac{dC_A}{dt} = -k_1 C_A C_S \tag{2.1.26}$$

is the rate of disappearance of A.

$$\frac{dC_Y}{dt} = k_2 C_X C_S \tag{2.1.27}$$

is the rate of formation of Y. Dividing Equation 2.1.25 by Equation 2.1.26 results in

$$\frac{dC_X}{dC_A} = -1 + \frac{k_2}{k_1} \frac{C_X}{C_A} \tag{2.1.28}$$

a linear first order differential equation.

For an integrating factor

$$\mu(C_A) = C_A^{-2}$$

the differential equation can be represented as

$$\left(C_X C_A^{-2}\right)' = -C_A^{-2}$$

which integrates to

$$C_X = C_A + m_1 C_A^2$$

subject to

$$C_A = 1, \quad C_X = 0 \quad \text{at} \quad t = 0$$

Therefore,

$$C_X = C_A - C_A^2$$

Similarly, dividing Equation 2.1.27 by Equation 2.1.26 gives

$$\frac{dC_Y}{dC_A} = -2 + 2C_A$$

which integrates to

$$C_Y = 1 - 2C_A + C_A^2$$

based on the initial condition

$$C_A = 1, \quad C_Y = 0 \quad \text{at} \quad t = 0$$

Finally, the mole fraction of X is

$$\frac{C_A - C_A^2}{C_A + C_S + C_X + C_Y} = \frac{1}{9}$$

when half of A is consumed.

The above examples demonstrate a technique to solve linear first order differential equations of the type given by Equation 2.1.5. While this method is straightforward, there are three things to note. First, the form given by Equation 2.1.5 is required; that is, the coefficient of the derivative term, $\rho'$, must be one. Second, the functions $a_1(t)$, $b_2(t)$, and $[g(t) - a_0(t)]/a_2(t)$ must be continuous. Third, Equation 2.1.7A provides a check as to whether the derivative of the product of $\mu$ and $\rho$ is in fact the appropriate left-hand side of Equation 2.1.7.

It should also be noted that each example problem was stated in prose and required transformation to mathematical symbolism. This transformation or problem setup is an important step and is usually where most students are left behind. However, in this book, whenever the demonstration involves physical phenomena such as those encountered in chemical engineering, the formats of Examples 2.1.1 and 2.1.2 will be followed. As an aid to this step, it is suggested that the student invest some time in reviewing the laws of conservation of mass and energy, as well as the unit operations principles discussed in undergraduate chemical engineering courses.

## 2.2 Additional Information on Linear Equations

In this section a very important fundamental theorem will be discussed. This theorem is important because it resolves two of the issues raised at the end of Section 1.1. Specifically, the theorem addresses existence and uniqueness of a solution.

An initial value problem for a first order linear equation will always have a unique solution if the conditions of the theorem stated below are satisfied (Boyce and DiPrima[1]).

*THEOREM 2.1*

*If the functions p and g are continuous on an open interval $\alpha < x < \beta$ containing the point $x = x_0$, then there exists a unique function $y = \phi(x)$ that satisfies the differential equation*

$$y' + p(x)y = g(x) \tag{2.2.1}$$

*for $\alpha < x < \beta$, and that also satisfies the initial condition*

$$y(x_0) = y_0 \tag{2.2.2}$$

*where $y_0$ is an arbitrary prescribed initial value.*

**PROOF** (nonrigorous)

We seek a function $\mu$ such that if Equation 2.2.1 is multiplied by $\mu$, then the left-hand side of Equation 2.2.1 can be written as the derivative of the single function $\mu(x)y$, i.e., $\mu(x)[y' + p(x)y] = [\mu(x)y]' = \mu(x)y' + \mu'(x)y$. ∎

Thus, $\mu(x)$ must satisfy

$$\mu(x)p(x)y = \mu'(x)y$$

or

$$\frac{\mu'(x)}{\mu(x)} = p(x), \quad \mu(x) > 0$$

then

$$ln\,\mu(x) = \int^x p(t)\,dt$$

or

$$\mu(x) = exp\left[\int^x p(t)\,dt\right] \tag{2.2.3}$$

Therefore,

$$\left[\mu(x)y\right]' = \mu(x)g(x) \tag{2.2.4}$$

following the multiplication of Equation 2.2.1 by $\mu(x)$.

Integrating both sides of Equation 2.2.4 with respect to x and solving for y gives

$$y = \frac{1}{\mu(x)}\left[\int^x \mu(s)g(s)\,ds + c\right] \tag{2.2.5}$$

Further, since p is continuous for $\alpha < x < \beta$, it follows that $\mu$ is defined in this interval and is a nonzero differentiable function. Thus, the conversion of Equation 2.2.1 into the form of Equation 2.2.4 is justified. Also, the function $\mu g$ has an antiderivative since $\mu$ and g are continuous and Equation 2.2.5 follows from Equation 2.2.4. The assumption that there is at least one solution of Equation 2.2.1 is verifiable by substituting Equation 2.2.5 into Equation 2.2.1. The initial condition, Equation 2.2.2, determines the integration constant c uniquely.

Sometimes nonlinear equations can be reduced to linear ones by a substitution. One example where such a substitution is helpful is in solving the Bernoulli equations. The form of the Bernoulli equations is

$$y' + p(x)y = q(x)y^n \tag{2.2.6}$$

and if $n \neq 0, 1$ then

$$v(x) = y^{1-n}(x) \tag{2.2.7}$$

reduces Equation 2.2.6 to a linear equation.

## Example 2.2.1

Solve:   $x^2 y' + 2xy - y^3 = 0$

*Solution*
By comparison to Equation 2.2.6, $n = 3$, i.e.,

$$x^2 y' + 2xy = y^3$$

or

$$y' + \frac{2}{x} y = \frac{1}{x^2} y^3 \tag{2.2.8}$$

let

$$v = y^{1-3} = y^{-2} = \frac{1}{y^2}$$

$$\frac{dv}{dx} = \frac{-2}{y^3} \frac{dy}{dx}$$

Solving for $\dfrac{dy}{dx}$ to get

$$\frac{dy}{dx} = \frac{-y^3}{2} \frac{dv}{dx} = -\frac{1}{2} v^{-3/2} \frac{dv}{dx}$$

Substituting for $\dfrac{dy}{dx}$ and y in Equation 2.2.8 gives

$$-\frac{1}{2} v^{-3/2} \frac{dv}{dx} + \frac{2}{x} v^{-1/2} = \frac{1}{x^2} v^{-3/2}$$

following simplification, the differential equation becomes

$$\frac{dv}{dx} - \frac{4}{x} v = -\frac{2}{x^2} \tag{2.2.9}$$

a linear first order differential equation with new dependent variable $v(x)$. Equation 2.2.9 can now be solved using the method of Section 2.1.

A more engineering-type example is demonstrated below.

### Example 2.2.2

Suppose that in a certain autocatalytic chemical reaction a compound A reacts to form a compound B. Further, suppose that the initial concentration of A is $C_{A0}$ and that $C_B(t)$ is the concentration of B at time t. Then $C_{A0} - C_B(t)$ is the concentration of A at time t. Determine $C_B(t)$ if $C_B(0) = C_{B0}$.

*Solution*

Note that in an autocatalytic reaction, the substance produced stimulates the reaction; thus, the reaction rate $\dfrac{dC_B}{dt}$ is proportional to both $C_B(t)$ and $C_{A0} - C_B(t)$, that is

$$\frac{dC_B(t)}{dt} = kC_B(t)\left(C_{A0} - C_B(t)\right) \tag{2.2.10}$$

subject to

$$C_B(0) = C_{B0}$$

where $k$ is the reaction rate coefficient.

Equation 2.2.10 can be recasted as

$$\frac{dC_B}{dt} - k\ C_B C_{A0} = -k\ C_B^2 \tag{2.2.11}$$

By comparison to Equation 2.2.6, $n = 2$.

Let $v(t) = C_B^{1-2} = \dfrac{1}{C_B}$

Then

$$\frac{dC_B}{dt} = -C_B^2 \frac{dv}{dt} = -v^{-2}\frac{dv}{dt}$$

Substitute $\dfrac{dC_B}{dt}$ and $v(t)$ into Equation 2.2.11 to give

$$-v^{-2}\frac{dv}{dt} - kv^{-1}C_{A0} = -kv^{-2}$$

Then multiplying both sides of the resulting equation by $-v^2$ gives

$$\frac{dv}{dt} + k\,C_{A0}\,v = k$$

$$\mu(t) = \exp\left(\int k\,C_{A0}\,dt\right)$$

Then

$$\left[v\exp\left(k\,C_{A0}\,t\right)\right]' = k\exp\left(k\,C_{A0}\,t\right)$$

Integrating both sides with respect to $t$ gives

$$v\exp\left(k\,C_{A0}\,t\right) = \frac{1}{C_{A0}}\exp\left(k\,C_{A0}\,t\right) + m_1$$

where $m_1$ is an arbitrary constant.

Therefore,

$$v = \frac{1}{C_{A0}} + m_1\exp\left(-k\,C_{A0}\,t\right) = \frac{1}{C_B}$$

such that

$$C_B(t) = \frac{1}{\dfrac{1}{C_{A0}} + m_1\exp\left(-k\,C_{A0}\,t\right)}$$

with

$$m_1 = \frac{C_{A0} - C_{B0}}{C_{A0}\,C_{B0}}$$

## 2.3   Nonlinear Equations

For those first order equations that cannot be expressed in polynomial form, there is no single analytical method to produce a solution as seen earlier in

Section 2.1. This difficulty increases the importance of the issues of existence and uniqueness of a solution. For a very lucid discussion on the existence and uniqueness theorem for nonlinear first order differential equations, many excellent texts are available (Boyce and DiPrima[1]).

In this section a few standard methods are presented for use on those first order nonlinear differential equations that can be solved analytically.

Even though the form

$$\frac{d\rho}{dt} = f(t,\rho) \tag{2.3.1}$$

is common, it is sometimes more convenient to rewrite Equation 2.3.1 in an alternate form

$$M(t,\rho) + N(t,\rho)\frac{d\rho}{dt} = 0 \tag{2.3.2}$$

### 2.3.1  Separable Equations

Suppose M is a function of $t$ only, and N is a function of $\rho$ only, then Equation 2.3.2 becomes

$$M(t) + N(\rho)\frac{d\rho}{dt} = 0 \tag{2.3.3}$$

which can be written as

$$N(\rho)d\rho = -M(t)dt \tag{2.3.4}$$

Whenever a first order differential equation can be written in either of the forms Equation 2.3.3 or Equation 2.3.4, the equation is said to be *separable*.

If we reconsider Equation 2.2.10:

$$\frac{dC_B}{dt} = kC_B(C_{A0} - C_B)$$

subject to

$$C_B(0) = C_{B0}$$

Then, this differential equation is separable and results in

$$\frac{dC_B}{C_B(C_{A0} - C_B)} = k\,dt \qquad (2.3.5)$$

In order to solve Equation 2.3.5, the left-hand side must first be simplified. Consider now the fraction

$$\frac{1}{C_B(C_{A0} - C_B)} = \frac{\alpha}{C_B} + \frac{\gamma}{C_{A0} - C_B} \qquad (2.3.6)$$

Where $\alpha$ and $\gamma$ are constants to be determined. Then

$$\alpha(C_{A0} - C_B) + \gamma C_B = 1$$

If we put

$$C_B = 0 : \alpha C_{A0} = 1$$

then

$$\alpha = \frac{1}{C_{A0}}$$

If we put

$$C_B = C_{A0} : \gamma C_{A0} = 1$$

then

$$\gamma = \frac{1}{C_{A0}}$$

Equation 2.3.6 can now be expressed as

$$\frac{1}{C_B(C_{A0} - C_B)} = \frac{\dfrac{1}{C_{A0}}}{C_B} + \frac{\dfrac{1}{C_{A0}}}{C_{A0} - C_B}$$

and Equation 2.3.5 becomes

$$\frac{1}{C_{A0}}\left[\frac{1}{C_B} + \frac{1}{C_{A0} - C_B}\right]dC_B = k\,dt \qquad (2.3.7)$$

which integrates to

$$\left[\frac{C_B}{C_{A0} - C_B}\right]^{\frac{1}{C_{A0}}} = m_1 \exp(kt)$$

where $m_1$ is an arbitrary constant to be determined with the given initial condition. At $t = 0$, $C_B = C_{B0}$, then

$$\left[\frac{C_{B0}}{C_{A0} - C_B}\right]^{\frac{1}{C_{A0}}} = m_1$$

$$\therefore \qquad C_B(t) = \frac{C_{B0}\,C_{A0}}{(C_{A0} - C_{B0})\exp(-kC_{A0}t) + C_{B0}}$$

following simplification.

As is evidenced in this later illustration, the potential difficulty in applying this separation of variable technique lies in one's ability to carry out the resulting integration that may arise, such as in Equation 2.3.5 above.

Another nonlinear problem that is not of the variable-separable type may be solvable if it is exact or can be made exact by use of an appropriate factor.

### 2.3.2 Exact Equations

Suppose Equation 2.3.2 is given, then if a function $w(t, \rho)$ exists such that

$$\frac{\partial w(t,\rho)}{\partial t} = M(t,\rho), \qquad \frac{\partial w(t,\rho)}{\partial \rho} = N(t,\rho) \qquad (2.3.8)$$

and such that $w(t, \rho)$ = constant defines $\rho = \varphi(t)$ implicitly as a differentiable function of $t$ (Boyce and DiPrima[1] and Thomas and Finney[3]), then

$$M(t,\rho) + N(t,\rho)\frac{d\rho}{dt} = \frac{\partial w}{\partial t} + \frac{\partial w}{\partial \rho}\frac{d\rho}{dt} = \frac{d}{dt}w(t,\rho(t)) \qquad (2.3.9)$$

By comparing Equation 2.3.9 with Equation 2.3.2, we get

$$\frac{d}{dt}w\big(t, \rho(t)\big) = 0 \tag{2.3.10}$$

When Equation 2.3.8 to Equation 2.3.10 holds, Equation 2.3.10 is called an *exact differential equation*. To determine whether an equation is exact in a given region $R$, the following criteria are essential:

1. The functions $M, N, \dfrac{\partial M}{\partial \rho}$ and $\dfrac{\partial N}{\partial t}$ must be continuous in the given region.

2. $\dfrac{\partial M}{\partial \rho} = \dfrac{\partial N}{\partial t}$ must hold at each point in the region.

3. The region $R$ must be simply connected; that is, a single closed curve which does not cross itself or a region without holes.

Sometimes criterion 2 is not immediately satisfied, and an adjustment can be made that will remedy such occurrences. Whenever such an adjustment is possible, the differential equation, Equation 2.3.2 will become exact. In order to determine this adjustment consider

$$\frac{\partial}{\partial \rho}(\mu M) = \frac{\partial}{\partial t}(\mu N) \tag{2.3.11}$$

where $\mu$ is the adjustment to be determined and can be a function of both $t$ and $\rho$.

Then

$$\frac{1}{\mu}\left[ N\frac{\partial \mu}{\partial t} - M\frac{\partial \mu}{\partial \rho} \right] = \frac{\partial M}{\partial \rho} - \frac{\partial N}{\partial t} \tag{2.3.12}$$

Equation 2.3.12 is not easy to solve in its present form; however, if either $\mu = \mu(t)$ or $\mu = \mu(\rho)$, then Equation 2.3.12 simplifies to

$$\frac{1}{\mu}\frac{d\mu}{dt} = \frac{1}{N}\left( \frac{\partial M}{\partial \rho} - \frac{\partial N}{\partial t} \right) \tag{2.3.13}$$

or

$$\frac{1}{\mu}\frac{d\mu}{d\rho} = \frac{1}{M}\left( \frac{\partial N}{\partial t} - \frac{\partial M}{\partial \rho} \right) \tag{2.3.14}$$

Either Equations 2.3.13 or 2.3.14 gives a formula to determine $\mu$. When $\mu = \mu(t, \rho)$, then Equation 2.3.12 must be solved directly.

Equation 2.2.10 can be solved by first finding $\mu = \mu(C_B)$ and multiplying both sides with $\mu$ to get an exact differential equation.

$$\mu(C_B)\frac{dC_B}{dt} - k\mu(C_B)C_B(C_{A0} - C_B) = 0 \qquad (2.3.15)$$

Homogeneous equations comprise a third group of nonlinear-type problems that usually do not yield to either the variable-separable or exact solution techniques. An equation of this type, however, may yield a solution if a new variable can be introduced.

### 2.3.3   Homogeneous Equations

Whenever Equation 2.3.1 can be rewritten in the form

$$\frac{d\rho}{dt} = h\left(\frac{\rho}{t}\right) \qquad (2.3.16)$$

then Equation 2.3.16 is said to be homogeneous. The quantity $\rho/t$ can now be treated as a new variable, and one of the solution techniques of the previous sections may now be applicable.

So far, the examples have assumed that the differential equations are given. However, as chemical engineers, we know that more often than not, the main problem is in the derivation of the differential equation and the associated conditions. To address that aspect of mathematical methods in this chapter, a problem setup section follows.

## 2.4   Problem Setup

The traditional approach of outlining the theory and presenting some supporting examples has been followed up to now. However, a needed deviation from tradition is a "how to" or a problem setup section. This section is included to demonstrate one approach to formulating a physically applicable first order ordinary differential equation.

*Problem Statement*
Consider the continuous extraction of benzoic acid from a mixture of benzoic acid and toluene, using water as the extracting solvent (Jenson and Jeffreys[4]). Both streams (acidic mixture and water) are fed into a tank where they are stirred efficiently and the mixture is then pumped into a second tank where

it is allowed to settle into two layers. The upper organic phase and the lower aqueous phase are removed separately, and the problem is to determine what proportion of the acid has passed into the solvent phase.

A list of simplifications for the idealized problem (model) follows:

1. Combine the two tanks into a single stage (see Figure 2.1).
2. Express stream-flow rates on solute-free basis.
3. Assume steady flow rate for each phase.
4. Assume that toluene and water are immiscible.
5. Assume that feed concentration is constant.
6. Assume that the mixing is efficient enough such that the two streams leaving the stage (Figure 2.1) are always in equilibrium with each other and can be expressed as

$$y = mx \qquad\qquad (2.4.1)$$

    where $m$ is the distribution coefficient, $x$ is the concentration of benzoic acid leaving the stage in the organic phase, and $y$ is the aqueous phase benzoic mass concentration.

7. Assume that the composition of a stream leaving the stage is the same composition as that phase in the stage.
8. Assume that the stage initially contains $V_1$ liter of toluene, $V_2$ liter of water, and no benzoic acid.

Then, using Equation 2.1.10, that is

      Rate of accumulation = Rate of input − Rate of output

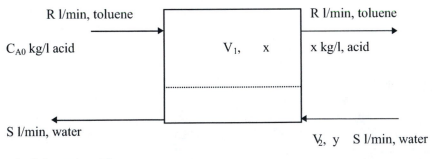

      R l/min, toluene                               R l/min, toluene

$C_{A0}$ kg/l acid                      $V_1$,    x            x kg/l, acid

    S l/min, water

                                              $V_2$, y    S l/min, water

    y kg/l, benzoic acid

**FIGURE 2.1**
Equilibrium stage.

the quantities for any time $t$ can be derived. A helpful procedure (Jenson and Jeffreys[4]) is to tabulate the quantities for any time $t$ and for a small change in time $\Delta t$. A typical table, Table 2.1, is given below.

**TABLE 2.1**

Quantities for $t$ and $\Delta t$

| System Property | $t$ | $t + \Delta t$ |
|---|---|---|
| Flow rate of organic phase | $R$ | $R$ |
| Flow rate of aqueous phase | $S$ | $S$ |
| Volume of organic phase in stage | $V_1$ | $V_1$ |
| Volume of aqueous phase in stage | $V_2$ | $V_2$ |
| Input acid concentration in organic phase | $C_{A0}$ | $C_{A0}$ |
| Output acid concentration in organic phase | $x$ | $x + \dfrac{dx}{dt}\Delta t$ |
| Output acid concentration in aqueous phase | $y$ | $y + \dfrac{dy}{dt}\Delta t$ |
| Amount (mass) of acid in organic phase | $V_1 x$ | $V_1 x + V_1 \dfrac{dx}{dt}\Delta t$ |
| Amount (mass) of acid in aqueous phase | $V_2 y$ | $V_2 y + V_2 \dfrac{dy}{dt}\Delta t$ |
| Input acid concentration in aqueous phase | $0$ | $0$ |

Adapted from Jenson and Jeffreys[4]

Then, during a time interval $\Delta t$, input of acid $= RC_{A0}\,\Delta t$ while output of acid

$$= R(x + \frac{dx}{dt}\,\Delta t)\Delta t + S(y + \frac{dy}{dt}\,\Delta t)\,\Delta t \text{ and accumulation of acid} = V_1 \frac{dx}{dt}\,\Delta t + V_2 \frac{dy}{dt}\,\Delta t. \text{ Therefore,}$$

$$\left( V_1 \frac{dx}{dt} + V_2 \frac{dy}{dt} \right)\Delta t = RC_{A0}\,\Delta t - \left\{ R\left( x + \frac{dx}{dt}\Delta t \right) + S\left( y + \frac{dy}{dt}\Delta t \right) \right\}\Delta t$$

which simplifies to

$$V_1 \frac{dx}{dt} + V_2 \frac{dy}{dt} = RC_{A0} - R\left( x + \frac{dx}{dt}\Delta t \right) - S\left( y + \frac{dy}{dt}\Delta t \right)$$

then

$$\lim_{\Delta t \to 0}\left[ V_1 \frac{dx}{dt} + V_2 \frac{dy}{dt} \right] = \lim_{\Delta t \to 0}\left\{ RC_{A0} - R\left( x + \frac{dx}{dt}\Delta t \right) - S\left( y + \frac{dy}{dt}\Delta t \right) \right\}$$

gives

$$V_1 \frac{dx}{dt} + V_2 \frac{dy}{dt} = RC_{A0} - Rx - Sy \qquad (2.4.2)$$

Equation 2.4.2 reduces to

$$V_1 \frac{dx}{dt} + V_2 m \frac{dx}{dt} = RC_{A0} - Rx - Smx$$

or

$$(V_1 + mV_2)\frac{dx}{dt} = RC_{A0} - (R + sm)x \qquad (2.4.3)$$

a linear first order ordinary differential equation subject to the initial condition ( state ) of the system (assumption 8 )

$$x = 0 \text{ when } t = 0 \qquad (2.4.4)$$

The solution of Equation 2.4.3 and Equation 2.4.4 will give the organic phase acid concentration profile as a function of time. Both the organic and aqueous phase acid concentration profiles can be used to forecast the behavior of a single stage liquid–liquid extraction unit during start-up.

In the derivation of Equation 2.4.3, eight assumptions were listed. No general rule governs the number of assumptions that will result in a perfect model. However, a balance between too many and too few assumptions must be found if a workable solution is to be expected. If too many assumptions are made, the result will be in gross error, while too few assumptions can result in a mathematical problem that is not tractable.

The above format may be modified according to the situation under consideration. The reader should keep in mind that there is no single way to set up problems. However, there are some key items to pay close attention to:

- Always check for the involvement of some physical law or principle (mass, momentum, or energy balances).
- Include a consistent set of units (same units in each term).
- Appropriately include given conditions (initial or boundary).
- Check the mode of operation (steady or unsteady process).

Below are some examples for which the straightforward application of mass or energy balance is sufficient to set up the differential equation.

**Example 2.4.1** Transient Behavior of an Air-Cooling System (Felder and
                 Rousseau[5])

Consider an engine that generates heat at a rate of 8530 Btu/min. Suppose
this engine is cooled with air, and the air in the engine housing is circulated
rapidly enough so that the air temperature can be assumed uniform and is the
same as that of the outlet air. The air is fed to the housing at 6.0 lb-mole/min
and 65°F. Also, an average of 0.20 lb-mole of air is contained within the engine
housing and its temperature variation can be neglected. If heat is lost from the
housing to its surroundings at a rate of $Q_l$(Btu/min) = 33.0(T – 65°F) and the
engine is started with the inside air temperature equal to 65°F:

a. Derive a differential equation for the variation of the outlet
   temperature with time.

b. Calculate the steady-state air temperature if the engine runs
   continuously for an indefinite period of time, using $C_v$ = 5.00
   Btu/lb mole °F.

*Solution*

The unsteady-state balance equation for this system (air within the engine
housing) is the first law of thermodynamics for open systems with changes
in kinetic and potential energies neglected. Also, the temperature and composition of the system contents are assumed independent of position and no
phase changes occur. This gives the equation

$$MC_V \frac{dT_{sys}}{dt} = mC_p\left(T_m - T_{sys}\right) + Q + W_s$$

where $W_s$ is the rate of transfer of shaft work, $Q$ is the rate of heat transfer, M
is the mass (or number of moles) of the system contents, while m is the mass
flow rate.

   Taking

$$C_p = C_v + R = 6.99 \text{ Btu / lb mole } °F$$

then

$$mC_p = (6.0 \text{ lb mole/min}) (6.99 \text{ Btu/lb mole } °F) = 41.9 \frac{\text{Btu}}{\text{min} °F}$$

For this problem, $W_s$ = 0, since there are no moving parts and $Q = Q_{gen} - Q_l$.
Therefore, the differential equation becomes

$$(0.2) (5.0) \frac{dT}{dt} = 41.9 (65 - T) + 8530 - 33 (T - 65)$$

which reduces to

$$\frac{dT}{dt} = -74.9T + 13398$$

subject to

$$t = 0, \quad T = 65°F$$

The solution to the differential equation and initial condition is

$$T(t) = 179 - 112e^{-74.9t}$$

The steady-state temperature is

$$\lim_{t \to \infty} T(t) = 179°F$$

**Example 2.4.2** Comminution Operation (Felder and Rousseau[5])
Suppose a copper ore is fed to a ball mill at a steady rate $Q(kg/h)$, and crushed ore is withdrawn at the same rate. At an initial time ($t = 0$) the total mass of ore in the mill is $M(kg)$. Further suppose that a series of ten particle size ranges is defined and $x_i$ represents the mass fraction of particles in the $i$th size range, where $i = 1$ is the largest size range and $i = 10$ is the smallest. The rate at which particles are broken out of the $i$th size range is

$$r_i = k_i m_i$$

where $m_i$ is the mass of particles in this size range. Further suppose that of the particles broken out of size range $i$ in a differential time interval, a fraction $b_{ij}$ go into size range $j$. Assume that the size distribution of particles is uniform throughout the mill and equals that of the product, and let $x_{if}$ be the mass fraction of the feed that falls in the $i$th size range. Show that a mass balance on the $j$th size fraction in the tank yields

$$\frac{dx_j}{dt} = \left(\frac{Q}{M}\right)\left(x_{jf} - x_j\right) - k_j x_j + \sum_{i=1}^{j-1} k_i x_i b_{ij}$$

*Solution*

Accumulation = Input + Generation – Output – Consumption

$$\text{Accumulation} = \frac{d}{dt}\left(Mx_j\right) = M\frac{dx_j}{dt}$$

Input $= Q\,x_{if}$

Generation: the rate at which particles enter the $j$th size fraction from the $i$th size fraction by breakage is $k_i\,m_i\,b_{if}$.

Output $= Q\,x_j$

Consumption $= k_j\,m_j$

Therefore, the balanced equation becomes

$$M\frac{dx_j}{dt} = Qx_{if} - Qx_j - k_jm_j + \sum_{i=1}^{10} k_im_ib_{ij}$$

Note that $b_{ij} = 0$ for $j \le i$ (particle size cannot increase, and breakage within a size range $i = j$ does not count as an event), then dividing by M gives

$$\frac{dx_j}{dt} = \frac{Q}{M}\left(x_{if} - x_j\right) + \sum_{i=1}^{j=1} k_ix_ib_{ij} - k_jx_j$$

subject to $t = 0$, $x_j = x_{j0}$ ($j$th size fraction of initial contents of the mill).

**Example 2.4.3** Semi-Batch Reacting System (Felder and Rousseau[5])
A liquid-phase chemical reaction with stoichiometry $A \to B$ takes place in a semi-batch reactor. The rate of consumption of A per unit volume of the reactor is given by the first order rate expression

$$r_A\left(\text{mol/liter}\,.\,\text{s}\right) = kC_A$$

where $C_A(\text{mol/liter})$ is the reactant concentration. The tank is initially empty. At time $t = 0$, a solution containing A at a concentration $C_{A0}(\text{mol/liter})$ is fed to the tank at a steady rate $\phi(\text{liters/s})$. Develop differential balances on the total volume of the tank contents, V, and on the moles of A in the tank, $n_A$.

*Solution*
Total volume balance: Accumulation = Input

$$\frac{dV}{dt} = \phi$$

$$t = 0, \quad V = 0$$

Moles of A balance: Accumulation = Input − Consumption

$$C_A = \frac{n_A}{V}$$

$$\frac{dn_A}{dt} = C_{A0}\phi - kn_A$$

$$t = 0, \quad n_A = 0$$

## Example 2.4.4

Water containing 2 oz of pollutant/gal flows through a treatment tank at a rate of 500 gal/min (West[7]). In the tank, the treatment removes 2% of the pollutant per min and the water is thoroughly stirred. The tank holds 10,000 gal of water. On the day the treatment plant opens, the tank is filled with pure water. Determine the concentration profile of the tank effluent.

*Solution*

Let P(t) be the amount of pollutant in the tank at any time t.
Then

$$\frac{dP}{dt} = \text{input} - \text{output}$$

$$= \underbrace{\left(2\frac{oz}{gal}\right)\left(500\frac{gal}{min}\right)}_{\text{input}} - \underbrace{\left(P\frac{oz}{10,000gal}\right)\left(500\frac{gal}{min}\right)}_{\text{effluent}} - \underbrace{0.02P\frac{oz}{min}}_{\text{treatment}}$$

subject to P = 0 at t = 0. Therefore,

$$P(t) = (100,000/7)\left(1 - e^{-0.07t}\right)$$

---

## 2.5　Problems

1.

    a. Derive the organic phase benzoic acid concentration profile as a function of time (Section 2.4).

    b. Determine the steady state solution to

$$\left(V_1 + mV_2\right)\frac{dx}{dt} = RC_{A0} - (R + sm)x$$

    i. directly

    ii. by taking the limit as $t \to \infty$.

   c. If E is the proportion of benzoic acid extracted and $\alpha = R/ms$, what is the relationship between E and $\alpha$ for the steady-state process?

2. Consider a two-stage solvent extraction of the benzoic acid with the previously made assumptions and

$$y_i = mx_i, \quad i = 1,2$$

where $i$ denotes the stage.[4]

   a. Develop a table similar to Table 2.1.

   b. Use your table to write the time dependent mass balance of acid for each stage.

   c. Find the steady state organic ($x_2$) and aqueous ($y_1$) acid concentration profiles.

   d. If E is the proportion of acid extracted and $\alpha = R/ms$, what is the relationship between E and $\alpha$ for the steady-state process?

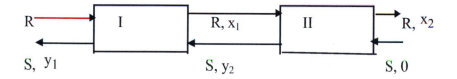

3. Give suitable initial conditions for the two first order differential equations and outline a solution.

4. Phosgene ($COCl_2$) is formed by reacting CO and $Cl_2$ in the presence of activated charcoal (Felder and Rousseau[5]):

$$CO + Cl_2 \to COCl_2$$

At a temperature of 303.8 K in the presence of 1 g of charcoal, the rate of formation of phosgene is

$$R_f(mol/min) = \frac{8.75[CO][Cl_2]}{\left(1 + 58.6[Cl_2] + 34.3[COCl_2]\right)^2}$$

where "[ ]" denotes concentration (mol/liter).

   a. Given that the input to a 3.00-liter batch reactor is 1.00 g of charcoal and a gas containing 60 mole % CO and 40 mole % $Cl_2$, and that the initial reactor conditions are 303.8 K and 1 atm, determine the initial concentrations (mol/liter) of both reactants. The volume occupied by the charcoal may be neglected.

b. Write a differential balance on phosgene and show that it simplifies to

$$\frac{d[COCl_2]}{dt} = \frac{2.92(0.02407 - [COCl_2])(0.01605 - [COCl_2])}{(1.941 - 24.3[COCl_2])^2}$$

5. A gas that contains $CO_2$ is contacted with liquid water in an agitated batch absorber (Felder and Rousseau[5]). Henry's Law gives the equilibrium solubility of $CO_2$ in water

$$C_A = \frac{P_A}{H_A}$$

where $C_A$ (mol/liter) is the $CO_2$ concentration in solution, $p_A$ (atm) is the partial pressure of $CO_2$ in the gas phase and $H_A$ is Henry's Law constant. The rate of transfer $CO_2$ from the gas to the liquid per unit area of gas-liquid interface is given by

$$r_a(mol/cm^2 \cdot s) = k(C_A^* - C_A)$$

where $C_A^*$ is the concentration of $CO_2$ that would be in equilibrium with the $CO_2$ in the gas phase ($C_A^* = P_A/H_A$). Suppose the gas phase total pressure is P(atm) and contains $y_A$ mol fraction of $CO_2$, and the liquid phase initially has V(cm$^3$) of pure water with the agitation of the liquid phase sufficient to neglect spatial composition dependency, and if the amount of absorbed $CO_2$ is low enough for P, V, and $y_A$ to be assumed constant, write a differential balance on $CO_2$ in the liquid phase and solve the differential equation to show that

$$C_A(t) = C_A^*\{1 - \exp(-kSt/V)\}$$

6. An iron bar 2 cm × 3 cm × 10 cm at a temperature of 95°C is dropped into a barrel of water at 25°C. The barrel is large enough that the water temperature rises negligibly as the bar cools. The rate at which heat is transferred from the bar to the water is given by

$$Q(J/min) = UA(T_b - T_w)$$

where $U(= 0.050 \text{ J/min} \cdot \text{cm}^2 \cdot {}°\text{C})$ is a heat-transfer coefficient, $A$ (cm$^2$) is the exposed area of the bar, $T_b$ is the surface temperature of the bar, and $T_w$ is the water temperature. Given that the heat capacity of the bar is 0.460 J/g $\cdot$ °C, and heat conduction in iron is fast enough to assume that the temperature $T_b(t)$ is uniform throughout the bar, write an energy balance on the bar and determine the steady state temperature of the bar. Also, calculate the time required for the bar to cool to 30°C.

7. A steam coil is immersed in a stirred heating tank (Felder and Rousseau[5]). Saturated steam at 7.50 bar condenses within the coil, and the condensate emerges at its saturation temperature. A solvent with a heat capacity of 2.30 kJ/kg.°C is fed to the tank at a steady rate of 12.0 kg/min and a temperature of 25°C, and the heated solvent is discharged at the same rate. The tank is initially filled with 760 kg of solvent at 25°C, at which point the flows of both steam and solvent are commenced. The rate at which heat is transferred from the coil to the solvent is given by

$$Q(kJ/min) = UA(T_{steam} - T)$$

where UA = 11.5 kJ/min·°C. The tank is well stirred such that the temperature can be considered as spatially uniform and is the same as the outlet temperature. Derive a differential energy balance on the tank contents and calculate the time required to heat the solvent to an arbitrary temperature $T_f$(°C).

# References

1. Boyce, W.E. and DiPrima, R.C., *Elementary Differential Equations and Boundary Value Problems*, 3rd ed., John Wiley & Sons, New York, 1977.
2. Giordano, F.R. and Weir, M.D., *Differential Equations, a Modeling Approach*, Addison-Wesley, New York, 1991.
3. Thomas, G.B. and Finney, R.L., *Calculus and Analytic Geometry*, 6th ed., Addison-Wesley, Reading, MA, 1984.
4. Jenson, V.G. and Jeffreys, G.V., *Mathematical Methods in Chemical Engineering*, Academic Press, London, 1963.

5. Felder, R.M. and Rousseau, R.W., *Elementary Principles of Chemical Processes*, 2nd ed., John Wiley & Sons, New York, 1976.
6. Rice, R.G. and Do, D.D., *Applied Mathematics and Modeling for Chemical Engineers*, John Wiley & Sons, New York, 1995.
7. West, B. H., Setting up differential equations from word problems, in *Modules in Applied Mathematics*, vol. 1, *Differential Equations*, Lucas, W.F., Ed., Springer-Verlag, New York, 1983, chap. 1.

# 3

## Linear Second Order Ordinary Differential Equations

## 3.1 Introduction

Chemical engineers working in the area of Transport Phenomena must frequently solve problems that involve linear second order differential equations. These may occur as boundary value problems in diffusional systems or initial value problems in process control or reacting systems. But most frequently, they are the result of a reduction of a partial differential equation.

In this chapter, linear second order ordinary differential equations will be reviewed. There will be examples on applications; however, Chapters 6 and 7 emphasize both mathematical and engineering applications.

Herein, the meaning of linearity is the same as that given in Chapter 2. Attention will now be focused on equations that can be written in the form:

$$\zeta = f(t, \zeta, \zeta') \tag{3.1.1}$$

Previously, it was observed that in the case of first order equations, the integrated solutions contained one arbitrary constant. This constant could be determined by a given condition at an initial time. In the case of second order equations, two constants of integration are expected to occur, and therefore two conditions at an initial time will be needed for the so-called initial value problems; or two conditions at separate locations for the so-called boundary value problems (Boyce and DiPrima,[1] O'Neil,[4] Mickley et al.[6]).

In the discussion of the theory of second order linear ordinary differential equations, the standard mathematical symbols will be employed. In chemical engineering applications the usual chemical engineering nomenclature will be used whenever there is no conflict.

The *general second order linear differential equation* is of the form

$$P(x)\frac{d^2y}{dx^2}+Q(x)\frac{dy}{dx}+R(x)y=G(x) \qquad (3.1.2)$$

where P, Q, R, and G are given functions.

Three important examples of second order linear differential equations that frequently occur in chemical engineering are Legendre's equation (Greenberg,[3] O'Neil[4]):

$$\left(1-x^2\right)y''-2xy'+\alpha(\alpha+1)y=0 \qquad (3.1.3)$$

of order $\alpha$, Bessel's equation (Mickley, Sherwood, and Reid,[6] Watson,[9] Gray and Matthews,[10] Tranter[11]) of order n:

$$x^2y''+xy'+\left(x^2-n^2\right)y=0 \qquad (3.1.4)$$

and the confluent hypergeometric equation (Hildebrand,[5] Loney,[7] Huang,[8] Slater[12]):

$$x\frac{d^2y}{dx^2}+(c-x)\frac{dy}{dx}-ay=0 \qquad (3.1.5)$$

where the quantities a, n, and c are constants.

In the discussion to follow, unless otherwise stated, the functions P, Q, R, and G in Equation 3.1.2 are taken to be continuous on some interval $a < x < b$, ( a may be $-\infty$ and/or b may be $+\infty$ ).

If the function $P(x) \neq 0$ everywhere on the interval, then Equation 3.1.5 can be rewritten as:

$$\frac{d^2y}{dx^2}+p(x)\frac{dy}{dx}+q(x)y=g(x) \qquad (3.1.6)$$

following division by $P(x)$.

Similar to first order equations, the issue of existence and uniqueness of solutions to second order equations must be dealt with. Below is a theorem (Boyce and DiPrima,[1] O'Neil[4]) that addresses existence and uniqueness of solutions of second order differential equations.

*THEOREM 3.1*

*If the functions p, q, and g are continuous on the open interval $a < x < b$, then there exists one and only one function $y = \phi(x)$ satisfying the differential Equation 3.1.6*

$$y'' + p(x)\,y' + q(x)\,y = g(x)$$

*on the entire interval a < x < b and the given initial conditions*

$$y(x_0) = y_0, \quad y'(x_0) = y'_0$$

*at a particular point $x_0$ in the interval.*

It is important to note that this theorem does not address the issue of existence and uniqueness of solutions for boundary value problems. Boundary value problems are discussed in the next chapter.

The following three elementary examples serve to demonstrate how the above-mentioned interval can be determined for a unique solution to exist.

### Example 3.1.1

Given:   $xy'' + 3y = x, \quad y(x_0) = y_0; \quad y'(x_0) = y'_0$

rewrite as:   $y'' + \dfrac{3}{x}y = 1$

then the interval consists of all points not including the origin.

### Example 3.1.2

Given:   $y'' + by' + 7y = 2\sin x, \quad y(x_0) = y_0; \quad y'(x_0) = y'_0$

Since the differential equation is defined on $-\infty < x < +\infty$, the whole real line is the interval.

### Example 3.1.3

Given:   $x(x-1)y'' + 3xy' + 4y = 2, \quad y(x_0) = y_0; \quad y'(x_0) = y'_0$

rewrite as:   $y'' + \dfrac{3x}{x(x-1)}y' + \dfrac{4}{x(x-1)}y = \dfrac{2}{x(x-1)}$

This differential equation is defined everywhere except at 0 and 1. Thus, the interval is any interval excluding the points 0 and 1.

Essentially, to determine the interval or domain of definition of the differential equation, one elementary procedure is to look for points where division by zero will occur or points where the equation becomes unbounded. These so-called singular points will be classified later, but for now, these are to be excluded from the interval over which Theorem 3.1 is applied.

In order to discuss the methods used to solve linear second order differential equations, it is necessary to reintroduce the term *homogeneous*, or *complementary*, but with a meaning unrelated to previous usage in this book.

When the forcing function g(x) is set to zero in Equation 3.1.6, the reduced equation

$$\frac{d^2y}{dx^2} + p(x)\frac{dy}{dx} + q(x)y = 0 \qquad (3.1.7)$$

results. This is the *homogeneous* form of the second order linear differential equation. Any equation of second order or higher with right-hand side identically zero is termed homogeneous as opposed to nonhomogeneous.

---

## 3.2   Fundamental Solutions of the Homogeneous Equation

Suppose p and q in Equation 3.1.7 are continuous on $a < x < b$, then for any twice differentiable function $\phi$ on $a < x < b$, the *linear differential operator* **L** is defined to mean

$$L[\phi] = \phi'' + p\phi' + q\phi \tag{3.2.1}$$

for

$$L \equiv D^2 + pD + q \tag{3.2.2}$$

where D is the derivative operator.
Then Equation 3.1.7 becomes

$$L[y] = y'' + p(x)y' + q(x)y = 0 \tag{3.2.3}$$

The use of the operator L here reduces the task of integrating the linear second order differential equation as will be seen below. However, before the integration process can be employed, a few more definitions are needed.

*THEOREM 3.2*
*If $y = y_1(x)$ and $y = y_2(x)$ are solutions of the differential Equation 3.2.3*

$$L[y] = y'' + p(x)y' + q(x)y = 0$$

*then the linear combination of $y = c_1y_1(x) + c_2y_2(x)$, with $c_1$ and $c_2$ being arbitrary constants is also a solution.*

**PROOF**
Since $y = y_1(x)$ is a solution set of Equation 3.2.3 then $L[y_1] = y_1'' + p(x)y_1' + q(x)y_1 = 0$ and since $y = y_2(x)$ is also a solution set of Equation 3.2.3, then $L[y_2] = y_2'' + p(x)y_2' + q(x)y_2 = 0$. But

$$L[C_1y_1 + C_2y_2] = L[C_1y_1] + L[C_2y_2]$$

$$= C_1 L[y_1] + C_2 L[y_2]$$

$$= 0 \qquad \blacksquare$$

Theorem 3.2 is a statement of the *Superposition Principle* (Boyce & Diprima,[1] O'Neil,[4] Giordano and Weir[2]), which is also applicable to higher order linear differential equations. The two solutions $y_1$ and $y_2$ form what is called a *fundamental set* of solutions for Equation 3.2.3.

In general, two solutions $y_1$ and $y_2$ of Equation 3.2.3 are said to form a fundamental set of solutions if every solution of Equation 3.2.3 can be expressed as a linear combination of $y_1$ and $y_2$. In particular

**THEOREM 3.3**
*If the functions p and q are continuous on the interval $a < x < b$ and if $y_1$ and $y_2$ are solutions of the differential Equation 3.2.3*

$$L[y] = y'' + p(x)y' + q(x)y = 0$$

*satisfying the condition:*

$$y_1(x)y_2'(x) - y'_1(x)y_2 \neq 0 \qquad (3.2.4)$$

*at every point in $a < x < b$, then any solution of Equation 3.2.3 on the interval $a < x < b$ can be expressed as a linear combination of $y_1$ and $y_2$.*

The condition described by Equation 3.2.4 is called the Wronskian and is commonly written in the determinant form as (Boyce and DiPrima[1])

$$W(y_1, y_2) = \begin{vmatrix} y_1 & y_2 \\ y'_1 & y'_2 \end{vmatrix} \qquad (3.2.5)$$

The linear combination mentioned in Theorem 3.3 is usually called the general, complementary or homogeneous solution of Equation 3.2.3. The following theorem should help to clarify when Equation 3.2.4 or Equation 3.2.5 is expected to be different from zero.

**THEOREM 3.4**
*If the function p and q are continuous on $a < x < b$ and the function $y_1$ and $y_2$ are linearly independent solutions of the differential Equation 3.2.3:*

$$L[y] = y'' + p(x)y' + q(x)y = 0$$

*the $W(y_1, y_2)$ is nonzero on $a < x < b$ and hence any solution of Equation 3.2.3 can be expressed as a linear combination of $y_1$ and $y_2$.*

Where linear independence is defined as two or more functions on the interval $a < x < b$ that are not linearly dependent (Boyce and DiPrima,[1] O'Neil,[4] Giordano and Weir[2]). For example, if f and g are two functions on $a < x < b$, such that

$$c_1 f(x) + c_2 g(x) = 0$$

for all x in $a < x < b$ is a true statement for $c_1$ and $c_2$ not both zero, then f and g are linearly dependent, otherwise they are linearly independent. In other words, if f and g are a constant multiple of each other on the interval, then they are linearly dependent; otherwise, they are linearly independent.

Finally, we come to the "how to" find solution for linear second order differential equations. Specifically, the case where p and q are constant in Equation 3.2.3.

## 3.3  Homogeneous Equations with Constant Coefficients

The most general constant coefficient, linear, second order, ordinary, homogeneous differential equation is:

$$ay'' + by' + cy = 0 \qquad (3.3.1)$$

where a, b, and c are real constants and $a \neq 0$. Equation 3.3.1 in operator notation is:

$$L[y] = ay'' + by' + c\,y = (a\,D^2 + b\,D + c)y = 0 \qquad (3.3.2)$$

Now we are interested in integrating this equation, Equation 3.3.2, and expressing our findings as $y = \phi(x)$ and in this case, include the two integration constants as implied by the presence of the second derivative term.

To solve Equation 3.3.2, we seek a function $\phi(x)$ such that a times its second derivative added to b times its first derivative added to c times the function itself results in zero. Among all possible candidate functions, the function $\phi(x) = e^{rx}$, where r is a constant, turns out to be the best candidate. Thus:

$$L[e^{rx}] = a(e^{rx})'' + b(e^{rx})' + c(e^{rx}) = e^{rx}(ar^2 + br + c) = 0 \qquad (3.3.3)$$

Since $e^{rx}$ is never zero, then

$$ar^2 + br + c = 0 \tag{3.3.4}$$

Equation 3.3.4 is called the *characteristic* or *auxiliary* equation and the resulting roots $r_1$ and $r_2$ are called the *characteristic roots* or *eigenvalues*. The roots $r_1$ and $r_2$ are given by

$$r_1 = \frac{-b + \sqrt{b^2 - 4ac}}{2a}, r_2 = \frac{-b - \sqrt{b^2 - 4ac}}{2a} \tag{3.3.5}$$

If the discriminant $b^2 - 4ac > 0$, then $r_1$ and $r_2$ are real and unequal. Then

$$y_1(x) = e^{r_1 x}, y_2(x) = e^{r_2 x} \tag{3.3.6}$$

and Theorem 3.2, the superposition principle tells us that the linear combination

$$y = c_1 e^{r_1 x} + c_2 e^{r_2 x} \tag{3.3.7}$$

with $c_1$ and $c_2$ being arbitrary (integration) constants is also a solution.

It is easy to check that Theorem 3.3 and Theorem 3.4 both hold; that is, $e^{r_1 x}$ and $e^{r_2 x}$ are linearly independent functions and thus form a fundamental set.

If the discriminant $b^2 - 4ac = 0$, then $r_1$ and $r_2$ are identical and there only results one solution, namely $e^{-\left(\frac{b}{2a}\right)x}$, that is

$$y_1 = e^{-\left(\frac{b}{2a}\right)x} \tag{3.3.8}$$

In order to find $y_2$, consider the following procedure.
Let

$$y = v(x) e^{-\left(\frac{b}{2a}\right)x} \tag{3.3.9}$$

then

$$y' = v'(x) e^{-\left(\frac{b}{2a}\right)x} - \frac{b}{2a} v(x) e^{-\left(\frac{b}{2a}\right)x} \tag{3.3.10}$$

$$y'' = \left[ v''(x) - \frac{b}{a} v'(x) + \frac{b^2}{4a^2} v(x) \right] e^{-\left(\frac{b}{2a}\right)x} \tag{3.3.11}$$

Substitute Equation 3.3.9 to Equation 3.3.11 into Equation 3.3.1 to get

$$a\left(v'' - \frac{b}{a}v' + \frac{b^2}{4a^2}v\right) + b\left(v' - \frac{b}{2a}v\right) + cv = 0 \tag{3.3.12}$$

and simplify further to get:

$$v'' = 0 \tag{3.3.13}$$

Integrating Equation 3.3.13 twice gives:

$$v(x) = (c_1 x + c_2) \tag{3.3.14}$$

Then Equation 3.3.9 becomes:

$$y(c_1 x + c_2)e^{-\left(\frac{b}{2a}\right)x} \tag{3.3.15}$$

Therefore,

$$y_2 = xe^{-\left(\frac{b}{2a}\right)x} \tag{3.3.16}$$

and Equation 3.3.15 is the general solution when $r_1 = r_2$, that is:

$$y = c_1 xe^{(r_1 x)} + c_2 e^{(r_1 x)} \tag{3.3.17}$$

If the discriminant $b^2 - 4ac < 0$ then $r_1$ and $r_2$ are complex numbers. Furthermore, since a, b, and c are real, then $r_1$ and $r_2$ will be a conjugate pair. That is:

$$r_1 = -\frac{b}{2a} + \frac{i\sqrt{b^2 - 4ac}}{2a} = \lambda + i\mu \tag{3.3.18}$$

$$r_2 = -\frac{b}{2a} - \frac{i\sqrt{b^2 - 4ac}}{2a} = \lambda - i\mu \tag{3.3.19}$$

Where the real numbers $\lambda$ and $\mu$ are introduced for convenience. Similar to the results obtained in Equation 3.3.7, the general solution is

$$y = k_1 e^{(\lambda + i\mu)x} + k_2 e^{(\lambda - i\mu)x} \tag{3.3.20}$$

where $k_1$ and $k_2$ are arbitrary constants.

Equation 3.3.20 can be rewritten as

$$
\begin{aligned}
y &= \left[ k_1 e^{i\mu x} + k_2 e^{-i\mu x} \right] e^{\lambda x} \\
&= \left[ k_1 \left( \cos \mu x + i \sin \mu x \right) + k_2 \left( \cos \mu x - i \sin \mu x \right) \right] e^{\lambda x} \\
&= \left[ \left( k_1 + k_2 \right) \cos \mu x + i \left( k_1 - k_2 \right) \sin \mu x \right] e^{\lambda x} \\
&= \left[ c_1 \cos \mu x + c_2 \sin \mu x \right] e^{\lambda x}
\end{aligned}
\tag{3.3.21}
$$

Where $c_1$ and $c_2$ are arbitrary constants since the sum or difference of two arbitrary constants is still an arbitrary constant.

In going from Equation 3.3.20 to Equation 3.3.21, the following concept was employed to define $e^{i\mu x}$ and $e^{-i\mu x}$ :

The function $e^{\mu x}$ has a Taylor series expansion of

$$
e^{\mu x} = 1 + \mu x + \frac{(\mu x)^2}{2!} + \frac{(\mu x)^3}{3!} + \frac{(\mu x)^4}{4!} + \frac{(\mu x)^5}{5!} + \cdots
$$

about the origin. Then

$$
\begin{aligned}
e^{i\mu x} &= 1 + i\mu x + \frac{(i\mu x)^2}{2!} + \frac{(i\mu x)^3}{3!} + \frac{(i\mu x)^4}{4!} + \frac{(i\mu x)^5}{5!} + \cdots \\
&= \left( 1 - \frac{(\mu x)^2}{2!} + \frac{(\mu x)^4}{4!} - \cdots \right) + i \left( \mu x - \frac{(\mu x)^3}{3!} + \frac{(\mu x)^5}{5!} - \cdots \right)
\end{aligned}
$$

It is also known that the functions $\cos \mu x$ and $\sin \mu x$ each have Taylor series expansion of

$$
\cos \mu x = 1 - \frac{(\mu x)^2}{2!} + \frac{(\mu x)^4}{4!} - \cdots
$$

and

$$
\sin \mu x = \mu x - \frac{(\mu x)^3}{3!} + \frac{(\mu x)^4}{5!} - \cdots
$$

about the origin. Therefore $e^{i\mu x}$ is defined as

$$
e^{i\mu x} = \cos \mu x + i \sin \mu x
\tag{3.3.22}
$$

and

$$e^{-i\mu x} = \cos \mu x - i \sin \mu x \qquad (3.3.23)$$

Equations 3.3.7, 3.3.17, and 3.3.21 give the general solutions of the second order, constant coefficient, homogeneous and linear differential equation for the respective cases of real unequal, repeated, and complex characteristic roots. However, the actual steps that are used in deriving a solution to the homogeneous problem are as follows:

**Step 1:** Put differential equation into operator form.
**Step 2:** Identify the characteristic equation and roots.
**Step 3:** Use Equations 3.3.7, 3.3.17 or 3.3.21 to express the general solution.

The following examples should clarify the three steps given above.

## Example 3.3.1
Given: $2y'' - 3y' + y = 0$

**Step 1:**

$$(2D^2 - 3D + 1)y = 0 \qquad\qquad \textit{Operator form}$$

**Step 2:**

$$(2r^2 - 3r + 1) = 0 \qquad\qquad \textit{Characteristic equation}$$

$$r = 1, {}^1\!/_2 \qquad\qquad \textit{Characteristic roots}$$

**Step 3:**
Since the roots are real and distinct, Equation 3.3.7 gives

$$y = c_1 e^x + c_2 e^{\frac{1}{2}x}$$

## Example 3.3.2
Given: $y'' + 2y' + y = 0$

**Step 1:**

$$(D^2 + 2D + 1)y = 0 \qquad\qquad \textit{Operator form}$$

**Step 2:**

$$r^2 + 2r + 1 = 0 \qquad\qquad \textit{Characteristic equation}$$

$$r = -1, -1 \qquad\qquad \textit{Characteristic roots}$$

**Step 3:**
Since roots are repeated, Equation 3.3.17 gives

$$y = c_1 x e^{-x} + c_2 e^{-x}$$

## Example 3.3.3
Given: $y'' + 6y' + 13y = 0$

**Step 1:**

$$(D^2 + 6D + 13)y = 0 \qquad\qquad \textit{Operator form}$$

**Step 2:**

$$r^2 + 6r + 13 = 0 \qquad\qquad \textit{Characteristic equation}$$

$$r = -3 + 2i, -3 - 2i \qquad \textit{Characteristic roots}$$

**Step 3:**
Since roots are complex, Equation 3.3.21 gives

$$y = [c_1 \cos 2x + c_2 \sin 2x]e^{-3x}$$

where $\lambda = -3$ and $\mu = 2$

## 3.4  Nonhomogeneous Equations

Previously, some brief theory and a few examples were given that demonstrate the approach necessary to construct solutions to the homogeneous problem. The summarized procedure for the constant coefficient case is very simple, and one would like to maintain that simplicity for the case of at least the nonhomogeneous constant coefficient problem.

The nonhomogeneous differential equation can be written as

$$L[y] = y'' + p(x)y' + q(x)y = g(x) \qquad (3.4.1)$$

where $p$, $q$, and $g$ are continuous on the interval of interest.

The following two theorems are needed to establish the procedure for solving Equation 3.4.1:

*THEOREM 3.5*
*The difference of any two solutions of the differential Equation 3.4.1*

$$L[y] = y'' + p(x)y' + q(x)y = g(x)$$

*is a solution of the corresponding homogeneous differential equation*

$$L[y] = y'' + p(x)y' + q(x)y = 0 \qquad (3.4.2)$$

**PROOF**  Suppose the two functions $u_1$ and $u_2$ are solutions of Equation 3.4.1 then

$$L[u_1] = g$$

and                                $$L[u_2] = g$$

Therefore,                  $$L[u_1] - L[u_2] = 0$$

but L is a linear operator

thus                $$L[u_1] - L[u_2] = L[u_1 - u_2] = 0 \qquad \blacksquare$$

*THEOREM 3.6*
*Given one solution $Y_p$ of the nonhomogeneous linear differential Equation 3.4.1*
*$L[y] = y'' + p(x)y' + q(x)y = g(x)$*
*then any solution $y = f(x)$ of this equation can be expressed as*

$$f(x) = Y_p(x) + c_1 y_1(x) + c_2 y_2(x) \qquad (3.4.3)$$

*where $y_1$ and $y_2$ are linearly independent solutions of the corresponding homogeneous equation.*

Equation 3.4.3 is the general solution of the nonhomogeneous Equation 3.4.1. That is, the general solution $y_g$, of the nonhomogeneous equation is

$$y_g = y_c + y_p \qquad (3.4.4)$$

where $y_c$ is the general solution of the associated homogeneous equation and is termed the *complementary solution*. The function $y_p$ is the particular solution satisfying the nonhomogeneous differential Equation 3.4.1.

A relatively simple procedure for finding the most suitable candidate for $y_p$ is demonstrated below for the constant coefficient case.

For example, given

$$y'' - 3y' - 4y = 2 \sin x \qquad (3.4.5)$$

then the associated homogeneous problem is

$$y'' - 3y' - 4y = 0$$

which can be put in the operator form

$$(D^2 - 3D - 4)y = 0$$

In fact, the nonhomogeneous problem can be put in the operator form as well

$$(D^2 - 3D - 4)y = 2 \sin x \tag{3.4.6}$$

Then one can see that if the RHS or nonhomogeneous part is differentiated two times, the result is $-2 \sin x$.

Further, if the nonhomogeneous part is added to the twice differentiated result, one gets a final result of zero. That is:

$$\frac{d^2}{dx^2}(2 \sin x) + 2 \sin x = 0$$

or

$$\left(D^2 + 1\right)2 \sin x = 0$$

Then, since the RHS of Equation 3.4.6 is operated on by $D^2 + 1$, the same operation must be carried out on the LHS, giving

$$\left(D^2 + 1\right)\left(D^2 - 3D - 4\right)y = 0 \tag{3.4.7}$$

a homogeneous equation of the fourth order. Following steps 2 and 3 of Section 3.3, the characteristic equation is

$$(r^2 + 1)\,(r^2 - 3r - 4) = 0$$

with characteristic roots $r_1 = -1$, $r_2 = 4$, $r_3 = +i$, $r_4 = -i$.

Then Equation 3.4.7 and Equation 3.4.21 give

$$y_g = c_1 e^{-x} + c_2 e^{4x} + c_3 \cos x + c_4 \sin x \tag{3.4.8}$$

where

$$y_c = c_1 e^{-x} + c_2 e^{4x}$$

which comes from the associated homogeneous equation and therefore

$$y_p = c_3 \cos x + c_4 \sin x$$

is the best candidate for the particular solution. The constants $c_3$ and $c_4$ are not arbitrary and are determined by substituting $y_p$ into Equation 3.4.5

$$y_p'' - 3y_p' - 4y_p = 2 \sin x$$

that is

$$\left(c_3 \cos x + c_4 \sin x\right)'' - 3\left(c_3 \cos x + c_4 \sin x\right)' - 4\left(c_3 \cos x + c_4 \sin x\right) = 2 \sin x$$

which results in an identity:

$$\left(-c_3 \cos x - c_4 \sin x\right) - 3\left(c_4 \cos x + c_3 \sin x\right)' - 4\left(c_3 \cos x + c_4 \sin x\right) = 2 \sin x$$

Following simplifications and equating coefficients, one gets

$$-\left(5c_3 + c_4\right)\cos x = 0$$

$$\left(3c_3 - 5c_4\right)\sin x = 2 \sin x$$

or

$$-5c_3 - 3c_4 = 0$$

and

$$3c_3 - 5c_4 = 2$$

giving $c_4 = -5/17$, $c_3 = 3/17$ such that

$$y_p = \tfrac{3}{17}\cos x - \tfrac{5}{17}\sin x$$

Therefore, the problem of finding the general solution to Equation 3.4.1 is reduced to finding a particular solution to Equation 3.4.1. It should be noted that the method works when the nonhomogeneous part can be differentiated to zero or when an appropriate number of differentiation will bring back some constant multiple of the original function. For example, this method would fail for RHS of the types: $\tan x$ or $x^{m/n}$, where the number $m/n$ is not an integer and for combinations of these. In other words, the method works in theory for g(x) of the type

$$g(x) = P_n(x) = a_o x^n + a_1 x^{n-1} + a_1 x^{n-1} + \cdots + a_n$$

$$= e^{\alpha x} P_n(x)$$

$$= e^{\alpha x} P_n(x) \, \sin \beta x$$

$$= e^{\alpha x} P_n(x) \, \cos \beta x$$

In general, given a linear ordinary differential equation:

$$P(D)y = R(x) \tag{3.4.9}$$

the general solution is

$$y = y_c + y_p$$

where $P(D)y_c = 0$ and $P(D)y_p = R(x)$. Suppose there is a differential operator, "Annihilator," A(D) that is linear with constant coefficients such that

$$A(D)R(x) = 0$$

If we operate on both sides of Equation 3.4.9 with A(D) we would get

$$A(D)P(D)y = A(D)R(x) = 0.$$

Now, consider this new equation

$$A(D)P(D)y = 0 \tag{3.4.10}$$

In order to find the general solution of Equation 3.4.9, we need the roots of the polynomial A(D)P(D); and they are: $r_1, r_2, \ldots, r_j; q_1, q_2, \ldots, q_k$ where the $r_1, r_2, \ldots, r_j$ come from P(D) and $q_1, q_2, \ldots, q_k$ come from A(D).
The general solution of Equation 3.4.10 can be written as

$$y = y_c + y_q$$

But

$$A(D)P(D)\left[y_c + y_p\right] = A(D)R(x) = 0$$

thus $y_c + y_p$ is also a solution of Equation 3.4.10. Since $y_c + y_q$ is the general solution of Equation 3.4.10, it contains $y_c + y_p$. Therefore

$$y_c + y_q \subset y_c + y_q \quad \text{or} \quad y_p \subset y_q$$

The function $y_q$ can be thought of as the best candidate for the particular solution, and the method of undetermined coefficients can be used to find $y_p$.

Below are three more elementary examples that demonstrate successful application of the method, and a fourth example in which the method fails.

## Example 3.4.1
Find the general solution of $y'' - y' - 6y = 12xe^x$

*Solution*

$$(D^2 - D - 6)y = 12 \; xe^x \qquad\qquad\qquad \textit{Put in operator form}$$

Determine the operator that will annihilate the nonhomogeneous part: *12xe^x*

$$\frac{d}{dx}(12xe^x) = 12e^x + 12xe^x$$

$$\qquad\qquad\qquad\qquad\qquad\qquad \textit{Annihilate RHS}$$

$$\frac{d^2}{dx^2}(12xe^x) = \frac{d}{dx}(12e^x + 12xe^x) = 24e^x + 12xe^x$$

then the combination of $\dfrac{d^2}{dx}(\cdot)$, $-2\dfrac{d}{dx}(\cdot)$, and the function itself adds to zero.

That is

$$(D^2 - 2D + 1)12xe^x = 0$$

operates on the LHS of the given equation with the newly found operator to get

$$(D^2 - 2D + 1)(D^2 - D - 6)y = 0$$

Solve using the procedure for the homogeneous problem.
$(r - 1)^2(r - 3)(r + 2) = 0$ is the characteristic equation
$r = 1, 1, -2, 3$ are the characteristic roots.
    Therefore,

$$y_g = c_1 e^x + c_2 x e^x + c_4 e^{-2x} + c_5 e^{3x}$$

$$y_c = c_4 e^{-2x} + c_5 e^{3x}$$

where $y_c$ is the solution to the associated homogeneous problem, and the candidate for $y_p$ is

$$y_p = c_1 e^x + c_2 x e^x$$

Determine $c_1$ and $c_2$ by substituting $y_p$ into the given differential equation

$$\left(c_1 e^x + c_2 x e^x\right)'' - \left(c_1 e^x + c_2 x e^x\right)' - 6\left(c_1 e^x + c_2 x e^x\right) = 12 x e^x$$

which results in

$$c_1 e^x + 2c_2 e^x + c_2 x e^x - c_1 e^x - c_2 e^x - c_2 x e^x - 6c_1 e^x - 6c_2 x e^x = 12 x e^x$$

an identity which simplifies to

$$c_2 - 6c_1 = 0$$

$$-6c_2 = 12$$

Therefore,

$$y_p = -\tfrac{1}{3} e^x - 2x e^x$$

and

$$y_g = c_4 e^{-2x} + c_5 e^{3x} - \tfrac{1}{3} e^x - 2x e^x$$

where $c_4$ and $c_5$ are the arbitrary constants.

## Example 3.4.2

Find the general solution of $y'' - 6y' + 8y = 3e^x$

$$(D^2 - 6D + 8) = 3e^x$$

becomes

$$(D - 1)(D^2 - 6D + 8)y = 0$$

$$(D - 1)(D - 4)(D - 2)y = 0$$

with characteristic roots: 1, 2, 4

$$y_g = c_1 e^{2x} + c_2 e^{4x} + c_3 e^x = y_c + y_p$$

where

$$y_p = c_3 e^x$$

$$y'_p = y''_p = c_3 e^x$$

That is

$$c_3 e^x - 6c_3 e^x + 8c_3 e^x = 3e^x$$

Therefore,

$$3c_3 = 3$$

giving

$$c_3 = 1$$

and

$$y_p = e^x$$

such that

$$y_g = c_1 e^{2x} + c_2 e^{4x} + e^x$$

## Example 3.4.3

Find the general solution of $y'' - 2y' + 10y = 20x^2 + 2x - 8$

*Solution*

$$(D^2 - 2D + 10)y = 20x^2 + 2x - 8$$

$$D^3 (D^2 - 2D + 10)y = 0$$

$0, 0, 0, 1 + 3i, 1 - 3i$: characteristic roots

$$\therefore \quad y_g = \underbrace{(c_1 \cos 3x + c_2 \sin 3x)e^x}_{y_c} + \underbrace{c_4 + c_5 x + c_6 x}_{y_p}$$

$$y'_p = 2c_6 x + c_5$$

$$y''_p = 2c_6$$

$$\therefore \quad 2c_6 - 2(2c_6 x + c_5) + 10c_4 + 10c_5 x + 10c_6 x^2 = 20x^2 + 2x - 8$$

Equating Coefficients:

$$x^0 : 2c_6 - 2c_5 + 10c_4 = -8$$

$$x :: - 4c_6 + 10c_5 = 2$$

$$x^2 : 10c_6 = 20$$

from which

$$c_6 = 2, c_5 = 1, c_4 = -1 \text{ and}$$

$$y_p = -1 + x + 2x^2$$

$$\therefore \quad y_g = \left[ c_1 \cos(3x) + c_2 \sin(3x) \right] e^x + 2x^2 + x - 1$$

## Example 3.4.4

Find the general solution of

$$y'' + y = \tan x, \quad 0 < x < \frac{\pi}{2}$$

$$\left(D^2 + 1\right) y = \tan x = \frac{\sin x}{\cos x}$$

There is no combination of derivatives that will give the appropriate multiple of sin x and cos x to drive the RHS to zero. Therefore, a more general method is needed.

## 3.4.1. Method of Variation of Parameters

Given the equation

$$y'' + p(x)y' + q(x)y = g(x) \tag{3.4.1}$$

a general method known as the method of *variation of parameters* where p, q, and g are continuous on the interval of interest can be used to find particular solutions of the nonhomogeneous differential equation.

To use this method we must know a fundamental set of solutions (for the second order case $y_1$ and $y_2$) for the homogeneous equation.

In the homogeneous case, we found that

$$y_H = c_1 y_1 + c_2 y_2$$

where $y_1$ and $y_2$ are the fundamental solutions. Then if we replace $c_1$ and $c_2$ with two functions, $u_1$ and $u_2$, such that a candidate for the particular solution

$$y_p = u_1(x)y_1 + u_2(x)y_2 \tag{3.4.11}$$

satisfies the nonhomogeneous Equation 3.4.1 (two conditions on $u_1$ and $u_2$ are required), then

$$y_p' = (u_1' \, y_1 + u_2' \, y_2) + (u_1 y_1' + u_2 y_2') \tag{3.4.12}$$

in order to simplify the computation, let

$$u_1' \, y_1 + u_2' \, y_2 = 0 \tag{3.4.13}$$

which is one condition on $u_1$ and $u_2$, thus

$$y_p' = u_1 y_1' + u_2 y_2' \text{ and } y_p'' = u_1' \, y_1' + u_1 y_1'' + u_2' \, y_2' + u_2 y_2''$$

substitute into Equation 3.4.1 gives

$$u_1' y_1' + u_1 y_1'' + u_2' y_2' + u_2 y_2'' + p(u_1 y_1' + u_2 y_2') + q(u_1 y_1 + u_2 y_2) = g$$

$$u_1(y_1'' + p \, y_1' + q \, y_1) + u_2(y_2'' + p \, y_2' + qy_2) + u_1' \, y_1' + u_2' \, y_2' = g$$

but

$$y_1'' + p \, y_1' + q \, y_1 = 0$$

$$y_2'' + p \, y_2' + qy_2 = 0$$

both of which satisfy the homogeneous equation, therefore

$$u_1' \, y_1' + u_2' \, y_2' = g$$

is the second condition on $u_1$ and $u_2$. That is, for second order the two conditions that must be met are:

$$u_1'y_1 + u_2'y_2 = 0$$

$$u_1' y_1' + u_2' y_2' = g$$

for third order the three conditions are:

$$u_1' y_1 + u_2' y_2 + u_3' y_3 = 0$$

$$u_1' y_1' + u_2' y_2' + u_3' y_3' = 0$$

$$u_1' y_1'' + u_2' y_2'' + u_3' y_3'' = g$$

As a demonstration of the procedure, consider the following examples.

## Example 3.4.5

$y'' - 5y' + 6y = 2e^x$
consider $y'' - 5y' + 6y = 0$
operator form $(D^2 - 5D + 6)y = 0$
then $(D - 3)(D - 2) = 0$
thus $y_c = c_1 e^{3x} + c_2 e^{2x}$

$$\text{let } y_p = u_1 e^{3x} + u_2 e^{2x}$$

$$u_1' e^{3x} + u_2' e^{2x} = 0$$

$$3u_1' e^{3x} + 2u_2' e^{2x} = 2e^x$$

$$\therefore u_1' = \frac{\begin{vmatrix} 0 & e^{2x} \\ 2e^x & 2e^{2x} \end{vmatrix}}{\begin{vmatrix} e^{3x} & e^{2x} \\ 3e^{3x} & 2e^{2x} \end{vmatrix}} = \frac{-2e^{3x}}{2e^{5x} - 3e^{5x}}$$

$$\therefore u_1' = 2e^{-2x}$$

$$u_1 = 2\int e^{-2x} dx = 2\left(-\frac{1}{2}e^{-2x}\right) = -e^{-2x}$$

and

$$u_2' = -\frac{u_1' e^{3x}}{e^{2x}} = \left(-2e^{-2x}\right)e^{3x}e^{-2x} = -2e^{-x}$$

such that $u_2 = 2e^{-x}$. Therefore

$$y_p = -e^{-2x}e^{3x} + 2e^{-x}e^{2x} = -e^x + 2e^x = e^x$$

and

$$y_g = c_1 e^{3x} + c_2 e^{2x} + e^x$$

## Example 3.4.6

Find a particular solution of $y'' + y = \tan x$

*Solution*

$$\left(D^2 + 1\right)y = \tan x$$

$$y_c = c_1 \cos x + c_2 \sin x$$

$$\therefore \quad y_1 = \cos x, \ y_2 = \sin x, \ y_1' = -\sin x, \ y_2' = \cos x$$

the candidate for $y_p$ is

$$y_p = u_1(x)y_1 + u_2(x)y_2$$

then

$$u_1' \cos x + u_2' \sin x = 0$$

$$-u_1' \sin x + u_2' \cos x = \frac{\sin x}{\cos x}$$

$$\therefore \quad u_1' = \frac{\begin{vmatrix} 0 & \sin x \\ \dfrac{\sin x}{\cos x} & \cos x \end{vmatrix}}{\begin{vmatrix} \cos x & \sin x \\ -\sin x & \cos x \end{vmatrix}} = \frac{-\sin^2 x}{\cos x}$$

and

$$u_1(x) = -\int \frac{\sin^2 x}{\cos x}\, dx$$

also

$$u_2' = \frac{-u_1' \cos x}{\sin x} = \sin x$$

such that

$$u_2(x) = -\cos x$$

$$\therefore \quad y_p - \cos x \int \frac{\sin^2 x}{\cos x}\, dx - \sin x \cos x$$

As is evident in Example 3.4.6, challenging integration problems can arise when the variation of parameter method is used.

---

## 3.5   Variable Coefficient Problems

Both of the previously discussed methods are applicable to constant coefficient as well as variable coefficient problems. However, for some variable coefficient problems, substitutions are possible that will reduce variable coefficient problems to constant coefficient ones.

One such class of variable coefficient problems is the *Euler* or *Equidimensional* differential equation

$$x^2 y'' + bxy' + cy = 0 \tag{3.5.1}$$

for the second order case in which b and c are constants. If one lets

$$x = e^z \tag{3.5.2}$$

then

$$\frac{dy}{dx} = e^{-z} \frac{dy}{dz} \tag{3.5.3}$$

and

$$\frac{d^2 y}{dx^2} = e^{-2z} \left( \frac{d^2 y}{dz^2} - \frac{dy}{dz} \right) \tag{3.5.4}$$

where Equation 3.5.3 and Equation 3.5.4 are derived by applying the chain rule. By substituting Equation 3.5.2, 3.5.3, and 3.5.4 into Equation 3.5.1 gives

$$\frac{d^2 y}{dz^2} + (b-1) \frac{dy}{dz} + cy = 0 \tag{3.5.5}$$

a constant coefficient equation.

For the case

$$y'' + p(x) y' + q(x) y = 0 \tag{3.5.6}$$

a change of independent variable can sometimes be made that will transform Equation 3.5.6 into a constant coefficient problem. That is, if

$$z = \int^{x} \left[ q(t) \right]^{1/2} dt \qquad (3.5.7)$$

then

$$\frac{dy}{dx} = \frac{dz}{dx}\frac{dy}{dz}$$

and

$$\frac{d^2y}{dx^2} = \left(\frac{dz}{dx}\right)^2 \frac{d^2y}{dz^2} + \frac{d^2z}{dx^2}\frac{dy}{dx}$$

transform Equation 3.5.6 into

$$\left(\frac{dz}{dx}\right)^2 \frac{d^2y}{dz^2} + \left(\frac{d^2z}{dx^2} + p(x)\frac{dz}{dx}\right)\frac{dy}{dz} + q(x)y = 0 \qquad (3.5.8)$$

provided that

$$\frac{\left[q'(x) + 2p(x)q(x)\right]}{2\left[q(x)\right]^{3/2}} = a\ cons\tan t \qquad (3.5.9)$$

If the procedure of changing the independent variable is too work intensive, then a simple substitution can be made for the Euler equation case. That
is, if

$$y = x^r \qquad (3.5.10)$$

where r is a constant to be determined, then

$$y' = rx^{r-1}$$

and

$$y'' = r(r-1)x^{r-2}$$

can be substituted into Equation 3.5.1 to get

$$\left[r(r-1) + br + c\right]x^r = 0 \qquad (3.5.11)$$

Then for any interval not containing the origin, the following statements hold:

$$y = c_1|x|^{r_1} + c_2|x|^{r_2} \tag{3.5.12}$$

for real and unequal roots

$$y = \left(c_1 + c_2 \ln(x)\right)|x|^{r_1} \tag{3.5.13}$$

for real and equal roots

$$y = |x|^{\lambda}\left[c_1 \cos\left(\mu \ln|x|\right) + c_2 \sin\left(\mu \ln|x|\right)\right] \tag{3.5.14}$$

where $r_1, r_2 = \lambda \pm i\mu$ are the complex roots.

For the class of problems for which convenient simplifying substitutions are not available, infinite series methods may be successfully applied. The next section discusses such class of variable coefficient problems.

### 3.5.1 Series Solutions Near a Regular Singular Point

Consider the variable coefficient, linear second order and homogeneous differential equation

$$P(x)y'' + Q(x)y' + R(x)y = 0 \tag{3.5.15}$$

in the neighborhood of a *regular singular* point $x = x_0$. By regular singular point (as opposed to irregular) we mean that both

$$\lim_{x \to x_0} (x - x_0) \frac{Q(x)}{P(x)} \text{ is finite} \tag{3.5.16}$$

and

$$\lim_{x \to x_0} (x - x_0)^2 \frac{R(x)}{P(x)} \text{ is finite} \tag{3.5.17}$$

Then one can assume a solution to Equation 3.5.15 of the form

$$y = \sum_{n=0}^{\infty} a_n x^{n+r} \tag{3.5.18}$$

where the values of r are to be determined and a relation for the $a_n$ is to be established. The singular points of Equation 3.5.15 are exactly those for which $P(x) = 0$, if P, Q, and R are polynomials without common factors. For a more in-depth discussion of this theory, standard differential equation text (see Boyce and DiPrima[1] and O'Neil[4]) should be consulted.

Presented below is an example that illustrates the essential steps needed to successfully solve Equation 3.5.15 using Equation 3.5.18.

Consider the differential equation

$$2x^2y'' - xy' + (1+x)y = 0$$

by comparison to Equation 3.5.15

$$P(x) = 2x^2, \quad Q(x) = -x, \quad R(x) = 1+x$$

then

$$\lim_{x \to 0} x \left( \frac{-x}{2x^2} \right) = -\tfrac{1}{2}$$

and

$$\lim_{x \to 0} x^2 \left( \frac{1+x}{2x^2} \right) = \tfrac{1}{2}$$

are both finite; therefore, $x = 0$ is a regular singular point.

Let

$$y = a_0x^r + a_1x^{1+r} + a_2x^{2+r} + \ldots + a_nx^{n+r} + \ldots = \sum_{n=0}^{\infty} a_nx^{n+r} \qquad (3.5.18)$$

then

$$y' = \sum_{n=0}^{\infty} (n+r)a_nx^{n+r-1}$$

and

$$y'' = \sum_{n=0}^{\infty} (n+r)(n+r-1)a_nx^{n+r-2}$$

thus, the given differential equation

$$2x^2y'' - xy' + (1+x)y = 0$$

becomes the identity

$$2x^2 \sum_{n=0}^{\infty}(n+r)(n+r-1)a_nx^{n+r-2} - x\sum_{n=0}^{\infty}(n+r)a_nx^{n+r-1} + (1+x)\sum_{n=0}^{\infty}a_nx^{n+r} = 0$$

and can be recasted as

$$\sum_{n=0}^{\infty}2(n+r)(n+r-1)a_nx^{n+r} - \sum_{n=0}^{\infty}(n+r)a_nx^{n+r} + \sum_{n=0}^{\infty}a_nx^{n+r}$$

$$(3.1.19)$$

$$+ \sum_{n=0}^{\infty}a_nx^{n+r+1} = 0$$

but

$$\sum_{n=0}^{\infty}a_nx^{n+r+1} = \sum_{n=1}^{\infty}a_{n-1}x^{n+r}$$

by shifting the index of summation (O'Neil[4]) or replacing n by n – 1 everywhere in the fourth term of Equation 3.5.19. Therefore, Equation 3.5.19 becomes

$$\sum_{n=0}^{\infty}\left[(n+r)(2n+2r-3)+1\right]a_nx^{n+r} + \sum_{n=1}^{\infty}a_{n-1}x^{n+r} = 0$$

after factoring. By expanding the summation for n = 0, we get

$$\left[(r)(2r-3)+1\right]a_0x^r + \sum_{n=1}^{\infty}\left\{\left[(n+r)(2n+2r-3)+1\right]a_n + a_{n-1}\right\}x^{n+r} =$$

for which $a_0$ is arbitrary and $x \neq 0$. Therefore,

$$r(2r-3)+1=0 \qquad\qquad (3.5.20)$$

and

$$\left[(n+r)(2n+2r-3)+1\right]a_n + a_{n-1} = 0 \quad \text{for } n \geq 1 \tag{3.5.21}$$

are the so-called *indicial equation* and *recurrence relationship,* respectively.

The roots of Equation 3.5.20 are $r_1 = 1$ and $r_2 = 1/2$.

Then the *recurrence relation* for $r_1$ is given by

$$a_n = \frac{-1}{n(2n+1)} a_{n-1}, \quad n \geq 1 \tag{3.5.22}$$

and for $r_2$, by

$$b_n = \frac{-1}{n(2n-1)} b_{n-1}, \quad n \geq 1 \tag{3.5.23}$$

where the switch from $a_n$ to $b_n$ is introduced to reduce confusion.

A few constants for Equation 3.5.22 and Equation 3.5.23 are evaluated as follows:

$$n = 1, \quad a_1 = \frac{-a_0}{1(3)}; \quad b_1 = \frac{-b_0}{1(1)}$$

$$n = 2, \quad a_2 = \frac{-a_1}{2(5)} = \frac{a_0}{1 \cdot 2 \cdot 3 \cdot 5}; \quad b_2 = \frac{-b_1}{2(3)} = \frac{b_0}{3 \cdot 2 \cdot 1}$$

$$n = 3, \quad a_3 = \frac{-a_2}{3(7)} = \frac{-a_0}{2 \cdot 3^2 \cdot 5 \cdot 7}; \quad b_3 = \frac{-b_2}{3(5)} = \frac{-b_0}{5 \cdot 3^2 \cdot 2 \cdot 1}$$

Both sets of constants exhibit a pattern that can be generalized to

$$a_n = \frac{(-1)^n}{\left[(2n+1)(2n-1)\cdots 5 \cdot 3\right]n!} a_0, \quad n \geq 1 \tag{3.5.24}$$

in the case of $r_1$, and

$$b_n = \frac{(-1)^n}{\left[(2n-1)(2n-3)\cdots 3 \cdot 1\right]n!} b_0, \quad n \geq 1 \tag{3.5.25}$$

in the case of $r_2$. Therefore, the two linearly independent solutions that are expected are:

$$y = a_0 x + a_1 x^2 + a_2 x^3 + \cdots + a_n x^{n+1} + \cdots = \sum_{n=0}^{\infty} a_n x^{n+1}$$

$$= a_0 x - \frac{a_0 x^2}{3} + \frac{a_0 x^3}{5 \cdot 3 \cdot 2} - \frac{a_0 x^4}{7 \cdot 5 \cdot 3^2 \cdot 2} + \cdots$$

$$= x \left\{ 1 + \sum_{n=0}^{\infty} \frac{(-1)^n x^n}{\left[ (2n+1)(2n-1) \cdots 5 \cdot 3 \right] n!} \right\} a_0$$

for the $r_1$ case, please elimate that is,

$$y_1 = x \left\{ 1 + \sum_{n=0}^{\infty} \frac{(-1)^n x^n}{\left[ (2n+1)(2n-1) \cdots 5 \cdot 3 \right] n!} \right\} a_0 \qquad (3.5.26)$$

and the $r_2$ case gives

$$y = b_0 x - b_0 x^2 + \frac{b_0 x^3}{3 \cdot 2 \cdot 1} - \frac{b_0 x^4}{5 \cdot 3^2 \cdot 2} + \cdots$$

$$= x^{\frac{1}{2}} \left\{ 1 + \sum_{n=1}^{\infty} \frac{(-1)^n x^n}{\left[ (2n-1)(2n-3) \cdots 3 \cdot 1 \right] n!} \right\} b_0$$

such that

$$y_2 = x^{\frac{1}{2}} \left\{ 1 + \sum_{n=1}^{\infty} \frac{(-1)^n x^n}{\left[ (2n-1)(2n-3) \cdots 3 \cdot 1 \right] n!} \right\} \qquad (3.5.27)$$

is the second linearly independent solution of the given differential equation.

Before the general solution can be stated, it is important to determine if the series obtained in Equation 3.5.26 and Equation 3.5.27 are convergent. One way is to apply a convergence test (Thomas and Finney[14]). For example, the *ratio test*

$$\lim_{n \to \infty} \left| \frac{a_{n+1} x^{n+1}}{a_n x^n} \right|$$

gives

$$\lim_{n\to\infty} \frac{|x|\left[(2n+1)(2n-1)\cdots5\cdot3\right]n!}{\left[2(n+1)+1(2(n+1)-1)\cdots5\cdot3\right](n+1)!}$$

$$= \lim_{n\to\infty} \frac{|x|}{(2n+3)(n+1)} = 0$$

Therefore, the series Equation 3.5.26 converges for all x. It can also be shown that Equation 3.5.27 converges for all x. Then the general solution is given by

$$y = a_0 y_1 + b_0 y_2$$

and the reader should verify that $y_1$ and $y_2$ are, in fact, linearly independent. There are exceptional cases ( O'Neil,[4] Mickley, Sherwood and Reid[6] ) that occur, and the following theorem is useful in dealing with those cases.

*THEOREM 3.7*
*Let $r_1$ and $r_2$ be roots of the indicial equation $x^2y'' + x[xq(x)]y' + x^2g(x)y = 0$, which is assumed to have a regular singular point at the origin. Then*

1. *If $r_1 \neq r_2$ and $r_1 - r_2$ is not an integer, then these two are linearly independent solutions*

$$y_1 = \sum_{n=0}^{\infty} a_n x^{n+r_1} \quad \text{and} \quad y_2 = \sum_{n=0}^{\infty} b_n x^{n+r_2}, \quad x>0$$

2. *If $r_1 - r_2$ is a positive integer, then there are linearly independent solutions of the form*

$$y_1 = \sum_{n=0}^{\infty} a_n x^{n+r_1} \quad \text{and} \quad y_2 = Ay_1 \ln(x) + \sum_{n=0}^{\infty} b_n x^{n+r_2}, x>0$$

*where A is a constant that may turn out to be zero.\**
3. *If $r_1 = r_2$, then there are linearly independent solutions*

$$y_1 = \sum_{n=0}^{\infty} a_n x^{n+r_1} \quad \text{and} \quad y_2 = y_1 \ln(x) + \sum_{n=0}^{\infty} b_n x^{n+r_1}, \quad x>0$$

\* Sometimes the smaller root will give the general solution or no solution.

## 3.6   Alternative Methods

So far, the methods discussed can be classified as standard. However, there are other techniques, some of which may even be applicable to a few nonlinear problems. For example, the problem

$$y'' + x(y')^2 = 0$$

can be solved by making the change of variable

$$y' = v, \ y'' = v'$$

such that the problem is reduced to

$$v' + xv^2 = 0$$

a separable first order differential equation. Here one takes advantage of the missing dependent variable, y, in

$$y'' = f(x, y')$$

Another example of an otherwise difficult problem is

$$yy'' + (y')^2 = 0$$

Here the independent variable, x, is missing from the equation. By making the substitutions $v = y'$ and using the chain rule

$$\frac{dv}{dx} = \frac{dv}{dy} \cdot \frac{dy}{dx} = v \frac{dv}{dy}$$

a new independent variable, y, can be defined. The problem is now reduced to

$$yv \frac{dv}{dy} + v^2 = y \frac{dv}{dy} + v = 0$$

a linear first order equation. Hence, by taking advantage of the missing independent variable, x, in

$$y'' = f(y, y')$$

can reduce the second order problem to a first order one that we know how to solve.

In engineering, prototype differential equations often occur as a result of some peculiarity of the system under investigation. Then the approach is to compare one's derived problem with a prototype and extract and modify the pertinent result. For example, the problem

$$\frac{1}{x}\frac{d}{dx}\left(x\frac{dy}{dx}\right)-\left(1-x^2\right)\lambda y = 0 \tag{3.6.1}$$

can be transformed into

$$y\frac{d^2f}{dy^2}+\left(1-y\right)\frac{df}{dy}-\left(\frac{1}{2}-\frac{\lambda}{4}\right)f = 0 \tag{3.6.2}$$

Equation 3.6.2 is the confluent hypergeometric equation (the prototype) with linearly independent solutions tabulated in the literature (Slater[12]) whereas Equation 3.6.1 resulted from a fluid flow problem with a parabolic velocity profile.

Another example of a prototype equation that should be familiar to most chemical engineers is Bessel's differential equation of order n

$$x^2y''+xy'+\left(x^2-n^2\right)y = 0 \tag{3.6.3}$$

There are tabulated solutions of Bessel's equation (Mickley, Sherwood, and Reid,[6] Watson,[9] Gray and Mathews,[10] and Tratner[11]), and the standard approach is to solve by comparison. Equations 3.6.1 to 3.6.3 could be solved by using the Frobenius series Equation 3.5.17, but that approach may be too work intensive.

In order to solve Bessel's differential equation by comparison one needs a standard form of the equation to compare to. For example, the form

$$y''-\left(\frac{2a-1}{x}\right)y'+\left(b^2c^2x^{2c-2}+\frac{a^2-\gamma^2c^2}{x^2}\right)y = 0 \tag{3.6.4}$$

has linearly independent solutions

$$y_1 = x^a J_\gamma\left(bx^c\right) \tag{3.6.5}$$

and

$$y_2 = x^a J_{-\gamma}\left(bx^c\right) \tag{3.6.6}$$

based on the fact that $J_\gamma$ is a solution of Bessel's equation of order $\gamma$. Thus, the problem

$$y'' + \frac{1}{x}y' + \left(4x^2 - \frac{4}{9x^2}\right)y = 0$$

can be compared to Equation 3.6.4 to give

$$2a - 1 = -1 \Rightarrow a = 0$$

$$2c - 2 = 2 \Rightarrow c = 2$$

$$b^2c^2 = 4 \Rightarrow b = \pm1$$

$$a^2 - \gamma^2 c^2 = -\frac{4}{9} \Rightarrow \gamma = \pm\frac{1}{3}$$

$$y_1 = J_{\frac{1}{3}}(x^2), \quad y_2 = J_{-\frac{1}{3}}(x^2)$$

$$y_g = c_1 J_{\frac{1}{3}}(x^2) + c_2 J_{-\frac{1}{3}}(x^2)$$

where $J_\gamma(\bullet)$ and $J_{-\gamma}(\bullet)$ are *Bessel functions of the first kind of order* $\gamma$.

*Summary:*
In this chapter, a few methods were presented for obtaining a solution to the linear second order (or higher) ordinary differential equations. To the inexperienced practitioner these many options could present a dilemma; that is, given a problem, which method should one use?

For example, consider the constant coefficient linear differential equation

$$y'' + y' = xe^{-x}\sin 3x$$

with the intent to determine a general solution. One could attempt to annihilate the nonhomogeneous portion, but that would require some clever algebra to be successful without a lot of labor. Therefore, it is not the recommended procedure for this problem.

A second alternative is,
let $v = y'$, then $v' = y''$ and we now have

$$v' + v = xe^{-x}\sin 3x$$

a linear first order differential equation, which can be solved by the method given in the previous chapter for such equation. That is

$$\mu(x) = e^x$$

$$\left(v\mu(x)\right)' = \left(ve^x\right)' x \sin 3x$$

such that

$$v = \frac{dy}{dx} = -\frac{x}{3}e^{-x}\cos 3x + \frac{1}{9}e^{-x}\sin 3x + c_1 e^{-x}$$

Therefore,

$$y = \frac{1}{3}\int xe^{-x}\cos 3x\,dx + \frac{1}{9}\int e^{-x}\sin 3x\,dx + c_1 e^{-x} + c_2$$

Where the integration (arbitrary) constant $c_1$ absorbs the change in sign resulting from an integration of $e^x$. With the aid of a good set of integral tables (Spiegel[13]), the first term on the right-hand side can be quickly evaluated. After some simplification the result is

$$y_g = \underbrace{\frac{1}{150}(5x-9)e^{-x}\cos 3x - \frac{1}{450}(45x+14)e^{-x}\sin 3x}_{y_p} + \underbrace{c_1 e^{-x} + c_2}_{y_c}$$

As a third alternative

$$y'' + y' = xe^{-x}\sin 3x$$

can be solved using the method of variation of parameters. That is

$$y_c = c_1 e^{-x} + c_2$$

where

$$y_1 = e^{-x} \quad \text{and} \quad y_2 = 1$$

$$y_1' = -e^{-x} \quad \text{and} \quad y_2' = 0$$

Then a candidate for the particular solution is

$$y_p = u_1(x)y_1 + u_2(x)y_2$$

subject to

$$u_1'y_1 + u_2'y_2 = 0$$

$$u_1'y_1' + u_2'y_2' = xe^{-x}\sin 3x$$

or

$$e^{-x}u_1' + u_2' = 0$$

$$-e^{-x}u_1' = xe^{-x}\sin 3x$$

such that

$$u_1 = -\int x\sin 3x\,dx = \frac{x}{3}\cos 3x - \frac{\sin 3x}{9}$$

and

$$u_2 = \int xe^{-x}\sin 3x\,dx = -\frac{e^{-x}}{100}(6\cos 3x - 8\sin 3x) + \frac{xe^{-x}}{10}(-\sin 3x - 3\cos 3x)$$

where this integral was evaluated using a set of tables (Spiegel[13]). Therefore

$$y_p = e^{-x}\left(\frac{x}{3}\cos 3x - \frac{\sin 3x}{9}\right) - \frac{xe^{-x}}{10}(\sin 3x + 3\cos 3x) + \frac{e^{-x}}{100}(8\sin 3x - 6\cos 3x)$$

which simplifies to

$$y_p - \frac{1}{150}(5x-9)e^{-x}\cos 3x - \frac{1}{450}(45x+14)e^{-x}\sin 3x$$

From a cursory glance, it appears that both the second and third alternates are equivalent in expediency; however, the underlying algebra is more work intensive for the second alternate. Therefore, one should make a choice based on all the factors, including the amount of time that can be spent on the overall problem. Bear in mind that the more steps that have to be executed, the greater the chance to introduce errors in an otherwise complicated analysis.

## 3.6.1 Initial Value Problems

Theorem 3.1 guarantees a unique solution to the initial value problem, and the subsequent methods that are discussed can be used to derive general solutions. Given a general solution, the integration constants can be evaluated with the use of the given initial conditions.

Another procedure, which transforms a differential equation into an algebraic equation, is given below. This is the Laplace transform method. For example, the problem

$$y'' - y' - 2y = 0 \quad y(0) = 1, \quad y'(0) = 0$$

can be solved using Laplace transform as follows:

$$L\{y'' - y' - 2y\} = L\{y''\} - L\{y'\} - 2L\{y\} = 0$$

giving

$$s^2 Y(s) - sy(0) - y'(0) - sY(s) + y(0) - 2Y(s) = 0$$

where

$$L\{y\} = Y(s)$$

Then, following simplification, we get

$$Y(s) = \frac{s-1}{(s-2)(s+1)}$$

Using partial fractions

$$= \frac{\frac{1}{3}}{s-2} + \frac{\frac{2}{3}}{s+1}$$

In order to recover the result in terms of the variables with which we started, an inversion of $Y(s)$ is needed.

The *inverse Laplace transform*, $L^{-1}\{y(s)\}$ is given as

$$\frac{1}{3} L^{-1}\left\{\frac{1}{s-2}\right\} = \frac{1}{3} e^{2t}$$

and

$$\frac{2}{3} L^{-1}\left\{\frac{1}{s+1}\right\} = \frac{2}{3} e^{-t}$$

Therefore, the complete solution to the given initial value problem is

$$y(t) = \frac{1}{3}e^{2t} + \frac{2}{3}e^{-t} = L^{-1}\{Y(s)\}$$

This method is very efficient when it is applicable, and is one of many *integral transformations* that are useful for solving initial value problems. Integral transforms are certain functions that are defined by an integral, such as

$$F(s) = \int_{\alpha}^{\beta} K(s,t)f(t)dt \tag{3.6.7}$$

where the function f(t) is transformed into F(s) and K(s, t) is called the *kernel* of the transform. In this case, the Laplace transform of f(t) is defined to be

$$L\{f(t)\} = F(s) = \int_{0}^{\infty} e^{-st} f(t)dt \tag{3.6.8}$$

where $e^{-st}$ is the kernel. Laplace transform is most efficient in solving problems with nonhomogeneous terms that are discontinuous or impulsive. These types of problems would at best be awkward if the previously discussed methods were attempted. When can this method be expected to work successfully? Below are some definitions that address this issue.

*THEOREM 3.8*
*Suppose that f is piecewise continuous on the interval $0 \leq t \leq A$ for any positive A and $|f(t)| \leq ke^{at}$ when $t \geq M$, where k, a, and M are real constants. Then the Laplace transform $L\{f(t)\} = F(s)$ exists for $s > a$.*

By *piecewise continuous* one means, given a function on an interval $\alpha \leq t$ $t \leq \beta$, where the interval can be subdivided by a finite number of points in the following way, $\alpha = t_0 < t_1 < \ldots < t_m = \beta$ such that f is continuous on each open subinterval $t_{i-1} < t < t_i$ and f approaches a finite limit as the endpoints of each subinterval are approached from inside. Essentially, Theorem 3.8 says that if a function is piecewise continuous and is exponentially bounded, then one can expect it to have a Laplace transform. The theory of Laplace transform is well documented (Boyce and Diprima,[1] Churchill and Brown,[16] Jenson and Jeffrey,[15] Mickley, Sherwood, and Reid,[6] and O'Neil[4]) and should be consulted for a deeper discussion. Also available are convenient tables of Laplace transforms (Spiegel[13]).

Equation 3.6.8 can be used to determine Laplace transforms of derivatives provided that the derivatives satisfy the appropriate conditions of continuity and boundedness as required by Theorem 3.8. For example

$$L\{y'(t)\} = \int_0^\infty e^{-st} y'(t)dt$$

By definition, the improper integral is

$$\lim_{B\to\infty} \int_0^B e^{-st} y'(t)dt$$

then applying integration by parts:

let

$$u = e^{-st} \quad \text{and} \quad dv = y'(t)dt$$

then

$$du = -se^{-st}dt \quad \text{and} \quad v = y(t)$$

therefore

$$\int_0^B e^{-st} y'(t)dt = \left[ e^{-st}y(t) + s\int e^{-st}y(t)dt \right]_0^B$$

$$= e^{-sB}y(B) - y(0) + s\int_0^B e^{-st}y(t)dt$$

then

$$\lim_{n\to\infty} \int_0^B e^{-st} y'(t)dt = \lim_{B\to\infty} \left\{ e^{-sB}y(B) - y(0) + s\int_0^B e^{-st}y(t)dt \right.$$

$$= -y(0) + s\int_0^B e^{-st}y(t)dt \qquad \text{(I)}$$

$$= -y(0) + sY(s)$$

where Equation 3.6.8 was used in Equation I. This integration procedure can be extended to finding the Laplace transform of an $n^{th}$ order derivative of a given function. That is

$$L\left\{f^{(n)}\left(t\right)\right\}=s^{n}L\left\{f(t)-s^{n-1}f(0)-\cdots-sf^{(n-2)}\left(0\right)-f^{(n-1)}\left(0\right)\right. \quad (3.6.9)$$

and for n = 2

$$L\left\{\frac{d^{2}f}{dt^{2}}\right\}=s^{2}F(s)-sf(0)-f'(0) \quad (3.6.10)$$

which is the Laplace transform of the second derivative that was used in the above example. As mentioned previously, Equation 3.6.9 can be found in tables of Laplace Transform in most mathematical handbooks (Spiegel[13]).

Equation 3.6.8 can also be used to derive the Laplace transforms of most elementary functions such as:

$$L\{1\}=\frac{1}{s}$$

$$L\{t\}=\frac{1}{s^{2}}=\frac{1}{s^{1+1}}$$

$$L\left\{t^{2}\right\}=\frac{2}{s^{3}}=\frac{2}{s^{2+1}}$$

$$L\left\{t^{n}\right\}=\frac{n!}{s^{n+1}};\quad n=1,2,\dots$$

Now suppose n is not a positive integer; for example, what is $L\left\{t^{-\frac{1}{2}}\right\}$? Since for any n,

$$L\left\{t^{n}\right\}=\int_{0}^{\infty}t^{n}e^{-st}dt,\quad s>0$$

Letting $\qquad\qquad u=st;\quad du=s\,dt$ gives

$$\int_{0}^{\infty}t^{n}e^{-st}dt=\int_{0}^{\infty}\left(\frac{u}{s}\right)^{n}e^{-u}\frac{du}{s}=\int_{0}^{\infty}\frac{u^{n}e^{-u}du}{s^{n+1}}$$

$$=\frac{1}{s^{n+1}}\int_{0}^{\infty}e^{-u}u^{n}du$$

If we denote $\displaystyle\int_{0}^{\infty}e^{-u}u^{n}\,du$ by $\Gamma(n+1)$, that is

$$\int_0^\infty e^{-u}u^n du = \Gamma(n+1),$$                                                           (3.6.11)

then L{$t^n$} becomes $\dfrac{\Gamma(n+1)}{s^{n+1}}$ where $\Gamma(n+1)$ is the so-called *gamma function* or

*factorial function*. Also, the quantity $\int_0^\infty e^{-u}u^n\, du$ can be evaluated using integration by parts to give

$$\int_0^\infty e^{-u}u^n du = -u^n e^{-u}\Big|_0^\infty + n\int_0^\infty e^{-u}u^{n-1}du$$

$$= n\Gamma(n)$$

Therefore

$$\Gamma(n+1) = n\Gamma(n); \quad n \neq 0.$$                                                            (3.6.12)

However, for n = 0, Equation 3.6.11 gives

$$\Gamma(1) = \int_0^\infty e^{-u}du = -e^{-u}\Big|_0^\infty = 1$$

Hence, 0! = 1.

The gamma function, or factorial function, can now be evaluated for other values of n using Equation 3.6.12. For example

$$n = 1: \ \Gamma(2) = 1\Gamma(1) = 1$$

$$n = 2: \ \Gamma(3) = 2\Gamma(2) = 2\bullet 1$$

$$n = 3: \ \Gamma(4) = 3\Gamma(3) = 3\bullet 2\Gamma(2) = 3\bullet 2\bullet 1$$

and in general

$$\Gamma(n+1) = n!$$

Now then, for n = $-1/2$,

$$\Gamma(-1/2+1)=\Gamma(1/2)=\int_0^\infty e^{-u}u^{-1/2}du \ = \ I$$

Thus, by substituting $u = x^2$, $du = 2x\,dx$, **I** becomes

$$I=\Gamma(1/2)=\int_0^\infty e^{-u}u^{-1/2}du \ = \int_0^\infty \frac{1}{x}e^{-x^2}2x\,dx = 2\int_0^\infty e^{-x^2}dx$$

In order to evaluate **I**, the following device is employed

$$I^2 = \left(2\int_0^\infty e^{-x^2}dx\right)\left(2\int_0^\infty e^{-y^2}dy\right)$$

$$= 4\int_0^\infty \int_0^\infty e^{-(x^2+y^2)}dx\,dy$$

But $x = r\cos\theta$, $y = r\sin\theta$ $dxdy = \dfrac{\partial(x,y)}{\partial(r,\theta)}dr\,d\theta = \begin{vmatrix} \dfrac{\partial x}{\partial r} & \dfrac{\partial y}{\partial r} \\ \dfrac{\partial x}{\partial \theta} & \dfrac{\partial y}{\partial \theta} \end{vmatrix} dr\,d\theta = r\,dr\,d\theta$

$$\therefore I^2 = 4\int_0^{\pi/2}\left(\int_0^\infty e^{-r^2}r\,dr\right)d\theta = \pi$$

such that

$$I=\Gamma(1/2)=\sqrt{\pi}\,,$$

and

$$L\{t^{-1/2}\}=\frac{\sqrt{\pi}}{s^{1/2}}=\sqrt{\frac{\pi}{s}}\,.$$

### 3.6.2  Some Useful Properties of Laplace Transforms

1. If the Laplace transform of $f(t)$ is $F(s)$, then

$$L\{e^{-at}f(t)\}=F(s+a)$$

where a is a constant. For example

$$L\{e^{-at}\cos bt\}=?$$

from a table of Laplace transforms

$$L\{\cos bt\}=\frac{s}{(s^2+b^2)}$$

therefore

$$L\{e^{-at}\cos bt\}=\frac{s+a}{\left[(s+a)^2+b^2\right]}$$

2. The transform of an integral is given by

$$L\left\{\int_0^t f(t)dt\right\}=\frac{1}{s}F(s)$$

3. The derivative of a transform is given by

$$\frac{d^n F(s)}{ds^n}=L\{(-t)^n f(t)\}$$

4. Transform of a step function: If U(t) is a unit step function

$$U(t)=0 \quad t<0$$

$$=1 \quad t>0$$

$$(3.6.13)$$

and if U(t – τ) is a unit step function starting at t = τ

$$U(t-\tau)=0 \quad t<\tau$$

$$=1 \quad t>\tau$$

$$(3.6.14)$$

then

$$L\{U(t)\}=\frac{1}{s}$$

and

$$L\{U(t-\tau)\}=\frac{e^{s\tau}}{s}$$

$$L\{f(t-\tau)\}=e^{-\tau s}F(s); \quad \text{if} \quad f(t-\tau)=0 \quad \text{for} \quad 0<t<\tau$$

and

$$L\{f(t-\tau)U(t)\}=e^{-\tau s}F(s)$$

5. Transform of an impulse function: The Dirac delta or unit impulse function is given as

$$\delta_\varepsilon(t-t_0)=\begin{cases} 0 \text{ if } t<t_0-\varepsilon \\ \dfrac{1}{2\varepsilon} \text{ if } t_0-\varepsilon \le t_0+\varepsilon \\ 0 \text{ if } t \ge t_0+\varepsilon \end{cases}$$

where $\varepsilon > 0$ is a small number and $t_0 > 0$.

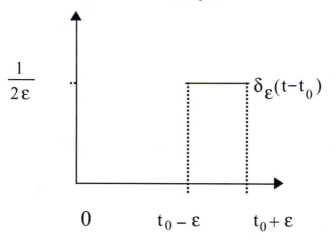

As can be seen from the sketch, as $\varepsilon \to 0$, the rectangular pulse $\delta_\varepsilon(t-t_0)$ gets taller and thinner, but the limit does not exist. Even though the limit does not exist in the usual sense, one can still derive useful properties of the impulse function. For example,

(1)
$$\int_{-\infty}^{\infty}\delta_\varepsilon(t-t_0)\ dt=\int_{t_0-\varepsilon}^{t_0+\varepsilon}\frac{1}{2\varepsilon}dt=1$$

since $\delta_\varepsilon(t - t_0) = 0$ for all t outside of the interval $t_0 - \varepsilon \le t \le t_0 + \varepsilon$. That is, if $\delta_\varepsilon(t - t_0)$ is considered as a force, then property (1) says the total impulse is unity.

(2) $$\int_{-\infty}^{\infty} \delta_\varepsilon(t - t_0) f(t)\, dt = \int_{t_0-\varepsilon}^{t_0+\varepsilon} \frac{1}{2\varepsilon} f(t)\, dt = \frac{1}{2\varepsilon} \int_{t_0-\varepsilon}^{t_0+\varepsilon} f(t)\, dt = f(\tau_\varepsilon)$$

for some $\tau_\varepsilon$ in $t_0 - \varepsilon \le t \le t_0 + \varepsilon$ and any continuous function f(t) defined on the interval $-\infty < t < \infty$. Further, if $\varepsilon \to 0$, then

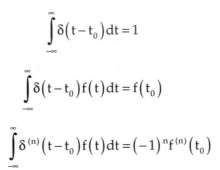

$$\int_{-\infty}^{\infty} \delta(t - t_0)\, dt = 1$$

$$\int_{-\infty}^{\infty} \delta(t - t_0) f(t)\, dt = f(t_0)$$

$$\int_{-\infty}^{\infty} \delta^{(n)}(t - t_0) f(t)\, dt = (-1)^n f^{(n)}(t_0)$$

(3) The Laplace transform of the impulse function is given by

$$L\{\delta_\varepsilon(t - t_0)\} = e^{-st_0} \text{ for } t_0 > 0$$

and                        $$L\{\delta(t)\} = 1 \text{ if } t_0 \to 0$$

### 3.6.3  Inverting the Laplace Transform

In order for this operational method to be useful to chemical engineers, there must be convenient ways to invert the transform. That is, it is necessary to be able to perform the operation

$$L^{-1}\{F(s)\} = f(t) \tag{3.6.15}$$

to complete a problem. There are three standard ways to invert the Laplace transform and each is outlined below. The Laplace transform of a function can be arranged to be of the form

$$F(s) = \frac{P(s)}{Q(s)} \tag{3.6.16}$$

where P and Q are polynomials in s and Q(s) is of higher degree than P(s). The function F(s) can then be expanded into its partial fractions (Thomas and Finney[14]) such that the form

$$F(s) = \sum_{i=1}^{n} \frac{A_1}{(s-r)^i} \tag{3.6.17}$$

is obtained, where the quantity $(s-r)$ is a linear factor of Q(s) and the $A_i$ are determined constants. Once the form Equation 3.6.17 is obtained, then Equation 3.6.15 can be applied to each term with the aid of a table of transforms (Spiegel[13]).

The second standard method of inverting the Laplace transform is by *convolution*. This method is most effective when F(s) is the product of two other transforms, H(s) and G(s), where transforms H(s) and G(s) are respectively transforms of the functions h(t) and g(t). The following theorem defines what is meant by convolution (Boyce and DiPrima[1] and Kreyzig[17]).

*THEOREM 3.9*
*If H(s) = L{h(t)} and G(s) = L{g(t)} both exist for s > a ≥ 0, then*

$$F(s) = H(s)G(s) = L\{f(t)\} \quad s > a \tag{3.6.18}$$

*where*

$$f(t) = \int_0^t h(t-\tau)g(\tau)d\tau = \int_0^t h(\tau)g(t-\tau)d\tau \tag{3.6.19}$$

*The function f is called the convolution of h and g and the integrals (Equation 3.6.19) are called convolution integrals.*

The convolution obeys the following rules:

$h*g = g*h$                                 Commutative law

$h*(g_1 + g_2) = h*g_1 + h*g_2$          Distributive law

$h*(g*h) = (h*g)*h$               Associative law

$h*0 = 0*h = 0$

However,

$$h*1 = h$$

is not generally a true statement. That is, if $h(t) = \sin t$, then

$$h * 1 = \int_0^t \sin(t - \tau)d\tau = \cos(t - \tau)\Big|_0^t = 1 - \cos t \neq \sin t$$

Also h*h is not necessarily nonnegative, that is

$$h * h = \int_0^t \sin(t - \tau)\sin \tau \, d\tau = \int_0^t (\sin t \, \cos \tau - \cos t \, \sin \tau)\sin \tau \, dt$$

$$= \sin t \int_0^t \sin \tau \, \cos \tau \, dt - \cos t \int_0^t \sin^2 \tau \, dt$$

$$= \frac{1}{2}(\sin t - t \, \cos t)$$

Essentially, the convolution integrals do not have all the properties of ordinary multiplication.

This example illustrates inversion using the convolution integral. Consider

$$F(s) = \frac{s}{(s+1)(s^2 + 4)} = \left(\frac{1}{s+1}\right)\left(\frac{s}{s^2 + 4}\right)$$

From a set of tables of Laplace transforms (Spiegel[13]) one can find

$$L^{-1} = \left\{\frac{1}{s+1}\right\} = e^{-t}$$

and

$$L^{-1} = \left\{\frac{s}{s^2 + 4}\right\} = \cos 2t$$

then

$$e^{-t} * \cos 2t = \int_0^t e^{-(t-\tau)} \cos 2\tau \, d\tau$$

$$= e^{-t} \int_0^t e^\tau \cos 2\tau \, d\tau$$

$$= e^{-t} \left[ \frac{e^{\tau} \left( \cos 2\tau + 2 \sin 2\tau \right)}{5} \right]_0^t$$

$$= \frac{1}{5} \left( \cos 2t + 2 \sin 2t - e^{-t} \right)$$

Therefore

$$L^{-1} \left\{ \frac{s}{(s+1)(s^2+4)} \right\} = \frac{1}{5} \left( \cos 2t + 2 \sin 2t - e^{-t} \right)$$

The third standard method of inverting a Laplace transform is by making use of the Residue theorem (Churchill and Brown,[16] Mickley, Sherwood, and Reid,[6] Jenson and Jeffreys,[15] Saff and Snider,[18] Dettman[22]). The transform function F(s) is analytic, except for singularities. In this discussion, when F(s) is analytic, the inverse transform of F(s) is given by

$$f(t) = L^{-1} \left\{ F(s) \right\} = \sum_1^\infty p_n(t) \tag{3.6.20}$$

Where $p_n(t)$ is termed the *residue* of F(s) at the singularities (poles) $S_n$. The residues of F(s) may be determined by employing the form given in Equation 3.6.16, that is

$$F(s) = \frac{P(s)}{Q(s)}$$

If $S_n$ is a simple pole of F(s), then $p_n(t)$ is given by

$$p_n(t) = \frac{P(S_n)}{Q'(S_n)} e^{S_n t} \tag{3.6.21}$$

where $Q'(S_n)$ is the value of $\dfrac{dQ}{ds}$ evaluated at the singular point of interest. Recall that

$$\frac{P(s_n)}{Q'(s_n)} = \lim_{s \to s_n} \frac{P(s)}{\left[ \dfrac{Q(s) - Q(s_n)}{s - s_n} \right]} = \lim_{s \to s_n} (s - s_n) \frac{P(s)}{Q(s)}$$

The form

$$\lim_{s \to s_n} (s - s_n) \frac{P(s)}{Q(s)}$$

is convenient for applications, especially when $s_n = 0$.

When $S_n$ is a multiple pole of order m of F(s), then

$$P_n(t) = e^{S_n t} \left[ A_1 + tA_2 + \frac{t^2}{2!} A_3 + \cdots + \frac{t^{m-1} A_m}{(m-1)!} \right]$$

$$= e^{S_n t} \sum_{i=1}^{m} A_i \frac{t^{i-1}}{(i-1)!}$$

(3.6.22)

where

$$A_i = \lim_{s \to s_n} \frac{1}{(m-i)!} \frac{d^{m-i}}{ds^{m-i}} \left[ (s - s_n)^m F(s) \right]$$

The following example illustrates the use of the residue theorem and should serve to clarify certain new terminology:

$$\text{Consider} \quad F(s) = \frac{1}{s(s-a)^2} = \frac{P(s)}{Q(s)}$$

then the polynomial Q(s) has singularities at s = 0 and s = a. These singularities are termed poles (Churchill and Brown[16] and Saff and Snider[18]) and in particular, s = 0 is termed a simple pole while s = a is a double pole (pole of order two). For the simple pole, Equation 3.6.21 gives

$$P_0(t) = \frac{1}{Q'(0)} e^{0t} = \frac{1}{Q'(0)} = \frac{1}{a^2}$$

since $Q'(s) = (s - a)^2 + 2s(s - a)$. Alternatively, we could first determine the quantity $P(0)/Q'(0)$ by taking the limit

$$\lim_{s \to 0} (s - 0) \frac{1}{s(s-a)^2} = \frac{1}{a^2}$$

Then, multiply the result by $e^{0t}$ to get the residue at s = 0.

The residue at the double pole, s = a, is given by

$$\rho_a(t) = e^{at}(A_1 + tA_2) \quad \text{since } m = 2$$

where

$$A_1 = \frac{1}{(2-1)!} \frac{d^{2-1}}{ds^{2-1}}\left[(s-a)^2 \frac{1}{s(s-a)^2}\right]_{s=a}$$

$$= \frac{d}{ds}\left[\frac{1}{s}\right]_{s=a} = -\frac{1}{a^2}$$

and

$$A_2 = \frac{1}{(2-2)!} \frac{d^{2-2}}{ds^{2-2}}\left[(s-a)^2 \frac{1}{s(s-a)^2}\right]_a$$

$$= \left[\frac{1}{s}\right]_{s=a} = \frac{1}{a}$$

then

$$\rho_a(t) = e^{at}\left(\frac{t}{a} - \frac{t}{a^2}\right)$$

and

$$f(t) = L^{-1}\{F(s)\} = \frac{1}{a^2} + e^{at}\left(\frac{t}{a} - \frac{t}{a^2}\right)$$

The previous example is very straightforward and could easily be inverted by use of partial fractions and a table of Laplace transforms. This next example may be tabulated. However, it is being used here to demonstrate more clearly how the residue theorem may be useful for similar or more complicated inversions. Consider

$$F(s) = \frac{\sinh x\sqrt{s}}{\sinh a\sqrt{s}}$$

Then

$$Q(s) = \sinh a\sqrt{s}$$

If we let $a\sqrt{s} = i\lambda; \ni s = -\lambda^2/a^2$ then F(s) transforms to $\dfrac{\sin(x\lambda/a)}{\sin\lambda}$ such that

$$\frac{dQ}{ds} = \frac{dQ}{d\lambda}\frac{d\lambda}{ds} = -a^2\frac{\cos\lambda}{2\lambda}$$

where $Q(\lambda)$ is given by $\sin \lambda$. Then application of Equations 3.6.20 and 3.6.21 result in

$$\sum_{n=0}^{\infty} -\frac{2\lambda_n \sin\left(x\lambda_n/a\right)}{a^2\cos\lambda_n}e^{-(\lambda_n^2/a^2)t}$$

But the poles of $\sinh a\sqrt{s}$ are the zeros of $\sin \lambda$ and these occur at $\lambda = n\pi$, $n = 0, 1, 2, \ldots$

Therefore replacing $\lambda_n$ by the quantity $n\pi$ and observing that $\cos(n\pi) = (-1)^n$, we get the result

$$\frac{2\pi}{a^2}\sum_{n=1}^{\infty}(-1)^{n+1}n\sin\left(xn\pi/a\right)e^{-(n^2\pi^2 t/a^2}$$

More detailed usage of the residue theorem to invert Laplace transforms is discussed in Chapter 6.

### 3.6.4   Taylor Series Solution of Initial Value Problems

In Section 3.5.1, the Frobenius series method was discussed in regard to differential equations with regular singular points. In this section, a method is given that effectively deals with differential equations with *ordinary* points. To illustrate the importance of this method consider the relatively harmless-looking first order initial value problem

$$y' + e^x y = x^2 \; ; \; y(0) = 4 \tag{3.6.23}$$

This problem has a unique solution (Theorem 2.1) for all x and is given by

$$y(x) = \frac{1}{e^{ex}}\left[\int_0^x \xi^2\, e^{e^\xi} d\xi + 4e\right]^{-x} \tag{3.6.24}$$

However, Equation 3.6.24 is not in *closed form*. By closed form we mean

$$y' + 2y = 1 \; ; \; y(0) = 3$$

has

$$y(x) = \frac{1}{2}\left(1 + 5e^{-2x}\right); \; -\infty < x < \infty$$

as its solution in closed form. Equation 3.6.24 is difficult to graph or to use in forecasting results, and these are important aspects of engineering practice.

Reconsider Equation 3.6.23, but this time, let

$$y(x) = y(0) + y'(0)x + \frac{1}{2!}y''(0)x^2 + \cdots + \frac{1}{n!}y^{(n)}(0)x^n + \cdots$$

$$= \sum_{n=0}^{\infty} \frac{1}{n!}y^{(n)}(0)x^n$$

(3.6.25)

then each derivative can be evaluated at x = 0 from

$$y'(0) + e^0 y(0) = 0$$

or

$$y'(0) = -4$$

to obtain the value of the second derivative at x = 0, one differentiates Equation 3.6.23 thus

$$y'' + e^x y' + e^x y = 2x$$

(3.6.26)

then at x = 0, Equation 3.6.26 gives

$$y''(0) + y'(0) + y(0) = 0 \quad \text{or} \quad y''(0) = 0$$

Differentiate Equation 3.6.26 to get

$$y^{(3)} + e^x y'' + 2e^x y' + e^x y = 2$$

(3.6.27)

then

$$y^{(3)}(0)+y''(0)+2y'(0)+y(0)=2 \quad \text{or} \quad y^{(3)}(0)=6$$

Therefore, the solution of Equation 3.6.23 using Equation 3.6.25 is

$$y(x)=4-4x+x^3+\cdots \tag{3.6.28}$$

Notice that one can take this process up to any desired amount of terms. Equation 3.6.28 is more manageable than Equation 3.6.24.

An appropriate question is when is this approach mathematically legal? The answer is whenever the initial value problem has analytic coefficients and forcing functions. The term *analytic* can be defined as any function having a Taylor series representation in some open interval about a given point.

*THEOREM 3.10*

*If p, q, and g are analytic at $x_0$, then the initial value problem*

$$y''+p(x)y'+q(x)y=g(x); \quad y(x_0)=A, \quad y'(x_0)=B$$

*has a unique solution which is also analytic at $x_0$.*

If the initial value is not given at the origin, a shift to the origin can be made as given in the example that follows.

Given

$$y''+xy=4 \quad ; \quad y(1)=2, \quad y'(1)=0 \tag{3.6.29}$$

let $t=x-1$ and $Y(t)=y(1+t)$

then

$$Y'(t)=\frac{dy}{dx}\frac{dx}{dt}$$

but

$$x=1+t, \quad \frac{dx}{dt}=1$$

$$\therefore \quad Y'(t)=\frac{dy}{dx}$$

similarly

$$Y''(t) = y''(x)$$

then Equation 3.6.29 becomes

$$Y''(t) + (1+t)Y(t) = 4; \quad Y(0) = 2, \quad Y'(0) = 0 \qquad (3.6.30)$$

Equation 3.6.30 can now be treated similar to the previous example to get

$$Y(t) = 2 + t^2 - \frac{1}{3}t^3 - \frac{1}{12}t^4 - \frac{1}{30}t^5 + \frac{1}{72}t^6 + L$$

such that

$$y(x) = 2 + (x-1)^2 - \frac{1}{3}(x-1)^3 - \frac{1}{12}(x-1)^4 - \frac{1}{30}(x-1)^5 + \frac{1}{72}(x-1)^6 + L$$

Even though the above discussion is focused on linear problems, the Taylor series approach may be applied to some nonlinear problems.

## 3.7 Applications of Second Order Differential Equations

In this section, a few applications of the theory and methods that were previously outlined will be illustrated. However, it should be noted that a substantial percentage of the application of second (and higher) order ordinary differential equations is in association with solving partial differential equations, a topic discussed in Chapter 6.

As in Section 2.4, some problem setup will be demonstrated as well as solution techniques. For example, consider the following problem in heat transfer through a cylindrical conductor (Jenson and Jeffreys[15]):

*Problem Statement:*
Two concentric cylindrical metallic shells are separated by a solid material. If the two metal surfaces are maintained at different constant temperatures, what is the steady state temperature distribution within the separating material?

In this problem both temperature, T, and the heat flow per unit area, Q, depend on the radius, r. Then if we consider the condition of the system about an element of thickness, $\Delta r$ and with the aid of Figure 3.1, the important quantities can be organized according to the procedure of Section 2.4.

| System Property | r | r + Δr |
|---|---|---|
| Temperature | T | $T + \dfrac{dT}{dr}\Delta r$ |
| Heat Transfer Area per Unit Length | $2\pi r$ | $2\pi(r + \Delta r)$ |
| Radial Heat Flux Density | Q | $Q + \dfrac{dQ}{dr}\Delta r$ |
| Total Radial Heat Flow | $2\pi r\, Q$ | $2\pi(r + \Delta r)\left(Q + \dfrac{dQ}{dr}\Delta r\right)$ |

Heat input to inner surface = $2\pi r\, Q$

Heat output from outer surface = $2\pi(r + \Delta r)\left(Q + \dfrac{dQ}{dr}\Delta r\right)$

Accumulation of heat = 0

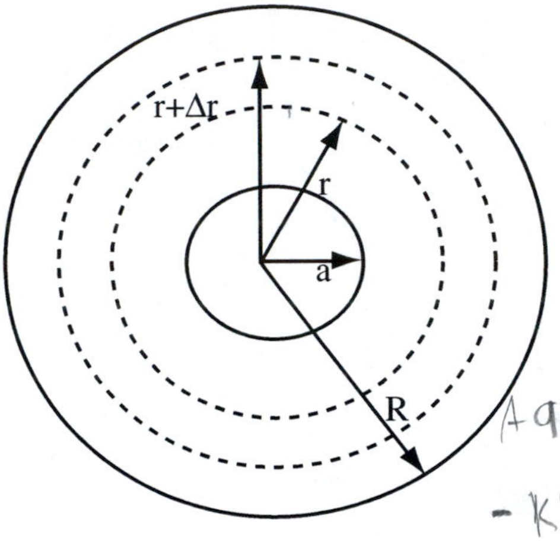

**FIGURE 3.1**
Radial heat flow through cylindrical conductor.

Therefore,

$$2\pi r Q - 2\pi(r + \Delta r)\left(Q + \frac{dQ}{dr}\Delta r\right) = 0$$

or

$$-r\frac{dQ}{dr} - Q - \frac{dQ}{dr}\Delta r = 0$$

Then, limit as $\Delta r \to 0$

$$r\frac{dQ}{dr} + Q = 0 \qquad (3.7.1)$$

*Handwritten annotations:*

$$+ q_{in} - q_{out} = 0.$$

$$- k\left.\frac{dT}{dr}\right|_r + k\left.\frac{dT}{dr}\right|_{r+\Delta r}$$

$$- 2\pi r\, k\left.\frac{dT}{dr}\right|_r + 2\pi(r + \Delta r)\,k\,\frac{d}{d\,}$$

$$- 2\pi r\, k\left.\frac{dT}{dr}\right|_r + 2\pi r\, k\left.\frac{dT}{dr}\right|_{r+\Delta r} + 2\pi\Delta r$$

$$r\left[\frac{\left.\frac{dT}{dr}\right|_{r+\Delta r} - \left.\frac{dT}{dr}\right|_r}{\Delta r}\right] + \frac{dT}{dr}$$

$$r\frac{d^2T}{dr^2} + \frac{dT}{dr} = 0$$

But Q is related to T by

$$Q = -k\frac{dT}{dr} \qquad (3.7.2)$$

where k is the thermal conductivity. Upon substituting Equation 3.7.2 into Equation 3.7.1 gives

$$r\frac{d^2T}{dr^2} + \frac{dT}{dr} = 0 \qquad (3.7.3)$$

*[handwritten annotations:]*
$T'' + \frac{1}{r}T' = 0$
$x^2 + \frac{1}{r}\cdot = 0$
$x(x + \frac{1}{r}) = 0$
$x = 0, -\frac{1}{r}$
$T = C_1 + C_2 e^{1/rt}$

for a constant thermal conductivity.

Since the two metal surfaces are maintained at different constant temperatures, say

$$\text{at } r = a, \ T = T_0 \qquad (3.7.4)$$

and

$$\text{at } r = R, \ T = T_1 \qquad (3.7.5)$$

one can solve the second order, linear, variable coefficient, and homogeneous differential equation subject to Equation 3.7.4 and Equation 3.7.5. That is,

$$\frac{d}{dr}\left(r\frac{dT}{dr}\right) = 0$$

gives

$$r\frac{dT}{dr} = c_1$$

and

$$T(r) = c_1 \ln r + c_2$$

then

$$T(a) = T_0 = c_1 \ln a + c_2$$
$$T(R) = T_1 = c_1 \ln R + c_2$$

therefore

$$T_1 - T_0 = c_1 \ln\left(\frac{R}{a}\right)$$

such that

$$c_1 = \frac{T_1 - T_0}{\ln\left(\frac{R}{a}\right)}$$

and

$$c_2 = \frac{T_0 \ln R - T_1 \ln a}{\ln\left(\frac{R}{a}\right)}$$

Finally

$$\frac{T(r) - T_0}{T_1 - T_0} = \frac{\ln\left(\frac{r}{a}\right)}{\ln\left(\frac{R}{a}\right)}$$

As can be expected, there are other ways to set up physical problems that may be of interest. One approach that is very prominent in chemical engineering is to use the equations of change (Bird, Stewart and Lightfoot[19] and Schlichting[20]). That is, to set up constant density, constant viscosity flow problems, one needs the equation of continuity and the equation of motion.

## Example 3.7.1
Consider the axial flow of an incompressible fluid in a circular tube of radius R. By considering a long tube and assuming that the $\theta$-component and the r-component of velocities are negligible, one can reduce the z-component of motion for constant $\rho$ and $\mu$ (Bird, Stewart and Lightfoot[19]) to

$$\rho v_z \frac{\partial v_z}{\partial z} = -\frac{\partial P}{\partial z} + \mu\left[\frac{1}{r}\frac{\partial}{\partial r}\left(r\frac{\partial v_z}{\partial r}\right) + \frac{\partial^2 v_z}{\partial z^2}\right] \tag{3.7.6}$$

Also, the equation of continuity ( Bird, Stewart, and Lightfoot[19]) reduces to

$$\frac{\partial v_z}{\partial z} = 0 \tag{3.7.7}$$

which further reduces Equation 3.7.6 to

$$0 = -\frac{dP}{dz} + \frac{1}{\mu}\frac{1}{r}\frac{d}{dr}\left(r\frac{dv_z}{dr}\right)$$

(3.7.8)

subject to

$$v_z \text{ is finite at } r = 0$$

(3.7.9)

and at the wall of the cylinder

$$v_z = 0 \text{ at } r = R$$

(3.7.10)

Equations 3.7.8, 3.7.9, and 3.7.10 integrate to

$$v_z = \frac{(P_0 - P_L)R^2}{4\mu L}\left[1 - \left(\frac{r}{R}\right)^2\right]$$

where $P_0$ and $P_L$ are the pressures at the entrance and exit of the cylinder respectively.

## Example 3.7.2

Derive the temperature profile $T(\cdot)$ in a solid cylinder with heat generation if the governing differential equation is

$$\frac{1}{r}\frac{\partial}{\partial r}\left(kr\frac{\partial T}{\partial r}\right) + \frac{1}{r^2}\frac{\partial}{\partial \phi}\left(k\frac{\partial T}{\partial \phi}\right) + \frac{\partial}{\partial z}\left(k\frac{\partial T}{\partial z}\right) + \dot{q} = \rho C_p \frac{\partial T}{\partial t}$$

(3.7.12)

where the coordinate system indicates the independent variables: $\rho$ is mass density and $C_p$ is specific heat. For a long solid cylinder as shown in Figure 3.2, with uniform heat generation $\dot{q}$ and at steady state conditions, the rate at which heat is generated within the cylinder will be equal to the rate at which heat is convected from the surface of the cylinder to a moving fluid. This condition allows the surface temperature $(T_s)$ to be maintained at a fixed value. Then, for constant thermal conductivity $k$, Equation 3.7.12 reduces to

$$\frac{1}{r}\frac{d}{dr}\left(r\frac{dT}{dr}\right) + \frac{\dot{q}}{k} = 0$$

or

$$d\left(r\frac{dT}{dr}\right) = -\frac{\dot{q}}{k}r\,dr$$

(3.7.13)

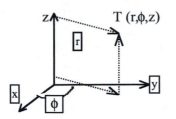

cold fluid

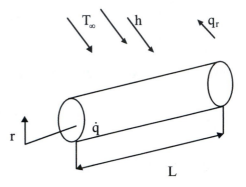

**FIGURE 3.2**
Solid cylinder with heat generation.

such that

$$T(r) = \frac{-\dot{q}}{4k}r^2 + C_1 \ln r + C_2 \qquad (3.7.14)$$

Along the centerline of the solid cylinder (r = 0) the temperature distribution is symmetric and the temperature gradient $\left(\dfrac{dT}{dr}\right)$ must be zero there. That is

$$\text{at } r = 0, \quad \frac{dT}{dr} = 0$$

therefore,

$$C_1 = 0$$

at

$$r = r_0, \ T = T_S$$

and

$$C_2 = T_S + \frac{\dot{q}}{4k} r_0^2$$

$$\therefore \quad T(r) - T_S = \frac{\dot{q}}{4k}\left(r_0^2 - r^2\right) \qquad (3.7.15)$$

From an overall energy balance

$$\dot{q}\left(\pi r_0^2 L\right) = h\left(2\pi r_0 L\right)\left(T_S - T_\infty\right)$$

which simplifies to

$$T_S = T_\infty + \frac{\dot{q} r_0}{2h} \qquad (3.7.16)$$

## Example 3.7.3

Consider a long solid tube, insulated at the outer radius $r_0$ and cooled at the inner radius $r_i$ with uniform heat generation $\dot{q}$ within the solid.

(a) Determine the general solution for the temperature profile in the tube.

(b) Suppose the maximum permissible temperature at the insulated surface $r_0$ is $T_0$. Identify appropriate boundary conditions that could be used to determine the arbitrary constants appearing in the general solution and find the temperature distribution.

(c) What is the heat removal rate per unit length of tube?

(d) If the coolant is available at a temperature $T_\infty$, obtain an expression for the convection coefficient that would have to be maintained at the inner surface to allow for operation at the prescribed values of $T_0$ and $\dot{q}$.

*Solution*
Assumptions:

1. Steady state conditions prevail.
2. One-dimensional radial conduction is reasonable.
3. Physical properties are constant.
4. Volumetric heat generation is uniform.
5. Outer surface is adiabatic.

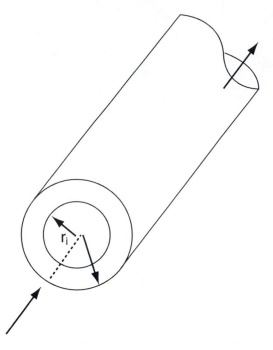

**FIGURE 3.3**
Solid tube with heat generation.

(a) Equation 3.7.12 in Example 3.7.2 is the governing differential equation and it is reducible to Equation 3.7.13 by using the reasoning per Example 3.7.2. Therefore, the general solution is as given in Equation 3.7.14

$$T(r) = \frac{-\dot{q}}{4k} r^2 + C_1 \ln r + C_2 \tag{3.7.14}$$

(b) Two boundary conditions are needed to determine $C_1$ and $C_2$. In this problem

$$T(r_0) = T_0 \tag{3.7.17}$$

and Fourier's law

$$Q_r = -kA \frac{dT}{dr} = -k(2\pi r L) \frac{dT}{dr}$$

expresses the rate at which heat is conducted across any cylindrical surface in the solid. In particular, at the adiabatic outer surface

$$Q_r = 0 \Rightarrow \text{at } r = r_0, \frac{dT}{dr} = 0 \qquad (3.7.18)$$

Therefore,

$$C_1 = \frac{\dot{q}}{2k} r_0^{\,2} \quad \text{and} \quad C_2 = T_0 + \frac{\dot{q}}{4k} r_0^{\,2} - \frac{\dot{q}}{2k} r_0^{\,2} \ln r_0$$

Finally, the temperature distribution is

$$T(r) = T_0 + \frac{\dot{q}}{4k}\left(r_0^{\,2} - r^2\right) - \frac{\dot{q}}{2k} r_0^{\,2} \ln\left(\frac{r_0}{r}\right) \qquad (3.7.19)$$

(c) The heat removal rate may be determined by obtaining the conduction rate at $r_i$ or by evaluating the total generation rate for the tube.

Since

$$Q_r = -2\pi kr \frac{dT}{dr}$$

represents the rate of conduction, then from Equation 3.7.19

$$\frac{dT}{dr} = \frac{\dot{q}}{2k}\left(\frac{r_0^2}{r} - r\right)$$

and at $r = r_i$

$$\frac{dT}{dr} = \frac{\dot{q}}{2k}\left(\frac{r_0^2}{r_i} - r_i\right)$$

Therefore,

$$Q_r(r_i) = -2\pi k r_i \frac{\dot{q}}{2k}\left(\frac{r_0^2}{r_i} - r_i\right) \qquad (3.7.20)$$

$$= -\pi\dot{q}\left(r_0^2 - r_i^2\right)$$

is the heat removal rate.

(d) Application of the energy conservation principle to the inner surface results in

$$Q_{r,\text{Cond}} = Q_{r,\text{Conv}} \qquad (3.7.21)$$

or

$$\pi\dot{q}\left(r_0^2 - r_i^2\right) = h2\pi r_i\left(T_i - T_\infty\right) \qquad (3.7.21)$$

where $-Q_r(r_i)$ accounts for heat flowing out of the wall, and $Q_{r,conv}$ is expressed as $(T_i - T_\infty)$ rather than $(T_\infty - T_i)$. Therefore

$$h = \frac{\dot{q}\left(r_0^2 - r_i^2\right)}{2r_i\left(T_i - T_\infty\right)}, \quad \text{where } T_i = T\left(r_i\right)$$

## Example 3.7.4

This example covers simultaneous diffusion and chemical reaction in a tubular reactor (Jenson and Jeffreys[15]).

A tubular reactor of length L and cross-sectional area $1.0 \text{ m}^2$ is used to carry out a first order chemical reaction of the type:

$$A \longrightarrow B$$

The rate coefficient is k ($\text{sec}^{-1}$). In a given feed rate of u $\text{m}^3/\text{sec}$, the initial feed concentration of species A is $C_0$ and the diffusivity of A is D $\text{m}^2/\text{sec}$. What is the concentration of A as a function of the reactor length? It may be assumed that during the reaction the volume remains constant and that steady-state conditions are established. Also there is no concentration variation in the section following the reactor.

*Solution*

Let x represent the distance of any point from the beginning of the reaction section ($0 < x < L$), C represent the concentration profile of species A in the entry section ($x < 0$), and y be the concentration profile of species A in the reaction section as shown in Figure 3.4.

Consider a material balance over the element of length $\Delta x$ at a distance x from the inlet. Then the bulk flow of A at x is (uy) and at $x + \Delta x$ is

$$\left(uy + u\frac{dy}{dx}\Delta x\right)$$

Also, the diffusion of A at x is

$$\left(-D\frac{dy}{dx}\right)$$

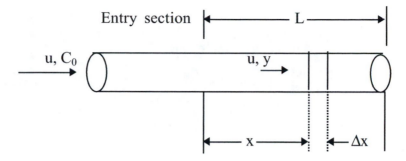

**FIGURE 3.4**
Tubular reactor.

and at $x + \Delta x$ is

$$\left( D\frac{dy}{dx} + \frac{d}{dx}\left( -D\frac{d}{dx} \right)\Delta x \right)$$

The accumulation rate is zero (assumed steady-state conditions given in problem statement). The rate of removal of species A by chemical reaction is given by:

$$ky\Delta x \text{ (cross-sectional area)}$$

Therefore,

$$uy - D\frac{dy}{dx} - \left( uy + u\frac{dy}{dx}\Delta x \right) - \left[ -D\frac{dy}{dx} + \frac{d}{dx}\left( -D\frac{dy}{dx} \right)\Delta x \right] = ky\Delta x. \quad (3.7.22)$$

Following simplification and division by $\Delta x$ we get:

$$D\frac{d^2y}{dx^2} - u\frac{dy}{dx} - ky = 0 \qquad (3.7.23)$$

Similarly, the material balance in the entry section gives:

$$D\frac{d^2C}{dx^2} - u\frac{dC}{dx} = 0 \qquad (3.7.24)$$

The characteristic equation for Equation 3.7.23 is

$$Dr^2 - ur - k = 0$$

such that

$$r_1 = u(1+m)/2D$$

and

$$r_2 = u(1-m)/2D, \text{ where } m = \sqrt{1 + \dfrac{4kD}{u^2}}$$

Therefore,

$$y = h_1 \exp\left[\frac{ux}{2D}(1+m)\right] + h_2 \exp\left[\frac{ux}{2D}(1-m)\right]$$

and the general solution for Equation 3.7.24 is

$$C = h_3 + h_4 \exp\left(\frac{ux}{D}\right)$$

The appropriate boundary conditions for Equation 3.7.23 are, at x = 0

$$C = y \text{ (continuity of composition) or } \frac{dc}{dx} = \frac{dy}{dx}$$

also at x = L, $\dfrac{dy}{dx} = 0$ (no concentration variation following the reaction section).
Appropriate boundary conditions for Equation 3.7.24 are, at x = −∞, C = $C_0$ and the condition at x = 0 can be reused (i.e., C = y). Following application of these boundary conditions we get

$$\frac{y}{C_0} = \frac{2}{K}\exp\left(\frac{ux}{D}\right)\left\{(m+1)\exp\left[\frac{um}{2D}(L-x)\right] + (m-1)\exp\left[-\frac{um}{2D}(L-x)\right]\right\}$$

where K is given by

$$K = (m+1)^2 \exp\left(uLm/2D\right) - (m-1)^2 \exp\left(\frac{-uLm}{2D}\right)$$

## Example 3.7.5
The Continuous Hydrolysis of Fat Compound in a Spray Column (Jenson and Jeffreys[15]).

A fat compound mixed with high-pressure hot water is fed to the bottom of a spray column. Water at the column operating conditions is sprayed into the top and descends in droplets through the rising fat phase. The hydrolysis reaction generates glycerine that is extracted by the descending water phase. Further, the hydrolysis reaction is first order with a specific rate constant of 0.17 sec$^{-1}$.

Estimate the concentration of glycerine in each phase as a function of column height. Also determine what fraction of the tower height H is required for the chemical reaction.

Data:
  Rate of fat compound input = 8070 lb/h
  Rate of high pressure hot water = 2270 lb/h
  Rate of water spray = 4120 lb/h
  Column operating conditions: 450°F and 600 psia
  Height = 72 ft, Diameter = 2 ft 2 in.
  Rate of final extract = 5560 lb/h containing 12.16% glycerine
  Rate of fatty acid raffinate = 8900 lb/h containing 0.24% raffinate

Glycerine distribution ratio between the water and the fat phase is 10.32 at the given column temperature and pressure.

*Solution*
As shown in Figure 3.5, L represents the mass flow of raffinate and G the mass flow of extract. In addition, the following symbols are used to designate the other quantities:

  $x$ = mass fraction of glycerine in raffinate
  $y$ = mass fraction of glycerine in extract
  $y^*$ = equilibrium mass fraction of glycerine in extract
  $z$ = mass fraction of hydrolyzable fat in raffinate
  $S$ = cross-sectional area of tower
  $a$ = interfacial area per unit volume of tower
  $K$ = overall mass transfer coefficient expressed in terms of extract compositions
  $m$ = distribution ratio
  $k$ = reaction rate coefficient
  $\rho$ = mass of fat per unit volume
  $h$ = distance coordinate from base of column
  $w$ = mass fat per mass glycerine
  $H$ = effective height of column

Then the changes occurring in the element of column of height $\Delta h$ are

- Glycerine transferred from fat to water phase: $KaS(y^* - y)\Delta h$.
- Rate of destruction of fat by hydrolysis: $k\rho Sz\Delta h$, thus giving the rate of production of glycerine as: $k\rho Sz\Delta h/w$.

A glycerine balance over the element $\Delta h$ (Figure 3.5) is

$$Lx + \frac{(k\rho Sz\Delta h)}{w} - L\left(x + \frac{dx}{dh}\Delta h\right)$$

$$= Gy - G\left(y + \frac{dy}{dh}\Delta h\right) = KaS(y^* - y)\Delta h \qquad (3.7.25)$$

where

$$y^* = mx. \qquad (3.7.26)$$

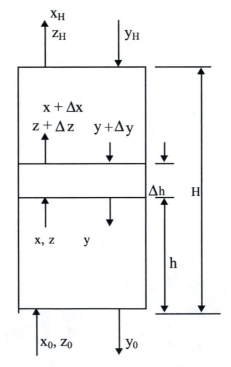

**FIGURE 3.5**
Continuously operating fat-hydrolyzing column under steady state.

Also a glycerine balance between the element and the base of the tower is

$$\frac{Lz_0}{w} + Gy = Lx + \frac{Lz}{w} + Gy_0 \tag{3.7.27}$$

Using Equation 3.7.26 and the last two parts of Equation 3.7.25 give

$$KaSmx = KaSy - G\frac{dy}{dh} \tag{3.7.28}$$

Substitution of Equation 3.7.27 into the first two parts of Equation 3.7.25 gives

$$k\rho S\left[\frac{z_0}{w} + \frac{G(y-y_0)}{L} - x\right]\Delta h - L\frac{dx}{dh}\Delta h = -G\frac{dy}{dh}\Delta h \tag{3.7.29}$$

Multiplication of Equation 3.7.29 by $(KaSm/LG)$ and substituting for x from Equation 3.7.28 gives

$$\frac{k\rho S^2 Ka}{LG}\left[\frac{mz_0}{w} + \frac{mG}{L}(y-y_0)\right] - \frac{k\rho S}{L}\left[\frac{KaS}{G}y - \frac{dy}{dh}\right]$$

$$-\left[\frac{KaS}{G}\frac{dy}{dh} - \frac{d^2y}{dh^2}\right] + \frac{KaSm}{L}\frac{dy}{dh} = 0$$

and using the following parameters:

$$r = \frac{mG}{L}; \quad p = \frac{k\rho S}{L}; \quad q = \frac{KaS}{G}(r-1)$$

reduce the above equation to

$$\frac{d^2y}{dh^2} + (p+q)\frac{dy}{dh} + pqy = \frac{pq}{r-1}\left[ry_0 - \frac{mz_0}{w}\right] \tag{3.7.30}$$

subject to the boundary conditions
   At

$$h = 0, \quad x = 0$$

and at

$$h = H, \quad y = 0$$

Equation 3.7.30 is a linear constant coefficient equation and can be solved in the following way:

$$\left(D^2 + [p+q]D + pq\right)y = \frac{pq}{r-1}\left[ry_0 - \frac{mz_0}{w}\right] \quad \text{Put in operator form.}$$

$$D\left(D^2 + [p+q]D + pq\right)y = 0 \quad \text{Annihilate in right-hand side.}$$

$$t\left(t^2 + [p+q]t + pq\right) = 0 \quad \text{Write down the characteristic equation.}$$

such that $t_1 = 0$, $t_2 = -p$, and $t_3 = -q$ are the characteristic roots. Then the general solution is

$$y = c_1 + c_2 e^{-ph} + c_3 e^{-qh}$$

where $c_2$ and $c_3$ are arbitrary and $c_1$ is a particular constant to be determined through substitution into Equation 3.7.30. That is, $y_p$ is given as

$$y_p = c_1, \quad y_p' = y_p'' = 0.$$

Then

$$pqc_1 = \frac{pq}{r-1}\left[ry_0 - \frac{mz_0}{w}\right]$$

or

$$c_1 = \frac{1}{r-1}\left[ry_0 - \frac{mz_0}{w}\right]$$

Finally, the general solution becomes

$$y_g = \frac{1}{r-1}\left[ry_0 - \frac{mz_0}{w}\right] + c_2 e^{-ph} + c_3 e^{-qh}$$

and substitution of the boundary conditions into the general solution gives

$$y = \frac{mz_0}{w(r-v)}\left[e^{-ph} + \left(\frac{e^{-pH}-v}{r-e^{-qH}}\right)e^{-qh} + \left(\frac{ve^{-qH}-re^{-pH}}{r-e^{qH}}\right)\right] \quad (3.7.31)$$

where

$$v = 1 + \frac{k\rho G}{KaL} = \frac{q + rp - p}{q}$$

$$y_0 = \frac{mz_0}{w\left(r - e^{-qH}\right)}\left[1 - \left(\frac{r-1}{r-v}\right)e^{-pH} + \left(\frac{v-1}{r-v}\right)e^{-qH}\right] \qquad (3.7.32)$$

The condition $y = y_0$ at $h = 0$ was used to derive Equation 3.7.32. Equation 3.7.31 gives the weight fraction of glycerine in the extract phase as a function of column height $h$. Taking the solubility of water in tallow into consideration and using mean flow rates together with the given data result in

$L = 8540$

$G = 3760$

$y_0 = 0.188$

$r = 4.544$

$p = 0.198$

$q = 0.00348Ka$

$v = 1 + 201.6/(Ka)$

Solving Equation 3.7.31 with $H = 72$ ft, gives $Ka = 14.2$ lb glycerine per hour per $ft^3$. With $Ka$ known, values of $y$, $y^*$ and $z$ can be determined as functions of column height using Equations 3.7.31, 3.7.28, 3.7.26, and 3.7.27, and the fraction of tower height principally required for chemical reaction can be determined.

## Example 3.7.6

This example covers Heat Loss through Pipe Flanges (Jenson and Jeffreys[15]).

Two thin wall metal pipes of 1 in outside diameter are connected by 1/2-in. thick and 4-in. diameter flanges that are carrying steam at 250°F. Determine the rate of heat loss from the pipe and the proportion that leaves the rim of the flange.

Thermal conductivity of the flange metal is $k = 220$ Btu/h $ft^2$°F $ft^{-1}$

The exposed surfaces of the flanges lose heat to the surroundings at $T_1 = 60$°F according to a heat transfer coefficient $h = 2$ Btu/h $ft^2$°F.

*Solution*

Consider one flange with one exposed circular face and an exposed rim with radial coordinate $r$ measured in inches from the axis of the pipe. Then the heat balance over an element of width $\Delta r$, as shown in Figure 3.6, give

$$\text{Input} = -2\pi 1/2r \frac{k}{12} \frac{dT}{dr}$$

$$\text{Output} = -\frac{\pi k r}{12} \frac{dT}{dr} + \frac{d}{dr}\left[-\frac{\pi k r}{12} \frac{dT}{dr}\right]\Delta r + \frac{2\pi r h \Delta r}{144}(T - T_1)$$

Accumulation = 0

Simplification of the heat balance leads to

$$r\frac{d^2T}{dr^2} + \frac{dT}{dr} - \frac{h}{6k}r(T - T_1) = 0$$

which can be reduced to

$$x^2 \frac{d^2y}{dx^2} + x\frac{dy}{dx} - x^2 y = 0 \tag{3.7.33}$$

for $y = T - T_1$, and $x = r\sqrt{h/6k}$

Equation 3.7.33 is a modified Bessel equation of zero order. By comparison with Equation 3.6.4, we get

$$y_1 = J_0(ix)$$

and

$$y_2 = Y_0(ix)$$

but $J_0(ix)$ is usually denoted as $I_0(x)$ and $Y_0(ix)$ as $K_0(x)$ where

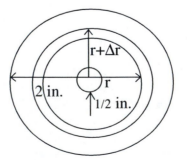

**FIGURE 3.6**
Pipe flange.

$$I_0(x) = 1 + \frac{x^2}{2^2} + \frac{x^4}{2^2 4^2} + \frac{x^6}{2^2 4^2 6^2} + \dots$$

and

$$K_0(x) = [\ln 2 - \gamma] I_0(x) - I_0(x) \ln(x) + \frac{1}{4} x^2 + \dots$$

are real forms. The number $\gamma$ is Euler's constant. Therefore, the general solution of Equation 3.7.33 is

$$y_g = c_1 I_0(x) + c_2 K_0(x)$$

Application of the boundary conditions:

$$\text{at} \quad r = \frac{1}{2}$$

$$T = 250 \quad \Rightarrow \quad x = 0.0195$$

$$y = 190$$

and at $r = 2$, $\quad -\dfrac{k}{12}\dfrac{dT}{dr} = \dfrac{h}{144}(T - T_1) \quad \Rightarrow \quad x = 0.078, \quad \dfrac{dy}{dx} = -0.0195y$

gives

$$T = 60 + 186.5 I_0(0.039r) + 0.8636 K_0(0.039r)$$

where the values of the pertinent Bessel function are taken from standard math tables (Spiegel[13]). The heat conducted from the pipe by the flange is given by

$$Q_1 = -\frac{\pi k}{24}\left(\frac{dT}{dr}\right)_{r=1/2} = 47.81 \text{ Btu/h}$$

and the heat lost through the rim is given by

$$Q_2 = -\frac{\pi k}{6}\left(\frac{dT}{dr}\right)_{r=2} = 16.62 \text{ Btu/h}$$

Therefore, the pipe loses 47.8 Btu/h through each flange of which 35% is lost through the rim.

## Example 3.7.7

This example covers Tubular Gas Preheater (Jenson and Jeffreys[15]).

1000 ft³/h of 70°F air is drawn through a 4-ft long and 4-in. diameter heated cylindrical pipe. The wall temperature, $T_w$, of the pipe is maintained at 600°F for the total length, and the overall heat transfer coefficient, h, as a function of x (distance measured from pipe inlet) is

$$h = 5x^{-1/2} \text{ Btu/h ft}^2\text{°F}$$

The air properties are:

Specific heat ($C_p$)                    = 0.24 Btu/lb°F
Thermal conductivity (k)          =0.020 Btu/h ft²°F ft⁻¹
Density (ρ)                              =0.050 lb/ft³

Assumptions:

Heat transfer occurs by conduction within the gas in an axial direction.
Mass flow of the gas is in an axial direction.

*Solution*
With the aid of Figure 3.7, the following heat balance can be established.

|  | **Input** | **Output** |
|---|---|---|
| Conduction | $-kA\dfrac{dT}{dx}$ | $-kA\dfrac{dT}{dx} + \dfrac{d}{dx}\left(-kA\dfrac{dT}{dx}\right)\Delta x$ |
| Mass flow | $u\rho C_p T$ | $u\rho C_p\left(T\dfrac{dT}{dx}\Delta x\right)$ |
| Wall heat transfer | $\pi Dh(T_w - T)\Delta x$ | |

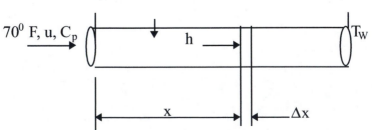

**FIGURE 3.7**
Gas preheater.

where A is the pipe cross-sectional area and D is its diameter. Then

$$-kA\frac{dT}{dx} + u\rho C_p T + \pi Dh\left(T_w - T\right)\Delta x$$

$$= -kA\frac{dT}{dx} + \frac{d}{dx}\left(-kA\frac{dT}{dx}\right)\Delta x + u\rho C_p\left(T + \frac{dT}{dx}\Delta x\right)$$

simplifies to

$$\pi Dh\left(T_w - T\right) = -kA\frac{d^2T}{dx^2} + u\rho C_p\left(\frac{dT}{dx}\right)$$

and rearranging gives

$$\frac{d^2T}{dx^2} - \frac{u\rho C_p}{kA}\frac{dT}{dx} + \frac{\pi Dh}{kA}\left(T_w - T\right) = 0 \qquad (3.7.34)$$

Inserting the numerical values and substituting $x = z^2$, and $t = T_W - T$, results in

$$z\frac{d^2t}{dz^2} - \left(1 + 13760z^2\right)\frac{dt}{dz} - 12000z^2t = 0$$

Attempting to solve this differential equation by the method of Frobenius gives $c = 0$ and $c = 2$ as roots of the indicial equation. Then, for $c = 2$ we get

$$a_1 = 0 \qquad\qquad a_3 = 800a_0$$

$$a_2 = 3440a_0 \qquad\qquad a_4 = 7.9 \times 10^6 a_0$$

The coefficients are increasing, even though a convergence test would suggest that this series is convergent. This is a case where a convergent series is not useful since more than 100 terms would be required to determine the first solution. This dilemma occurs due to the much larger coefficients of $\frac{dt}{dz}$ and $t$ in comparison to that of $\frac{d^2t}{dz^2}$. By neglecting the gas conduction term, we get

$$\frac{dt}{dx} + 0.436x^{-1/2}t = 0$$

which can be solved to give

$$t = \alpha \exp\left(-0.872x^{1/2}\right)$$

Then at $x = 0$, $t = 530$, such that

$$t = 530 \exp\left(-0.872x^{1/2}\right) \tag{3.7.35}$$

or

$$T = 600 - 530 \exp\left(-0.872x^{1/2}\right)$$

resulting in an exit gas temperature of 507°F.

A check on the accuracy of this approximation shows an error of 7.5% for an $x = 10^{-3}$. That is, Equation 3.7.35 can be differentiated twice to show that

$$\frac{\dfrac{d^2t}{dx^2}}{\left(688\dfrac{dt}{dx}\right)} = \frac{1}{13760x} + \frac{1}{15780x^{1/2}}$$

which is small, except when x is small.

### Example 3.7.8

A control valve of the type shown in Figure 3.8 is actuated by air pressure ranging from 3 psig to 15 psig operating on a 16-in. diameter diaphragm whose effective area is equivalent to 100 in.². The effective dead weight of the moving parts of the valve, allowing for the friction in the gland, etc., is estimated to be 300 $lb_f$, the stiffness of the spring is 600 $lb_f$/in., and the damping constant is estimated to be 17 $lb_f$-s/in. If the total lift of the valve is 2.0 in., predict the response of the valve if the controlling air pressure suddenly changes from 6 psig to 12 psig.

*Solution*
A force balance on the valve is Input force = $\Delta PAU(t)$, $U(t)$ is the unit step function. Output force consists of:

- force to overcome inertia of movable parts $= m\dfrac{d^2x}{dt^2}$
- force to overcome resistance of the spring $= kx$

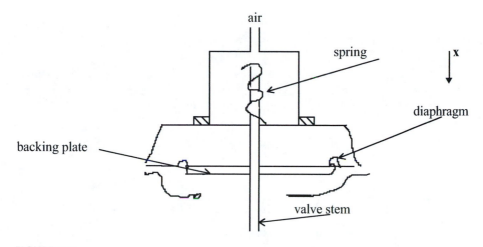

**FIGURE 3.8**
Control valve.

- force to overcome damping resistance = $c\dfrac{dx}{dt}$

Therefore, at equilibrium

$$m\frac{d^2x}{dt^2} + c\frac{dx}{dt} + kx = \Delta PAU(t)$$

By letting $\alpha = c/2m\omega$ where $\omega = \sqrt{k/m}$, simplify the differential equation to

$$\frac{d^2x}{dt^2} + 2\alpha\omega\frac{dx}{dt} + \omega^2 x = \frac{\Delta P}{m}U(t) \qquad (3.7.36)$$

Laplace transform of Equation 3.7.36 gives

$$\left(S^2 + 2\alpha\omega S + \omega^2\right)\bar{x} = \frac{\Delta PA}{mS} + Sx(0) + 2\alpha\omega x(0) + x'(0)$$

where

$$\bar{x} = L\{x(t)\}$$

Then, assuming that the origin of the displacement of the moving parts of the valve corresponds to an air pressure of 6 psig, gives

$$x(0) = x'(0) = 0$$

and $\bar{x}$ bcomes

$$\bar{x} = \frac{\Delta PA}{mS\left(S^2 + 2\alpha\omega S + \omega^2\right)} \qquad (3.7.37)$$

By the method of partial fractions we get

$$\frac{\Delta PA}{mS\left(S^2 + 2\alpha\omega S + \omega^2\right)} = \frac{\Delta PA}{m\omega^2 S} - \frac{\Delta PAS/m\omega^2}{S^2 + 2\alpha\omega S + \omega^2} - \frac{2\alpha\Delta PA/m\omega}{S^2 + 2\alpha\omega S + \omega^2}$$

Then, inverting each term on the right-hand side and simplifying the algebra results in

$$L^{-1}\{\bar{x}(s)\} = x(t)$$

$$= \frac{\Delta PA}{k}\left\{1 - \left(\cos\omega\sqrt{1-\alpha^2}\,t + \frac{\alpha}{\sqrt{1-\alpha^2}}\sin\omega\sqrt{1-\alpha^2}\,t\right)e^{-\alpha\omega t}\right\}$$

Alternatively, the inversion can be carried out in the following way: Consider

$$\frac{\Delta PA}{mS\left(S^2 + 2\alpha\omega S + \omega^2\right)} = \frac{P(s)}{Q(s)}$$

where the polynomial $Q(s)$ has singularities at $S_0 = 0$, $S_1 = -\alpha\omega + i\omega\sqrt{1-\alpha^2}$ and $S_2 = -\alpha\omega - i\omega\sqrt{1-\alpha^2}$, then

$$P_0(t) = \frac{\Delta PA}{Q'(0)}e^{0t} = \frac{\Delta PA}{m\omega^2}$$

$$P_{S_1}(t) = \frac{\Delta PA}{Q'(S_1)}e^{S_1 t} = \frac{\Delta PAe^{-\alpha\omega t}e^{-i\omega\sqrt{1-\alpha^2}\,t}}{2m\omega^2\left(\alpha^2 - 1 - i\alpha\sqrt{1-\alpha^2}\right)}$$

$$P_{S_2}(t) = \frac{\Delta PA}{Q'(S_2)}e^{S_2 t} = \frac{\Delta PAe^{-\alpha\omega t}}{2m\omega\left(\alpha^2 - 1 + i\alpha\sqrt{1-\alpha^2}\right)}e^{-i\omega\sqrt{1-\alpha^2}\,t}$$

Then application of Equation 3.6.20 gives

$$x(t) = \frac{\Delta PA}{m\omega^2} + \frac{\Delta PA}{2m\omega^2} \left\{ \frac{e^{i\omega\sqrt{1-\alpha^2}t}}{\alpha^2 - 1 - i\alpha\sqrt{1-\alpha^2}} + \frac{e^{-i\omega\sqrt{1-\alpha^2}t}}{\alpha^2 - 1 - i\alpha\sqrt{1-\alpha^2}} \right\} e^{-\alpha\omega t}$$

$$= \frac{\Delta PA}{k} \left\{ 1 - \left( \cos\omega\sqrt{1-\alpha^2}t + \frac{\alpha}{\sqrt{1-\alpha^2}} \sin\omega\sqrt{1-\alpha^2}t \right) e^{-\alpha\omega t} \right\}$$

where the Euler formulas

$$\left\{ e^{i\theta} = \cos\theta + i\sin\theta \right.$$

$$\left\{ e^{-i\theta} = \cos\theta - i\sin\theta \right.$$

have been used. Therefore

$$x(t) = 1 - (\cos 25.5t + 0.427 \sin 25.5t)e^{-10.9t}$$

## Example 3.7.9
This example covers an application to Process Control.
   Consider the differential equation

$$ay'' + by' + cy = u(t); \quad y(0) = y_0 \quad \text{and} \quad y'(0) = y_0' \qquad (3.7.38)$$

where a, b, and c are constants and u(t) is a function that has a Laplace transform. Then in terms of process control nomenclature, u(t) is the input, while the solution y(t) is the output or response function. Further,

$$y(t) = y_c(t) + y_p(t)$$

where $y_c(t)$ is the solution to the associated homogeneous equation

$$ay'' + by' + cy = 0$$

and $y_p$ is chosen such that $y_p(0) = y'_p(0) = 0$, while $y_c(0) = y_0$ and $y'_{c(0)} = y'_0$. Then

$$L\{ay'' + by' + cy = 0\} = 0$$

gives

$$Y_c(s) = \frac{1}{as^2 + bs + c} \left[ a(sy_0 + y_0') + by_0 \right]$$

Also

$$L\left\{ ay_p'' + by_p' + cy_p = (t) \right\} \text{ with } y_p(0) = y_p'(0) = 0 \quad \text{gives}$$

$$Y_p(s) = \frac{1}{as^2 + bs + c} U(s), \quad U(s) = L\left\{ u(t) \right\}$$

The factor $\dfrac{1}{as^2 + bs + c}$ is denoted as H(s) and is called the transfer function of Equation 3.7.38. Notice here that H(s) and the initial conditions $y(0) = y_0$ and $y'(0) = y_0'$ together produce the complementary solution

$$y_c(t) = L^{-1}\left\{ Y_c(s) \right\}$$

while the transfer function and the input function u(t) produce the particular solution

$$y_p(t) = L^{-1}\left\{ Y_p(s) \right\}$$

Therefore, u(t) is said to control the response $y_p(t)$; that is, u(t) is the *control function*.

From Figure 3.9, if H is considered fixed, then u(t) controls the output $y_p(t)$. Further, since H(s) = L{h(t)}, where h(t) is some function, then the transfer function is the Laplace transform of the impulse response. That is

$$ay'' + by' + cy = \delta(t), \quad y(0) = y'(0) = 0$$

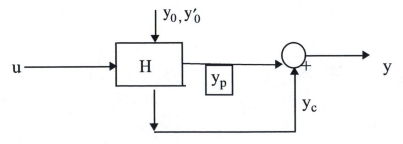

**FIGURE 3.9**
A control system with transfer function H(s).

transforms to

$$as^2Y(s) + bsY(s) + cY(s) = 1$$

or

$$Y(s) = \frac{1}{as^2 + bs + c}$$

If we recall the following three cases for positive coefficients of Equation 3.7.38, that is a, b, and c are all positive, then

*Case I:*
for $b^2 - 4ac > 0$

$$m_1 = \frac{-b + \sqrt{b^2 - 4ac}}{2a} \text{ and } m_2 = \frac{-b - \sqrt{b^2 - 4ac}}{2a}$$

such that

$$y_c = c_1 e^{m_1 t} + c_2 e^{m_2 t}$$

*Case II:*
for $b^2 - 4ac = 0$

$$m_1 = m_2 = \frac{-b}{2a} \text{ and } y_c = c_1 e^{m_1 t} + c_2 t e^{m_1 t}$$

and if

$$b2 - 4ac < 0$$

$$m_1 = \alpha + i\beta, \ m_2 = \alpha - i\beta$$

and

$$y_c = e^{\alpha t}(c_1 \cos\beta t + c_2 \sin\beta t)$$

In each of the above three cases, the solution $y_c(t) \to 0$ as $t \to \infty$. Therefore, $y_c(t)$ and the initial conditions $y_0$ and $y'_0$ become less important to the response function ($y = y_c + y_p$) as t gets larger and larger. That is, $y \to y_p$ as t $\to \infty$, thus $y_p$ is the *steady-state* solution. Also,

$$y_p = L^{-1}\{H(s)U(s)\} = h(t)*u(t)$$

That is, for positive coefficients, the response tends to the steady state given by the convolution of the impulse response with the control function.

## Example 3.7.10
Consider the steady flow between parallel planes (Walas[24]). Two planes are a distance 2b apart in the y direction and extend to infinity in the z direction. Assume that the fluid flow between them is steady. Derive the average velocity, $\bar{u}$, in terms of the impressed pressure gradient, the distance between the plates, and the fluid viscosity. Also write down the ratio of the velocity to that of the average.

*Solution*
Starting with the general equation in Cartesian coordinates for incompressible fluid with viscosity independent of position (Bird, Stewart and Lightfoot[19]), $\partial u / \partial z = 0$, $\partial^2 u / \partial z^2 = 0$, and the pressure gradient can be taken as constant, that is

$$\frac{dp}{dx} = A. \qquad\qquad (\textit{The reader should verify this statement.})$$

Then

$$\frac{d^2 u}{dy^2} = \frac{A}{\mu}$$

which integrates to

$$u = \frac{Ay^2}{2\mu} + c_1 y + c_2$$

Applying the conditions $u = 0$ when $y = \pm b$ (*the reader should verify these conditions*) results in

$$u = \frac{1}{2\mu}\frac{dP}{dx}\left(y^2 - b^2\right)$$

(*What is the sign of the pressure gradient when flow is to the right?*). The average velocity is determined by integrating over the cross-section:

$$\bar{u} = \frac{1}{2b}\int_{-b}^{b} u\, dy = -\frac{2Ab^2}{3\mu}, \quad\text{and}\quad \frac{u}{\bar{u}} = \frac{3}{2}\left[1 - \left(\frac{y}{b}\right)^2\right]$$

## 3.8 Problems

1. Two thin wall pipes of 1 in. outside diameter have flanges 1/2-in. thick and 4-in. diameter on the ends joining them together. If the conductivity of the flange metal is k Btu/h.ft².°F ft⁻¹, and the exposed surfaces of the flange lose heat to the surroundings, which are at $T_1$°F by means of a heat transfer coefficient, h Btu/h.ft²°F, show that the equation giving the temperature distribution in the flange is

$$6k\left(r\frac{d^2T}{dr^2}+\frac{dT}{dr}\right)=hr\left(T-T_1\right)$$

where r is the radial distance coordinate in inches.

If the circular faces of the flanges are thermally insulated and the flanges only lose heat through the rim where h = 20, solve the differential equation and determine the temperature of the rim for a pipe temperature of 200°F, $T_1$ = 60°F, and k = 200.

2. Mass transfer. As liquid flows across any plate of a distillation column its composition, x, changes from the entry concentration, $x_0$, to the exit concentration, $x_1$ (Jenson and Jeffreys[15]). The composition at any point on the plate is influenced by the passage of the stripping gas at a rate G, the bulk flow at a rate L, and the mixing on the plate by the eddy diffusivity, $D_E$. If a constant Murphree point efficiency

$$E^*_{mv}=\frac{y-y_1}{y^*-y_1}$$

can be assumed and a straight line equilibrium curve given by

$$y^*=mx+b$$

show that the liquid composition satisfies the equation

$$D_E\frac{d^2x}{dz^2}-V\frac{dx}{dz}-\frac{mGV}{Lz_1}E^*_{mv}\left(x-x^*\right)=0$$

where z is the distance measured along the plate from the inlet weir, $z_1$ is the distance between the weirs, V is the linear velocity of the liquid, and $x^*$ is the liquid composition in equilibrium with the entering gas, which is constant across the plate.

3. Consider the problem of heat loss from the surface of an oven wall due to "through metal," which conducts heat from the inside. The heat is dissipated to the air from the sheet-metal protective covering of the insulated housing. The metal covering consists of 0.005-ft thick steel having thermal conductivity of 25 Btu/h ft². °F/ft. The surface coefficient of heat transfer is 2.5 Btu/h.ft². °F and the head of the bolt is 5/8-in. diameter. The temperature of the room is 70°F and the bolt head temperature is constant at 150°F. Neglecting heat loss, except by conduction along the bolt, determine temperatures of the outer metal wall at several points up to 1 ft from the bolt. *Hint: Since the temperature is symmetrical about the bolt, then T can be assumed to be a function of r only.*

4. A copper fin L ft long is triangular in cross-section (Mickley, Sherwood and Reid[6]). It is w ft thick at the base and tapers off to a line (see following sketch). The base of this wedge-shaped piece of metal is maintained at a constant temperature $T_A$, and the fin loses heat by convection to the surrounding air, which is at a temperature $T_B$. The surface coefficient of heat transfer is h Btu/h.ft².°F. Derive the relationship between the temperature, T, of the metal fin and the distance from the base, $L-x$. *Hint: Assume T is a function of x only.*

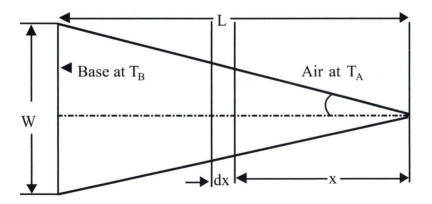

Wedge-shaped fin.

Answer   $\dfrac{d}{dx}\left(k\,\dfrac{w}{L}\times\dfrac{dT}{dx}\right)=h\,2\sec\theta\left(T-T_A\right)$

5. Steady Flow between a Fixed and a Moving Plate (White[25]). Assume that (1) Two infinite plates are 2h apart, and the upper plate moves at speed U relative to the lower. (2) The pressure p is assumed constant. (3) The upper plate is held at temperature $T_1$ and the lower plate at $T_0$. Derive the velocity and temperature

profiles assuming that u = u(y) and T = T(y). Also, find the shear stress.

Answer: $\dfrac{\mu U}{2h}$

$$y = +h,\ u = U,\ T = T_1 \qquad \text{moving}$$

$$\rho = \text{const.} \qquad y$$

$$y = -h,\ u = 0,\ T = T_0 \qquad z$$

Couette flow between parallel plates.

Continuity: $\dfrac{\partial u}{\partial x} = 0$

Momentum: $0 = \mu \dfrac{d^2 u}{dy^2}$

Energy: $0 = k \dfrac{d^2 T}{dy^2} + \mu \left( \dfrac{du}{dy} \right)^2$

6. Diffusion with Chemical Reaction. A gas is absorbed by a solution with which it reacts chemically (Mickley, Sherwood and Reid[6]). The rate of diffusion in the liquid can be assumed proportional to the concentration gradient, and the diffusing gas is eliminated as it diffuses by a first order chemical reaction. The reaction rate is propor-

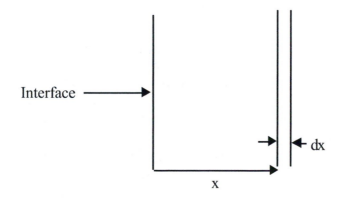

Interface

dx

x

Diffusion with chemical reaction.

tional to the concentration of the solute gas in the liquid. Obtain an expression for the concentration in the liquid as a function of the distance from the gas liquid interface (see sketch).

Answer: $\dfrac{d}{dx}\left(D\dfrac{dc}{dx}\right) = kc$

7. Consider the set of reversible reactions ( Mickley, Sherwood and Reid[6]).

$$A \xrightarrow{k_1} B$$
$$B \xrightarrow{k_2} A$$
$$B \xrightarrow{k_3} C$$
$$C \xrightarrow{k_4} B$$

Suppose the initial amount of A is 1 mole and $N_A$, $N_B$, and $N_C$ denote the moles of A, B, and C, respectively, at any time t. Derive a second order constant coefficient linear differential equation relating $N_A$, t, and the rate constants.

8. Oxygen dissolves into and reacts irreversibly with aqueous sodium sulfite solutions (Rice and Do[23]). If the gas solubility is denoted as $C_A^*$ at the liquid-gas interface, develop a differential equation that describes the steady state composition profiles of $O_2$ in the liquid phase when $KC_A^n$ is the $O_2$ reaction rate, and $J_A = -D_A\, dC_A/dz$ describes the local $O_2$ diffusion flux.

Answer: $D_A \dfrac{d^2 C_A}{dz^2} = kC_A^n$

9. If one of the planes of Example 3.7.10 is moving with speed V, what are the boundary conditions? Derive the velocity profile for this case.

Answer: $u = \dfrac{V}{2}\left(\dfrac{y}{b}+1\right) + \dfrac{A}{2\mu}\left(y^2 - b^2\right)$

10. Consider the steady flow of a fluid through a circular pipe of radius R. Suppose the fluid flows with a velocity $u_z = U$ in the z direction, starting with the continuity equation (Bird, Stewart and Lightfoot[19]) in polar coordinates and the Navier-Stokes equations (Bird, Stewart and Lightfoot[19]) show that:

$$0 = -\frac{A}{\rho} + \frac{\mu}{\rho}\left(\frac{1}{r}\frac{d}{dr}\left(r\frac{du_z}{dr}\right)\right)$$

where A is the pressure gradient, which is taken as constant. Why is the quantity $\partial^2 u_z/\partial\theta^2 = 0$. Also, show that

$$dP = -\frac{8\mu\bar{u}_z}{R^2}dz = -32\frac{\mu\bar{u}_z}{D^2}dz$$

11. Reduction of Linear Boundary Value Problems to Initial Value Problems (Na[26]). Consider the problem given by the differential equation

$$\frac{d^2y}{dx^2} + f_1(x)\frac{dy}{dx} + f_2(x)y = r(x) \tag{1}$$

and the boundary conditions

$$y(a) = y_a, \quad y(b) = y_b \tag{2}$$

This problem can be transformed into a set of two initial value problems (IVP).

Transform Equation 1 into a set of two differential equations by assuming that

$$y(x) = y_1(x) + \mu y_2(x) \tag{3}$$

is a linear combination of solutions to the desired differential equations, where the quantity $\mu$ is a constant. Use the boundary conditions given by Equation 2 to derive the initial conditions

$$y_1(a) = y_a, \quad y_2(a) = 0 \tag{4}$$

and set

$$\frac{dy_1(a)}{dx} = 0, \quad \frac{dy_2(a)}{dx} = 1 \tag{5}$$

to identify the missing slope. Transform the boundary condition at the second point to define $\mu$.

Answer: IVP-1:

$$\frac{d^2y_1}{dx^2} + f_1(x)\frac{dy_1}{dx} + f_2(x)y_1 = r(x)$$

$$y_1(a) = y_a, \quad \frac{dy_1(a)}{dx} = 0$$

$$\frac{d^2y_2}{dx^2} + f_1(x)\frac{dy_2}{dx} + f_2(x)y_2 = 0$$

IVP-2:  $y_2(a) = 0, \quad \dfrac{dy_2(a)}{dx} = 1$

$$\mu = \left[y_b - y_1(b)\right]\big/y_2(b)$$

---

# References

1. Boyce, W.E. and DiPrima, R.C., *Elementary Differential Equations and Boundary Value Problems*, 3rd ed., John Wiley & Sons, New York, 1977.
2. Giordano, F.R. and Weir, M.D., *Differential Equations, a Modeling Approach*, Addison-Wesley, Reading, MA, 1988.
3. Greenberg, M.D., *Advanced Engineering Mathematics*, Prentice-Hall, Englewood Cliffs, NJ, 1988.
4. O'Neil, P.V., *Advanced Engineering Mathematics*, 4th ed., PWS-Kent, Boston, 1995.
5. Hildebrand, F.B., *Advanced Calculus for Applications*, 2nd ed., Prentice-Hall, Englewood Cliffs, NJ, 1976.
6. Mickley, H.S., Sherwood, T.K., and Reid, C.E., *Applied Mathematics in Chemical Engineering*, McGraw-Hill, New York, 1957.
7. Loney, N.W., Analytical solution to mass transfer in laminar flow in hollow fiber with heterogeneous chemical reaction, *Chemical Eng. Sci.*, 51, 3995, 1995.
8. Huang, C.R., Heat transfer to a laminar flow fluid in a circular tube, *AIChE J.*, 30, 833, 1984.
9. Watson, G.N., *A Treatise on the Theory of Bessel Functions*, 2nd ed., Cambridge University Press, London, 1966.
10. Gray, A. and Matthews, G.B., *A Treatise on Bessel Functions and Their Applications to Physics*, 2nd ed., Dover, New York, 1966.
11. Tranter, C. J., *Bessel Functions with Some Physical Applications*, English University Press, London, 1968.
12. Slater, L.C.J., *Confluent Hypergeometric Functions*, Cambridge University Press, London, 1960.
13. Spiegel, M.R., *Mathematical Handbook*, McGraw-Hill, New York, 1968.
14. Thomas, G.B. and Finney, R.L., *Calculus and Analytic Geometry*, 6th ed., Addison-Wesley, Reading, MA, 1984.
15. Jenson, V.G. and Jeffreys, G.V., *Mathematical Methods in Chemical Engineering*, Academic Press, London, 1963.
16. Churchill, R.V. and Brown, J.W., *Complex Variables and Applications*, 4th ed., McGraw-Hill, New York, 1984.
17. Kreyszig, E., *Advanced Engineering Mathematics*, 7th ed., John Wiley & Sons, New York, 1993.
18. Saff, E.B. and Snider, A.D., *Fundamentals of Complex Analysis for Mathematics, Science and Engineering*, Prentice-Hall, Englewood Cliffs, NJ, 1976.

19. Bird, R.B., Stewart, W.E., and Lightfoot, E.N., *Transport Phenomena*, John Wiley & Sons, New York, 1960.
20. Schlichting, H., *Boundary-Layer Theory*, 7th ed., McGraw-Hill, New York, 1979.
21. Incropera, F.P. and DeWitt, D.P., *Fundamentals of Heat and Mass Transfer*, 2nd ed., John Wiley & Sons, New York, 1985.
22. Dettman, J.W., *Applied Complex Variables*, Dover, New York, 1984.
23. Rice, R.G. and Do, D.D., *Applied Mathematics and Modeling for Chemical Engineers*, John Wiley & Sons, New York, 1995.
24. Walas, S.M., *Modeling with Differential Equations in Chemical Engineering*, Butterworth-Hinemann, Boston, 1991.
25. White, F.M., *Viscous Fluid Flow*, McGraw-Hill, New York, 1991.
26. Na, T.Y., *Computational Methods in Engineering Boundary Value Problems*, Academic Press, New York, 1979.

# 4

## Sturm-Liouville Problems

### 4.1 Introduction

In the previous chapters, the discussion was primarily about initial value problems, those types for which one seeks a solution to a differential equation subject to conditions on the dependent variable and its derivatives specified at only one value of the independent variable.

In this chapter, the discussion will be about the class of problems, for which one seeks a solution to a differential equation subject to conditions on the dependent variable specified at two or more values of the independent variable, namely *boundary value problems*.

Among the many problems encountered in chemical engineering science, those that appear very frequently involve the Sturm-Liouville boundary value problems. Following is an illustration of what is meant by a Sturm-Liouville problem.

Consider the problem

$$y'' + \frac{2}{x}y' + \left(x^2 - \lambda x\right)y = 0 \qquad (4.1.1)$$

then an integrating factor for Equation 4.1.1 is

$$\mu(x) = e^{\int \frac{2}{x}dx} = x^2$$

If Equation 4.1.1 is multiplied by $x^2$ we get

$$x^2 y'' + 2xy' + \left(x^4 - \lambda x^3\right)y = 0$$

or

$$\left(x^2 y'\right)' + \left(x^4 - \lambda x^3\right)y = 0 \qquad (4.1.2)$$

Equation 4.1.2 is considered to be in the Sturm-Liouville form. That is

$$(p(x)y')' + [q(x) + \lambda r(x)]y = 0 \tag{4.1.3}$$

is the Sturm-Liouville differential equation.

It is also helpful to distinguish between two types of linear boundary value problems, those that are homogeneous and those that are not. The equation

$$p(x)y'' + Q(x)y' + R(x)y = G(x) \tag{4.1.4}$$

is a general, linear, second order, and nonhomogeneous differential equation and was discussed in the previous chapter. The condition

$$a_1 y(x_0) + a_2 y'(x_0) = c \tag{4.1.5}$$

is a general, linear, and nonhomogeneous boundary condition. Both Equation 4.1.4 and Equation 4.1.5 are considered homogeneous if their right-hand sides are identically zero. Therefore, we are using the same definition of "homogeneous" as was introduced in the discourse on second order differential equations. However, it is very important to note what is meant by a *homogeneous boundary value problem* as opposed to a differential equation.

*A boundary value problem is homogeneous when both the differential equation and all boundary conditions are homogeneous (Boyce and Diprima,[1] P.V.O'Neil,[2] Humi and Miller,[3] and Myint-U and Debnath [4]) and nonhomogeneous otherwise.*

For example, the problem

$$P(x)y'' + Q(x)y' + R(x)y = 0, 0 < x < 1$$

$$a_1 y(0) + a_2 y'(0) = 0$$

$$b_1 y(0) + b_2 y'(0) = 0$$

is a second order linear homogeneous boundary value problem.

Sturm-Liouville problems are categorized according to the type of boundary conditions that the differential equation must satisfy.

## 4.2. Classification of Sturm-Liouville Problems

A general homogeneous second order differential equation of the form

$$a_1(x)y'' + a_2(x)y' + [a_3(x) + \lambda]y = 0 \tag{4.2.1}$$

can be recasted in the Sturm-Liouville form

$$\frac{d}{dx}\left(p(x)\frac{dy}{dx}\right) + [q(x) + \lambda r(x)]y = 0 \tag{4.2.2}$$

with the aid of the following device

$$p(x) = \exp\left[\int^x \frac{a_2(t)}{a_1(t)}dt\right] \tag{4.2.3}$$

$$q(x) = \frac{a_3(x)}{a_1(x)}p(x) \tag{4.2.4}$$

and

$$r(x) = \frac{p(x)}{a_1(x)} \tag{4.2.5}$$

where p, q, and r are real-valued functions of x and $\lambda$ is a parameter. Further, to ensure the existence of solutions on a closed finite interval [a, b], q and r are to be continuous on [a, b], while p is to be continuously differentiable on [a, b] (Myint-U and Debnath[4] and O'Neil[2]).

Generally, there are three classes of Sturm-Liouville problems,

1. Regular
2. Periodic
3. Singular

Each class is discussed below with an illustrative problem. Again, the boundary conditions are very influential as to what class of boundary value problem one has to solve.

Class 1. *The Regular Sturm-Liouville Problem on [a, b]*

If p(x) > 0 and r(x) > 0 on a ≤ x ≤ b, one seeks numbers $\lambda$ and nontrivial solutions of

$$(p(x)y')' + [q(x) + \lambda r(x)]y = 0 \tag{4.2.6}$$

subject to:

$$A_1 y(a) + A_2 y'(a) = 0 \tag{4.2.7}$$

$$B_1 y(b) + B_2 y'(b) = 0 \tag{4.2.8}$$

in which $A_1$ and $A_2$ are given and not both zero, and $B_1$ and $B_2$ are given and not both zero. Equations 4.2.6, 4.2.7, and 4.2.8 constitute a *regular Sturm-Liouville* boundary value problem. For example, the problem

$$y'' + \lambda y = 0 \qquad 0 \le x \le \frac{\pi}{2}$$

$$y(0) = 0$$

$$y\left(\frac{\pi}{2}\right) = 0$$

where $p(x) = 1$, $r(x) = 1$ and $q(x) = 0$ is a regular Sturm-Liouville problem with $A_1 = B_1 = 1$ and $A_2 = B_2 = 0$.

There are three cases to consider, depending on the parameter $\lambda$, i.e., $\lambda = 0$, $\lambda > 0$, $\lambda < 0$.

*Case I:*

Suppose $\lambda = 0$, then $y'' = 0$ and

$$y_g(x) = k_1 x + k_2$$

$$g(0) = 0 = k_2$$

$$y\left(\frac{\pi}{2}\right) = 0 = k_1 \frac{\pi}{2} \quad \therefore \quad k_1 = 0$$

giving the trivial solution. Therefore, $\lambda \ne 0$.

*Case II:*

Suppose $\lambda > 0$, say $\lambda = \beta^2$ (for convenience), then

$$y'' + \beta^2 y = 0$$

$$y(0) = 0, \qquad y(\pi/2) = 0$$

has general solution

$$y(x) = k_1 \cos \beta x + k_2 \sin \beta x$$

and

$$y(0) = 0 = k_1$$

$$y\left(\pi/2\right) = 0 = k_2 \sin \beta \frac{\pi}{2}$$

therefore, either $k_2 = 0$ or $\sin \beta \frac{\pi}{2} = 0$.

If $k_2 = 0$, then we get the trivial solution. Then, for nontrivial solution to exist

$$\sin \beta \frac{\pi}{2} = 0, \quad \Rightarrow \quad \beta \frac{\pi}{2} = n\pi$$

or

$$\beta = 2n, \quad n = 1, 2, 3, K$$

Therefore, $\lambda_n = 4n^2$ defines the so-called *eigenvalues* of this problem. Corresponding to each eigenvalue is an *eigenfunction*

$$\phi_n(x) = k_{2n} \operatorname{Sin}(2nx)$$

with $k_{2n}$ an arbitrary constant.

*Case III:*

Suppose $\lambda < 0$, say $\lambda = -\alpha^2$, $\alpha > 0$, then

$$y'' - \alpha^2 y = 0$$

$$y(0) = y(\pi/2) = 0$$

Therefore, $y_g(x) = c_1 e^{\alpha x} + c_2 e^{-\alpha x}$ and

$$y(0) = 0 = c_1 + c_2 \quad \Leftrightarrow \quad c_2 = -c_1$$

$$y\left(\frac{\pi}{2}\right) = 0 = c_1 e^{\alpha\frac{\pi}{2}} + c_2 e^{-\alpha\frac{\pi}{2}} \quad \Rightarrow \quad c_1 \left( e^{\alpha\frac{\pi}{2}} - e^{-\alpha\frac{\pi}{2}} \right) = 0$$

from which one can conclude that $c_1 = 0 = c_2$, and there are no negative eigenvalues.

In summary, the regular Sturm-Liouville problem

$$y'' - \lambda y = 0, \quad y(0) = y\left(\frac{\pi}{2}\right) = 0$$

has eigenvalues

$$\lambda_n = 4n^2, \quad n = 1,2,3,\ldots$$

and corresponding eigenfunctions

$$\phi_n(x) = k_{2n} \sin(2nx)$$

with $k_{2n}$ nonzero but otherwise arbitrary.

Class 2. *The Periodic Sturm-Liouville Problem on [a, b]*

If $p(x) > 0$ and $r(x) > 0$ on $a \le x \le b$, one seeks number $\lambda$ and nontrivial solution of

$$(p(x)y')' + [q(x) + \lambda r(x)]y = 0 \tag{4.2.6}$$

subject to:

$$y(a) = y(b) \tag{4.2.9}$$

$$y'(a) = y'(b) \tag{4.2.10}$$

For example,

$$y'' + \lambda y = 0 \quad \text{on} \quad [-\pi, \pi]$$

$$y(-\pi) = y(\pi)$$

$$y'(-\pi) = y'(\pi)$$

is a periodic Sturm-Liouville problem. Again there are three cases to consider.

*Case I:*

Suppose $\lambda = 0$, then

$$y_g(x) = k_1 x + k_2 \text{ and } y(-\pi) = -k_1\pi + k_2 = y(\pi) = k_1\pi + k_2$$

which results in

$$y = k_2$$

Thus, $\lambda = 0$ is an eigenvalue for this problem with corresponding eigenfunction $y = k_2$, an arbitrary constant.

*Case II:*

Suppose $\lambda > 0$, say $\lambda = \beta^2$
then

$$y_g = c_1 \cos\beta x + c_2 \sin\beta x$$

and

$$y(-\pi) = c_1 \cos\beta x - c_2 \sin\beta x = y(\pi) = c_1 \cos\beta\pi + c_2 \sin\beta\pi$$

and

$$y'(-\pi) = c_1\beta \sin\beta\pi + c_2\beta \cos\beta\pi = -c_1\beta \sin\beta\pi + c_2 \cos\beta\pi = y'(\pi)$$

Therefore,

$$2c_2 \sin\beta\pi = 0$$

and

$$2c_1 \sin\beta\pi = 0 \implies \sin\beta\pi = 0 \implies \beta\pi = n\pi, \ n = 1, 2, 3, \ldots$$

for nontrivial solution. Therefore, $\lambda = n^2$ is an eigenvalue with corresponding eigenfunctions

$$\phi_n(x) = c_{1n} \cos(nx) + c_{2n} \sin(nx)$$

in which $c_{1n}$ and $c_{2n}$ are arbitrary constants which cannot both be zero but are otherwise arbitrary.

For the case with $\lambda < 0$, it can be shown that only trivial solutions result. Therefore, the periodic Sturm-Liouville problem

$$y'' + \lambda y = 0, \quad y(-\pi) = y(\pi) \quad \text{and} \quad y'(-\pi) = y'(\pi)$$

has eigenvalues

$$\lambda = n^2, \quad n = 0, 1, 2, \ldots$$

and eigenfunction

$$\phi_n(x) = c_{1n} \cos(nx) + c_{2n} \sin(nx)$$

with not both $c_{1n}$ and $c_{2n}$ zero.

Class 3. *The Singular Sturm-Liouville Problem on [a, b]*

If $p(x) > 0$ and $r(x) > 0$ on $a \leq x \leq b$, one seeks number $\lambda$ and nontrivial solution of

$$\left(p(x)y'\right)' + \left[q(x) + \lambda r(x)\right]y = 0 \tag{4.2.6}$$

satisfying one of the following three types of boundary conditions

1. If $p(a) = 0$, then

$$B_1 y(b) + B_2 y'(b) = 0 \tag{4.2.8}$$

   with $B_1$ and $B_2$ given and not both zero, also *solutions must be bounded at a.*

2. If $p(b) = 0$, then

$$A_1 y(a) + A_2 y'(a) = 0 \tag{4.2.7}$$

   with $A_1$ and $A_2$ given and not both zero, also *solutions must be bounded at b.*

3. If $p(a) = p(b) = 0$, we have no boundary conditions specified at either a or b but require that solutions be bounded on [a, b].

For example, Legendre's equation (O'Neil[2])

$$(1 - x^2)y'' - 2xy' + \ell(\ell + 1)y = 0$$

can be recasted as

$$\left[(1-x^2)y'\right]' + \lambda y = 0 \qquad -1 \leq x \leq 1$$

with

$$\lambda = \ell(\ell + 1), \quad p(x) = 1 - x^2 \quad \text{such that } p(-1) = p(1) = 0$$

This results in the class of singular Sturm-Liouville problems having type 3 boundary condition. Further, since

$$(1-x^2)y'' - 2xy' + \ell(\ell+1)y = 0$$

can be rewritten as

$$y'' - \frac{2x}{1-x^2}y' + \frac{\ell(\ell+1)}{1-x^2}y = 0$$

with singular points at $x = \pm 1$, but both $\dfrac{-2x}{1-x^2}$ and $\dfrac{\ell(\ell+1)}{1-x^2}$ have Taylor Series expansion about the origin in $-1 < x < 1$ ($x = 0$ is an ordinary point), then

$$y = \sum_{n=0}^{\infty} a_n x^n$$

can be assumed as a solution of the given differential equation. Proceeding formally, as in the Frobenius series method, results in

$$2a_2 + \ell(\ell+1)a_0 + \left[3 \cdot 2a_3 + \ell(\ell+1)a_1 - 2a_1\right]x$$

$$+ \sum_{n=2}^{\infty} \left\{(n+2)(n+1)a_{n+2} + \left[\ell(\ell+1) - n(n-1) - 2n\right]a_n\right\}x^n = 0$$

from which

$$a_2 = -\frac{\ell(\ell+1)}{2}a_0$$

$$a_3 = -\frac{(\ell+2)(\ell+1)}{3 \cdot 2}a_1$$

$$a_{n+2} = -\frac{(n+\ell+1)(\ell-n)}{(n+2)(n+1)}a_n, \quad n \geq 2$$

where the latter equation is the *recurrence relation*.

Following the determination of a few more constants, one can observe the patterns:

$$a_{2n} = \frac{(-1)^n(\ell+2n-1)(\ell+2n-3)...(\ell+1)\ell(\ell-2)...(\ell-2n+4)(\ell-2n+2)}{(2n)!}a_0$$

for even indexed coefficients and

$$a_{2n+1} = \frac{(-1)^n(\ell+2n)...(\ell+4)(\ell+2)(\ell-1)(\ell-3)...(\ell-2n+1)}{(2n+1)!}a_1$$

for odd indexed coefficients.

Therefore, one can obtain two linearly independent solutions of Legendre's equation. Say for $a_0 = 1$ and $a_1 = 0$

$$y_1 = \sum_{n=0}^{\infty}(-1)^n\frac{(\ell+2n-1)(\ell+2n-3)...(\ell+1)\ell(\ell-2)...(\ell-2n+4)(\ell-2n+2)}{(2n)!}x^{2n}$$

and for $a_0 = 0$ and $a_1 = 1$

$$y_2 = \sum_{n=0}^{\infty}(-1)^n\frac{(\ell+2n)...(\ell+4)(\ell+2)(\ell-1)(\ell-3)...(\ell-2n+1)}{(2n+1)!}x^{2n+1}$$

By making appropriate choices of $\ell$, the linearly independent solutions can be reduced to polynomials (Legendre's polynomials).

Knowledge of the properties of the eigenvalues and eigenfunctions can dramatically reduce the labor typically required to solve a Sturm-Liouville problem. These properties also provide a check on whether or not one's solution is reasonable.

## 4.2.1. Properties of the Eigenvalues and Eigenfunctions of a Sturm-Liouville Problem

The following theorem summarizes the important properties of the Sturm-Liouville problem and its eigenvalues (Boyce and Diprima[1] and O'Neil[2]).

*THEOREM 4.1*

1. *For the regular and periodic Sturm-Liouville problems there exist an infinite number of eigenvalues. Further, these eigenvalues can be labeled $\lambda_1, \lambda_2, \cdots$ so that $\lambda_n < \lambda_m$ if $n < m$ and $\lim\limits_{n \to \infty} \lambda_n = \infty$.*

2. *If $\lambda_n$ and $\lambda_m$ are distinct eigenvalues of any of the three types of Sturm-Liouville problems, with corresponding eigenfunctions $\phi_n$ and $\phi_m$, then $\phi_n$ and $\phi_m$ are orthogonal on $[a, b]$ with weight function $r(x)$. That is*

$$\int_a^b r(x)\phi_n(x)\phi_m(x)dx = 0 \quad \text{if} \quad n \neq m.$$

3. *For all three classes of Sturm-Liouville problems, all eigenvalues are real.*

4. *For a regular Sturm-Liouville problem, any two eigenfunctions corresponding to a given eigenvalue are linearly dependent.*

5. *The Laplace transform, $F(s)$, of a solution to a Sturm-Liouville equation is analytic for all finite s except for poles, which correspond to the eigenvalues of the system.*

Sometimes it is helpful to approximate the value of the smallest eigenvalue when one is faced with a computationally difficult engineering problem. An iterative procedure known as the method of Stodola and Vianello (Hildebrand[5]) is outlined as follows.

Consider the regular Sturm-Liouville problem

$$\frac{d^2y}{dx^2} + \lambda y = 0 \tag{4.2.11}$$

subject to

$$y(0) = 0, \quad y(L) = 0 \tag{4.2.12}$$

This problem can be easily shown to have eigenvalues

$$\lambda_n = \frac{n^2\pi^2}{L^2}, \quad n = 1, 2, \ldots \tag{4.2.13}$$

with corresponding eigenfunctions

$$\phi_n(x) = \sin\frac{n\pi x}{L} \tag{4.2.14}$$

However, the method can be illustrated using this familiar problem:

1. Rewrite Equation 4.2.11 as

$$\frac{d^2y}{dx^2} = -\lambda y \tag{4.2.15}$$

2. Replace the unknown function y on the right-hand side of Equation 4.2.15 by a conveniently chosen first approximation $y_1(x)$ giving

$$\frac{d^2y}{dx^2} = -\lambda y_1(x) \tag{4.2.16}$$

3. Solve Equation 4.2.16, and Equation 4.2.12 formally to get

$$y = \lambda f_1(x) \tag{4.2.17}$$

If $y_1(x)$ were actually an eigenfunction of the problem, then

$$y = y_1$$

and $\lambda$ would be the eigenvalue given by the constant ratio

$$\lambda = \frac{y_1(x)}{f_1(x)}$$

However, $y_1(x)$ is generally not an eigenfunction and therefore $y_1(x)$ and $f_1(x)$ generally will not be in a constant ratio. To illustrate further, suppose $y_1(x)$ is chosen to be

$$y_1(x) = x(L - x)$$

then

$$\frac{d^2y}{dx^2} = -\lambda x(L - x) = \lambda(x^2 - Lx)$$

whose solution satisfying Equation 4.2.12 is

$$y = \frac{\lambda}{12}\left(x^4 - 2Lx^3 + L^3x\right) \equiv \lambda f_1(x)$$

Observe that $y \neq y_1$, but a first approximation of the smallest eigenvalue, $\lambda_1$, can be obtained from

$$\int_0^L y_1(x)dx = \lambda \int_0^L f_1(x)\ dx \tag{4.2.18}$$

That is,

$$\int_0^L \left(Lx - x^2\right)dx = \frac{\lambda}{12}\int_0^L \left(x^4 - 2Lx^3 + L^3x\right)\ dx$$

or

$$\frac{\lambda}{12} = \frac{\displaystyle\int_0^L \left(Lx - x^2\right)\ dx}{\displaystyle\int_0^L \left(x^4 - 2Lx^3 + L^3x\right)\ dx}$$

such that

$$\lambda_1^{(1)} = \frac{10}{L^2}$$

is the first approximation to the smallest eigenvalue. By taking

$$y_2(x) = x^4 - 2Lx^3 + L^3x$$

as the next approximation, and formally solving

$$\frac{d^2y}{dx^2} = -\lambda y_2(x)$$

subject to Equation 4.2.12 gives

$$y = \lambda f_2(x) = -\frac{\lambda}{30}\left(x^6 - 3Lx^5 + 5L^3x^3 - 3L^5x\right)$$

then Equation 4.2.18 becomes

$$\int_0^L y_2(x)\ dx = \lambda \int_0^L f_2(x)\ dx$$

such that the second approximation to $\lambda_1$ is

$$\lambda_1^{(2)} = \frac{168}{17L^2} = \frac{9.882}{L^2}$$

In general, successive estimates of the eigenvalues $\lambda_1$ may be obtained after each iteration by requiring that the functions $y_n(x)$ and $y = \lambda f_n(x)$ agree as well as possible over the interval $(0, L)$. In particular, the $n^{th}$ approximation to be the smallest value of $\lambda$ is given by

$$\lambda_1^{(n)} = \frac{\displaystyle\int_0^L y_n(x)\ dx}{\displaystyle\int_0^L f_n(x)\ dx} \qquad (4.2.19)$$

and the corresponding eigenfunctions is $y_n(x)$. An improved version of Equation 4.2.19 is given by

$$\lambda_1^{(n)} = \frac{\displaystyle\int_a^b r(x)f_n(x)y_n(x)\ dx}{\displaystyle\int_a^b r(x)\left[f_n(x)\right]^2\ dx} \qquad (4.2.20)$$

Equation 4.2.20 is a consequence of point 2 of Theorem 4.1.

The next section briefly introduces an application of property 2 given in the theorem. This is a very important property that will be used repeatedly in solving applications problems in both Chapters 6 and 7.

## 4.3 Eigenfunction Expansion

Consider a Sturm-Liouville problem on [a, b], that is

$$\left(p(x)y'\right)' + \left[q(x) + \lambda r(x)\right]y = 0 \tag{4.3.1}$$

subject to

$$A_1 y(a) + A_2 y'(a) = 0 \tag{4.3.2}$$

$$B_1 y(b) + B_2 y'(b) = 0 \tag{4.3.3}$$

Then, if the eigenvalues $\lambda_1, \lambda_2, \ldots$, and corresponding eigenfunctions $\phi_1(x)$, $\phi_2(x)$, ... have been found, one can sometimes expand a given function $f(x)$ in a series of these eigenfunctions. That is

$$f(x) = \sum_{n=1}^{\infty} c_n \phi_n(x), \quad a \le x \le b \tag{4.3.4}$$

The series in Equation 4.3.4 is called an *eigenfunction expansion* of $f(x)$ on [a, b]. In order to determine the $c_n$'s, property 2 of Theorem 4.1 will be used in the following way:

Multiply Equation 4.3.4 by $r(x)\phi_k(x)$ and integrate the result from a to b.

$$\int_a^b r(x)f(x)\phi_k(x)dx = \sum_{n=1}^{\infty} c_n \int_a^b r(x)\phi_n(x)\phi_k(x)dx \tag{4.3.5}$$

assuming that term-by-term integration of the series is permissible. Then property 2 of Theorem 4.1 gives

$$\int_a^b r(x)\phi_n(x)\phi_k(x)dx = 0, \quad \text{if } n \ne k$$

Therefore, Equation 4.3.5 reduces to

$$\int_a^b r(x)f(x)\phi_k(x)dx = c_k \int_a^b r(x)[\phi_k(x)]^2 dx, \quad \text{for} \quad n = k$$

Finally,

$$c_k = \frac{\int_a^b r(x)f(x)\phi_k(x)dx}{\int_a^b r(x)[\phi_k(x)]^2 dx} \tag{4.3.6}$$

These $c_k$ are the so-called *Fourier coefficients* of $f(x)$ with respect to the eigen-functions of the given Sturm-Liouville problem. A more comprehensive discussion on Fourier coefficients will be given in the next chapter. Also, the conditions that a function must satisfy in order to have a series expansion, as given in Equation 4.3.4 will be discussed in Chapter 5.

Following is an example demonstrating the computation of Fourier coefficients.

### Example 4.3.1

Consider $y'' + \lambda y = 0$ subject to $y(0) = y(\pi/2) = 0$, a regular Sturm-Liouville problem on $[0, \pi/2]$ with eigenvalues $\lambda_n = 4n^2$, $n \geq 1$ and corresponding eigen-functions $\phi_n(x) = \text{Sin}2nx$.

Then, if $f(x) = x^2$ on $[0, \pi/2]$ satisfies all the required conditions to be given in Chapter 5, one can proceed to find the $c_n$'s as follows:

$$\int_0^{\pi/2} r(x)f(x)\phi_n(x)dx = \int_0^{\pi/2} x^2 \text{Sin}2nx\,dx$$

$$= \frac{-\pi^2}{8n}\cos n\pi + \frac{1}{4n^3}[\cos n\pi - 1]$$

and

$$\int_0^{\pi/2} r(x)[\phi_n(x)]^2 dx = \int_0^{\pi/2} x^2 \text{Sin}^2 2nx\,dx$$

$$= \pi/4$$

Such that

$$c_n = \frac{4}{\pi}\left[\frac{-\pi^2}{8n}\cos n\pi + \frac{1}{4n^3}(\cos n\pi - 1)\right]$$

$$= \frac{-\pi}{2n}\cos n\pi + \frac{1}{\pi n^3}(\cos n\pi - 1)$$

Finally,

$$x^2 = \sum_{n=1}^{\infty}\left[\frac{1}{\pi n^3}(\cos n\pi - 1) - \frac{\pi}{2n}\cos n\pi\right]\text{Sin}2nx$$

Again, eigenfunction expansion will be useful in solving certain types of boundary value problems to be discussed in Chapter 6.

## 4.4 Problems

1. Find all the values of $\lambda$, satisfying the boundary value problem

$$\frac{X''}{X} = \lambda$$

$$X(0) = 0, \quad X(1) = 0$$

2. Find all the values of $\lambda$, satisfying the boundary value problem

$$\frac{X''}{X} = \lambda$$

$$X(0) = 0, \quad X'(L) = 0$$

3. Find all the values of $\lambda$, satisfying the boundary value problem

$$\frac{X''}{X} = \lambda$$

$$X(-L) = X(L), \quad X'(-L) = X'(L)$$

4. Find all the values of $\lambda$, satisfying the boundary value problem

$$\frac{X''}{X} = \lambda$$

$$X'(0) = 0, \quad X'(L) = 0$$

5. Use property 2 of Theorem 4.1 to determine $B_n$ in the series

$$\sum_{n=1}^{\infty} B_n \sin \frac{n\pi x}{L} = f(x)$$

$$f(x) = \begin{cases} 1, & 0 < x < 0.5, \\ 0, & 0.5 < x < L \end{cases}$$

6. Use property 2 of Theorem 4.1 to determine $A_n$ in the series

$$\sum_{n=0}^{\infty} A_n \cos \frac{n\pi x}{L} = f(x)$$

$$f(x) = \begin{cases} 1, & 0 < x < 0.5, \\ 0, & 0.5 < x < L \end{cases}$$

7. Use property 2 of Theorem 4.1 to determine $A_n$ and $B_n$ in the series

$$\sum_{n=0}^{\infty} \left( A_n \cos \frac{n\pi x}{2} + B_n \sin \frac{n\pi x}{2} \right) = f(x)$$

$$f(x) = \begin{cases} -x, & -2 < x < 0 \\ 2, & 0 < x < 2 \end{cases}$$

8. Derive the equation that is required to yield the values of $\lambda$ in the singular Sturm-Liouville problem

$$\rho R''(\rho) + R'(\rho) + \lambda^2 \rho R(\rho) = 0$$

subject to

$$R(c) = 0.$$

*Hint:* Apply a boundedness condition after using Equation 3.6.4 to reduce labor.

# References

1. Boyce, W.E. and DiPrima, R.C., *Elementary Differential Equations and Boundary Value Problems*, 3rd ed., John Wiley, New York, 1977.
2. O'Neil, P.V., *Advanced Engineering Mathematics*, 4th ed., PWS, Boston, 1995.
3. Humi, M. and Miller, W.B., *Boundary Value Problems and Partial Differential Equations*, PWS, Boston, 1988.
4. Myint-U, T. and Debnath, L., *Partial Differential Equations for Scientists and Engineers*, 3rd ed., Prentice-Hall, Englewood Cliffs, NJ, 1987.
5. Hildebrand, F.B., *Advanced Calculus for Applications*, 2nd ed., Prentice-Hall, Englewood Cliffs, NJ, 1976.

# 5

## Fourier Series and Integrals

### 5.1   Introduction

One of the most common solution techniques applicable to linear homogeneous partial differential equation problems involves the use of Fourier series. A discussion of the methods of solution of linear partial differential equations will be the topic of the next chapter. In this chapter, a brief outline of Fourier series is given. The primary concerns in this chapter are to determine when a function has a Fourier series expansion and then, does the series converge to the function for which the expansion was assumed? Also, the topic of Fourier transforms will be briefly introduced, as it can also provide an alternative approach to solving certain types of linear partial differential equations.

In order to establish the conditions for a function to have a Fourier series expansion, the following definitions are necessary.

A function is said to be *piecewise continuous* in an interval $a \leq x \leq b$ if there exist finitely many points $a = x_1 < x_2 < \ldots < x_n = b$, such that $f(x)$ is continuous in $x_j < x < x_{j+1}$ and the one-sided limits $f(x_j^+)$ and $f(x_{j+1}^-)$ exist for all $j = 1, 2, 3, \ldots, n - 1$ (Myint-U and Debnath,[4] Churchill and Brown,[1] O'Neil,[2] and Spiegel[3]). Note that a function is piecewise continuous on the closed interval $a \leq x \leq b$, however continuity on the open interval $a < x < b$ does not imply piecewise continuity there. For example,

$$f(x) = \frac{1}{x}$$

is continuous on $0 < x < 1$, but it is not piecewise continuous since $f(0^+)$ fails to exist.

When a function $f(x)$ is piecewise continuous on an interval $a < x < b$, the integral of $f(x)$ from $x = a$ to $x = b$ always exists. That integral is the sum of the integrals of $f(x)$ over the open subintervals on which f is continuous, that is

$$\int_a^b f(x) \; dx = \int_a^{x_1} f(x) \; dx + \int_{x_1}^{x_2} f(x) \; dx + \cdots + \int_{x_{n-1}}^b f(x) \; dx \qquad (5.1.1)$$

For example

$$f(x) = x, \quad 0 \le x < 1$$

$$= -1, \quad 1 < x < 2$$

$$= 1, \quad 2 < x < 3$$

$$\int_0^3 f(x)dx = \int_0^1 xdx + \int_1^2 -dx + \int_2^3 dx = \frac{1}{2}$$

If two functions $f_1$ and $f_2$ are piecewise continuous on an internal $a < x < b$, then there is a finite subdivision of the interval such that both functions are continuous on each closed subinterval whenever the functions are given their limiting values from the interior at the endpoints. This means that a linear combination $c_1 f_1 + c_2 f_2$ or the product $f_1 f_2$ has the continuity on each subinterval and is, itself, piecewise continuous on $a < x < b$. As a consequence

$$\int [c_1 f_1 + c_2 f_2] dx, \quad \int f_1 f_2 \, dx, \quad \text{and} \quad \int [f_1(x)]^2 dx$$

all exist on that interval (Churchill and Brown[1]).

If $f(x)$ is piecewise continuous on $a \le x \le b$ and if the first derivative $f'(x)$ is continuous on each of the subintervals $x_j < x < x_{j+1}$ and the limits $f(x_j^+)$ and $f(x_j^-)$ exist, then f is said to be *piecewise smooth*.

The functions of a sequence $\{\phi_n(x)\}$ are said to be *orthogonal* with respect to the weight function $r(x)$ on $a \le x \le b$ if

$$\int_a^b r(x)\phi_n(x)\phi_m(x)dx = 0, \quad m \ne n \tag{5.1.2}$$

and if $m = n$ then

$$\left[ \int_a^b \phi_n^2 r(x)dx \right] = \|\phi_n\| \tag{5.1.3}$$

which is called the *norm* of the orthogonal system $\{\phi_n\}$.

For example, the sequence of function

$$1, \cos x, \sin x, \ldots, \cos nx, \sin nx$$

form an orthogonal system on $-\pi \le x \le \pi$ since

$$\int_{-\pi}^{\pi} \sin mx \sin nx\, dx = 0 \quad m \ne n$$

$$\int_{-\pi}^{\pi} \sin mx \sin nx\, dx = \pi \quad m = n$$

$$\int_{-\pi}^{\pi} \sin mx \cos nx\, dx = 0 \quad \text{for all } m, n \qquad (5.1.4)$$

$$\int_{-\pi}^{\pi} \cos mx \cos nx\, dx = 0 \quad m \ne n$$

$$\int_{-\pi}^{\pi} \cos mx \cos nx\, dx = \pi \quad m = n$$

for positive integers m and n. Note that an *orthonormal system,* one that is both orthogonal and normalized, for the given sequence is

$$\frac{1}{\sqrt{2\pi}}, \frac{\cos x}{\sqrt{\pi}}, \frac{\sin x}{\sqrt{\pi}}, \cdots, \frac{\cos nx}{\sqrt{\pi}}, \frac{\sin nx}{\sqrt{\pi}}$$

where each element of the sequence is divided by its norm.

It is also important to note that the elements of the sequence 1, cos x, sin x, ..., cos nx, sin nx are *periodic*. In general, a piecewise continuous function f(x) in an interval $a \le x \le b$ is said to be periodic if there exists a real positive number p such that

$$f(x + np) = f(x) \qquad (5.1.5)$$

for any integer n. Further if $f_1, f_2, \ldots, f_k$ have the period p, then the linear combination

$$f = c_1 f_1 + c_2 f_2 + \cdots + c_k f_k \qquad (5.1.6)$$

has the period p. Also, a constant function is a periodic function with an arbitrary period p.

## 5.2  Fourier Coefficients

Since the linear independent functions 1, cos x, sin x, cos 2x, sin 2x, … are mutually orthogonal to each other $-\pi \le x \le \pi$, one can form the series

$$f(x) \sim \frac{A_0}{2} + \sum_{n=1}^{\infty}\left(A_n \cos nx + B_n \sin nx\right) \tag{5.2.1}$$

where the symbol $\sim$ indicates an association of $A_0$, $A_n$, and $B_n$ to f(x) in some unique manner. The series may or may not be convergent.

Suppose f(x) is a Riemann integrable function that is defined on $-\pi \le x \le \pi$. Then one can define the $k^{th}$ partial sum

$$S_k(x) = \frac{A_0}{2} + \sum_{n=1}^{k}\left(A_n \cos nx + B_n \sin nx\right) \tag{5.2.2}$$

to represent f(x) on $-\pi \le x \le \pi$. We now seek the numbers $A_0$, $A_n$ and $B_n$ such that $S_k(x)$ is the best approximation to f(x) in the sense of least squares; that is, we need to minimize the integral

$$I(A_0, A_n, B_n) = \int_{-\pi}^{\pi}\left[f(x) - S_k(x)\right]^2 dx \tag{5.2.3}$$

The necessary condition for $I(A_0, A_n, B_n)$ to be a minimum is

$$\frac{\partial I}{\partial A_0} = 0 = \frac{\partial I}{\partial A_n} = \frac{\partial I}{\partial B_n} \tag{5.2.4}$$

Thus

$$\frac{\partial I}{\partial A_0} = -\int_{-\pi}^{\pi}\left[f(x) - \frac{A_0}{2} - \sum_{j=1}^{k}\left(A_j \cos jx + B_j \sin jx\right)\right]dx \tag{5.2.5}$$

$$\frac{\partial I}{\partial A_n} = -2\int_{-\pi}^{\pi}\left[f(x) - \frac{A_0}{2} - \sum_{j=1}^{k}\left(A_j \cos jx + B_j \sin jx\right)\right]\cos nx\, dx \tag{5.2.6}$$

$$\frac{\partial I}{\partial B_n} = -2 \int\limits_{-\pi}^{\pi} \left[ f(x) - \frac{A_0}{2} - \sum_{j=1}^{k} \left( A_j \cos jx + B_j \sin jx \right) \right] \sin nx \, dx \qquad (5.2.7)$$

From the orthogonality of the trigonometric function (Equation 5.1.4) and noting that

$$\int\limits_{-\pi}^{\pi} \cos mx \, dx = \int\limits_{-\pi}^{\pi} \sin mx \, dx = 0, \qquad m \text{ an integer}$$

Equations 5.2.5, 5.2.6, and 5.2.7 become

$$\frac{\partial I}{\partial A_0} = \pi A_0 - \int\limits_{-\pi}^{\pi} f(x) \, dx \qquad (5.2.8)$$

$$\frac{\partial I}{\partial A_n} = 2\pi A_n - 2 \int\limits_{-\pi}^{\pi} f(x) \cos nx \, dx \qquad (5.2.9)$$

$$\frac{\partial I}{\partial A_n} = 2\pi B_n - 2 \int\limits_{-\pi}^{\pi} f(x) \sin nx \, dx \qquad (5.2.10)$$

Based on Equation 5.2.4, there results

$$A_0 = \frac{1}{\pi} \int\limits_{-\pi}^{\pi} f(x) \, dx \qquad (5.2.11)$$

$$A_n = \frac{1}{\pi} \int\limits_{-\pi}^{\pi} f(x) \cos nx \, dx \qquad (5.2.12)$$

$$B_n = \frac{1}{\pi} \int\limits_{-\pi}^{\pi} f(x) \sin nx \, dx \qquad (5.2.13)$$

Further,

$$\frac{\partial^2 I}{\partial A_0^{\,2}} = \pi$$

$$\frac{\partial^2 I}{\partial A_n^{\,2}} = \frac{\partial^2 I}{\partial B_n^{\,2}} = 2\pi$$

and

$$\frac{\partial^2 I}{\partial A_0 A_n} = \frac{\partial^2 I}{\partial A_0 B_n} = \frac{\partial^{(m)} I}{\partial A_n^{\,(m)}} = \frac{\partial^{(m)} I}{\partial B_n^{\,(m)}} = 0, \quad m \geq 3, \quad n = 1, 2, \cdots$$

By expanding $I(A_0, A_n, B_n)$ in a Taylor series about $(A_0, A_1, \ldots, A_n, B_1, B_2, \ldots, B_n)$ one gets

$$I\left(A_0 + \Delta A_0, \cdots, B_n + \Delta B_n\right) = I\left(A_0, \cdots, B_n\right) + \Delta I \qquad (5.2.14)$$

where $\Delta I$ represents the remaining terms. But $\Delta I$ can be written as

$$\Delta I = \frac{1}{2!}\left[ \frac{\partial^2 I}{\partial A_0^2} \Delta A_0^2 + \sum_{n=1}^{k} \frac{\partial^2 I}{\partial A_n^2} \Delta A_n^2 + \frac{\partial^2 I}{\partial B_n^2} \Delta B_n^2 \right] \qquad (5.2.15)$$

since the first derivative, all mixed second derivatives, and all remaining higher order derivatives are zero. Then

$$\Delta I = \frac{1}{2!}\left[ \pi \Delta A_0^2 + \sum_{n=1}^{k} \left( 2\pi \Delta A_n^2 + 2\pi \Delta B_n^2 \right) \right] > 0$$

Therefore, in order for I to have a minimum value, the coefficients $A_0$, $A_n$, and $B_n$ must be given by Equations 5.2.11, 5.2.12, and 5.2.13, respectively. These coefficients are called the *Fourier Coefficients* of $f(x)$, and the series given by Equation 5.2.1 represent the Fourier series.

According to Equation 5.1.6, the Fourier series Equation 5.2.1, is periodic with period $2\pi$, if it converges.

The following are a few examples on calculating Fourier series of a given function.

**Example 5.2.1**
Find the Fourier series expansion for the function $f(x) = x + x^2$ on $-\pi < x < \pi$.

*Solution*

$$A_0 = \frac{1}{\pi}\int\limits_{-\pi}^{\pi} f(x)\,dx = \frac{1}{\pi}\int\limits_{-\pi}^{\pi}\left(x+x^2\right)dx = \frac{2\pi^2}{3}$$

$$A_n = \frac{1}{\pi}\int\limits_{-\pi}^{\pi} f(x)\cos nx\,dx = \frac{1}{\pi}\int\limits_{-\pi}^{\pi}\left(x+x^2\right)\cos nx\,dx = \frac{4}{n^2}\cos n\pi$$

$$= \frac{4}{n^2}(-1)^n, \quad n \ge 1$$

$$B_n = \frac{1}{\pi}\int\limits_{-\pi}^{\pi} f(x)\sin nx\,dx = \frac{1}{\pi}\int\limits_{-\pi}^{\pi}\left(x+x^2\right)\sin nx\,dx = \frac{2}{n}\cos n\pi$$

$$= \frac{2}{n}(-1)^n, \quad n \ge 1$$

Therefore, the Fourier series expansion for f is

$$f(x) = \frac{\pi^2}{3} + \sum_{n=1}^{\infty}\left[\frac{4}{n^2}(-1)^n\cos nx - \frac{2}{n}(-1)^n\,\mathrm{Sin}\,nx\right]$$

## Example 5.2.2
Find the Fourier series expansion of the function $f(x) = x$ in the interval $-\pi < x < \pi$

*Solution*

$$A_0 = \frac{1}{\pi}\int\limits_{-\pi}^{\pi} f(x)\,dx = \frac{1}{\pi}\int\limits_{-\pi}^{\pi} x\,dx = 0$$

Also, for $n \ge 1$

$$A_n = \frac{1}{\pi}\int\limits_{-\pi}^{\pi} f(x)\cos nx\,dx = \frac{1}{\pi}\int\limits_{-\pi}^{\pi} x\cos nx\,dx = 0$$

$$B_n = \frac{1}{\pi} \int_{-\pi}^{\pi} f(x) \sin nx \, dx = \frac{1}{\pi} \int_{-\pi}^{\pi} x \sin nx \, dx = \frac{2}{n}(-1)^{n+1}$$

Therefore

$$f(x) = 2 \sum_{n=1}^{\infty} \frac{(-1)^{n+1}}{n} \sin nx$$

## Example 5.2.3

Find the Fourier series of the periodic function shown in Figure 5.1

$$f(x) = -\pi, \quad -\pi < x < 0$$

$$= x, \quad 0 < x < \pi$$

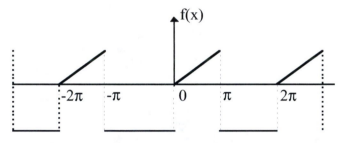

**FIGURE 5.1**

*Solution*

$$A_0 = \frac{1}{\pi} \int_{-\pi}^{\pi} f(x) \, dx = \frac{1}{\pi} \left[ \int_{-\pi}^{0} -\pi \, dx + \int_{0}^{\pi} x \, dx \right] = -\pi/2$$

$$A_n = \frac{1}{\pi} \int_{-\pi}^{\pi} f(x) \cos nx \, dx = \frac{1}{\pi} \left[ \int_{-\pi}^{0} -\pi \cos nx \, dx + \int_{0}^{\pi} x \cos nx \, dx \right]$$

$$= \frac{1}{\pi n^2} \left[ (-1)^n - 1 \right], \quad n \geq 1$$

$$B_n = \frac{1}{\pi} \int_{-\pi}^{\pi} f(x) \sin nx \, dx = \frac{1}{\pi} \left[ \int_{-\pi}^{0} -\pi \sin nx \, dx + \int_{0}^{\pi} x \sin nx \, dx \right] = \frac{1}{n} \left[ 1 - 2(-1)^n \right]$$

Therefore, the Fourier series is

$$f(x) = -\frac{\pi}{4} + \sum_{n=1}^{\infty} \left\{ \frac{1}{n^2 \pi} \left[ (-1)^n - 1 \right] \cos nx + \frac{1}{n} \left[ 1 - 2(-1)^n \right] \sin nx \right\}$$

---

## 5.3  Arbitrary Interval

In the previous two sections, the discussion was primarily based on the interval $[-\pi, \pi]$. However, in many applications, as will be observed in the next chapter, this interval is restrictive.

Suppose the interval of interest is $[a, b]$. Then the interval $a \le x \le b$ can be transformed to $-\pi \le x \le \pi$ by using

$$x = \frac{1}{2}(b + a) + \frac{b - a}{2\pi} t \qquad (5.3.1)$$

Therefore, the function

$$f\left[ \frac{(b + a)}{2} + \frac{b - a}{2\pi} t \right] = F(t)$$

has period $2\pi$. When $F(t)$ is expanded in a Fourier series, one obtains

$$F(t) = \frac{A_0}{2} + \sum_{n=1}^{\infty} \left( A_n \cos nt + B_n \sin nt \right) \qquad (5.3.2)$$

where for $n \ge 0$

$$A_n = \frac{1}{\pi} \int_{-\pi}^{\pi} F(t) \cos nt \, dt$$

and

$$B_n = \frac{1}{\pi} \int_{-\pi}^{\pi} F(t) \sin nt \, dt$$

Upon re-substituting $x$, the expansion for $f(x)$ in $a \le x \le b$ is

$$f(x) = \frac{A_0}{2} + \sum_{n=1}^{\infty} \left[ A_n \cos n\pi \left( \frac{2x-b-a}{b-a} \right) + B_n \sin n\pi \left( \frac{2x-b-a}{b-a} \right) \right] \quad (5.3.3)$$

where

$$A_n = \frac{2}{b-a} \int_a^b f(x) \cos n\pi \left( \frac{2x-b-a}{b-a} \right) dx, \quad n \geq 0 \quad (5.3.4)$$

$$B_n = \frac{2}{b-a} \int_a^b f(x) \sin n\pi \left( \frac{2x-b-a}{b-a} \right) dx, \quad n \geq 1 \quad (5.3.5)$$

To illustrate the change of interval, consider the function $f(x) = x, -2 < x < 2$ then

$$A_0 = \frac{1}{2} \int_{-2}^{2} x\, dx = 0$$

also for $n \geq 1$, $A_n = 0$ and

$$B_n = \frac{1}{2} \int_{-2}^{2} x \sin \frac{n\pi x}{2} dx$$

$$= \frac{-4}{n\pi} (-1)^n, \quad n \geq 1$$

Therefore, the Fourier series for $f(x)$ is

$$f(x) = \sum_{n=1}^{\infty} \frac{4}{n\pi} (-1)^{n+1} \sin \frac{n\pi x}{2}$$

a sine series.

An alternate representation of Equations 5.3.4 and 5.3.5 can be achieved if the convenient interval $[-L, L]$ is used. That is, let $a = -L$, and $b = L$, then

$$f(x) = \frac{A_0}{2} + \sum_{n=1}^{\infty} \left[ A_n \cos \frac{n\pi x}{L} + B_n \sin \frac{n\pi x}{L} \right] \quad (5.3.3A)$$

where

$$A_n = \frac{1}{L} \int_{-L}^{L} f(x) \cos \frac{n\pi x}{L} dx, \quad n \geq 0 \tag{5.3.6}$$

$$B_n = \frac{1}{L} \int_{-L}^{L} f(x) \sin \frac{n\pi x}{L} dx, \quad n \geq 1 \tag{5.3.7}$$

## 5.4 Cosine and Sine Series

Suppose $f(x)$ is an even function defined on the interval $[-\pi, \pi]$, then since cos $nx$ is an even function and sin $nx$ is an odd function, the product $f(x)\cos nx$ and $f(x)\sin nx$ are even and odd functions, respectively. Then the Fourier coefficients of $f(x)$ are

$$A_n = \frac{1}{\pi} \int_{-\pi}^{\pi} f(x) \cos nx dx = \frac{2}{\pi} \int_{0}^{\pi} f(x) \cos nx\, dx, \quad n \geq 0 \tag{5.4.1}$$

and

$$B_n = \frac{1}{\pi} \int_{-\pi}^{\pi} f(x) \sin nx\, dx = 0, \quad n \geq 1 \tag{5.4.2}$$

Therefore, the Fourier series representation of $f(x)$ is

$$f(x) \sim \frac{A_0}{2} + \sum_{n=1}^{\infty} A_n \cos nx \tag{5.4.3}$$

On the other hand, if $f(x)$ is an odd function defined on $[-\pi, \pi]$, then $f(x)$ cos $nx$ and $f(x)$ sin $nx$ are odd and even functions, respectively. Then

$$A_n = \frac{1}{\pi} \int_{-\pi}^{\pi} f(x) \cos nx\, dx = 0, \quad n \geq 0 \tag{5.4.4}$$

and

$$B_n = \frac{1}{\pi} \int_{-\pi}^{\pi} f(x) \sin nx \, dx = \frac{2}{\pi} \int_{0}^{\pi} f(x) \sin nx \, dx \qquad (5.4.5)$$

Therefore, the Fourier series representation of $f(x)$ becomes

$$f(x) \sim \sum_{n=1}^{\infty} B_n \sin nx \qquad (5.4.6)$$

In practice, one frequently encounters problems in which a function is defined on the interval $[-\pi, \pi]$. In these cases, a periodic extension as shown in Figure 5.2 can be made, and those functions can be represented by the Fourier series expansion even though one is only interested in the expansion on $[-\pi, \pi]$.

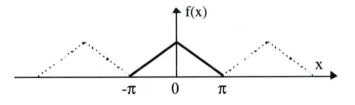

**FIGURE 5.2**
Periodic extension.

However, if a function $f(x)$ is defined only on $[0, \pi]$ then $f(x)$ may be extended in either an *even extension* or an *odd extension*.

By even extension of $f(x)$, one means

$$F_e(x) = \begin{array}{ll} f(x), & 0 < x < \pi \\ f(-x), & -\pi < x < 0 \end{array}$$

as shown in Figure 5.3.

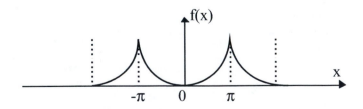

**FIGURE 5.3**
Even extension.

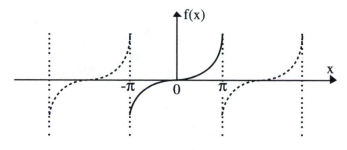

**FIGURE 5.4**
Odd extension.

By odd extension of $f(x)$, one means

$$F_0(x) = f(x), \quad 0 < x < \pi$$
$$f(-x), \quad -\pi < x < 0$$

as shown in Figure 5.4.

These functions $F_e(x)$ and $F_0(x)$ with period $2\pi$, both have Fourier series expansions given by

$$F_e(x) \sim \frac{A_0}{2} + \sum_{n=1}^{\infty} A_n \cos nx \qquad (5.4.7)$$

$$F_0(x) \sim \sum_{n=1}^{\infty} B_n \sin nx \qquad (5.4.8)$$

with the Fourier coefficients as given by Equations 5.4.1 and 5.4.5.

Following are 4 examples of the expediency that is achievable when the concept of odd or even function is used in the context of Fourier series.

## Example 5.4.1

Show that an even function can have no sine terms in its Fourier expansion.

*Solution*

No sine terms if $B_n = 0$, $n \geq 1$.

Consider $f(x)$ on $[-L, L]$, then

$$B_n = \frac{1}{L} \int_{-L}^{L} f(x) \sin \frac{n\pi x}{L} dx$$

$$= \frac{1}{L} \int_{-L}^{0} f(x) \sin \frac{n\pi x}{L} dx + \frac{1}{L} \int_{0}^{L} f(x) \sin \frac{n\pi x}{L} dx$$

But if $f(x)$ is even, then $f(x) = f(-x)$. Specifically, let $x = -y$, $dx = -dy$ then

$$\frac{1}{L}\int_{-L}^{0} f(x)\sin\frac{n\pi x}{L}dx = \frac{1}{L}\int_{L}^{0} f(-y)\sin\frac{n\pi y}{L}dy$$

$$= -\frac{1}{L}\int_{L}^{0} f(-y)\sin\frac{n\pi y}{L}dy$$

Therefore

$$B_n = -\frac{1}{L}\int_{L}^{0} f(-y)\sin\frac{n\pi y}{L}dy + \frac{1}{L}\int_{0}^{L} f(x)\sin\frac{n\pi x}{L}dx = 0$$

In which case the knowledge that the function is even would eliminate the need to calculate $B_n$.

### Example 5.4.2

Expand $f(x) = \sin x$, $0 < x < \pi$ in a Fourier cosine series.

*Solution:*   $B_n = 0$, $n \geq 1$

Since a Fourier series consisting of cosine terms alone is obtained only for an even function, then $f(x)$ is redefined as an even extension.

$$A_n = \frac{2}{\pi}\int_{0}^{\pi} \sin x \cos x \, dx, \quad n \geq 0$$

for $n = 0$,

$$A_0 = \frac{2}{\pi}\int_{0}^{\pi} \sin x \, dx = \frac{4}{\pi}$$

$f(x)$

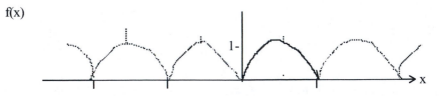

**FIGURE 5.5**
Even extension of sin x.

for n = 1,  $\qquad$  $A_1 = 0$ (why?)

for n > 1,

$$A_n = -\frac{2(1+\cos n\pi)}{\pi(n^2-1)}$$

Therefore

$$\sin x = \frac{2}{\pi} - \frac{2}{\pi}\sum_{n=2}^{\infty}\frac{(1+\cos n\pi)}{n^2-1}\cos nx$$

## Example 5.4.3

Expand $f(x) = x$, $0 < x < 2$ in a sine series.

*Solution*
We need the odd periodic extension of $f(x)$ shown in Figure 5.6, which follows.

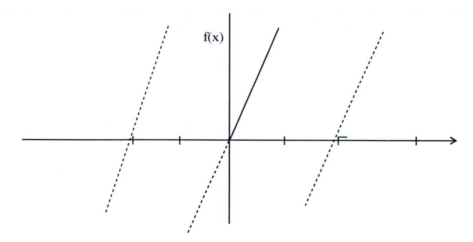

**FIGURE 5.6**
Odd extension of $f(x) = x$.

$A_n = 0$

$$B_n = \frac{2}{2}\int_0^2 x\sin\frac{n\pi x}{2}\,d$$

$$= \frac{-4}{n\pi}(-1)^n, \quad n \geq 1$$

Therefore,

$$f(x) = \sum_{n=1}^{\infty} \frac{4}{n\pi}(-1)^{n+1}\sin\frac{n\pi x}{2}$$

## Example 5.4.4

Given $f(x) = \begin{cases} 1, & 0 < x < 1/2 \\ 0, & 1/2 < x < 1 \end{cases}$

what is the cosine series expansion of f(x)?

*Solution*

An even extension of f(x) is needed for a period $2l = 2$. For an even extension

$$B_n = 0, \quad n \geq 1$$

$$A_0 = \frac{2}{L}\int_0^L f(x)dx = 2\int_0^{1/2} dx = 1$$

also

$$A_n = \frac{2}{L}\int_0^L f(x)\cos\frac{n\pi x}{L}dx = \int_0^{1/2} \cos n\pi x dx = \frac{2}{n\pi}\sin\frac{n\pi}{2}$$

Therefore,

$$f(x) = \frac{1}{2} + \sum_{n=1}^{\infty} \frac{2}{(2n-1)\pi}(-1)^{n-1}\cos(2n-1)\pi x$$

---

## 5.5    Convergence of Fourier Series

If f(x) is piecewise continuous and periodic with period $2\pi$, then

$$\int_{-\pi}^{\pi} \left[f(x) - S_k(x)\right]^{-2} dx \geq 0 \qquad (5.5.1)$$

where

$$S_k(x) = \frac{A_0}{2} + \sum_{n=1}^{k} \left( A_n \cos nx + B_n \sin nx \right)$$

Upon expansion of Equation 5.5.1, one obtains

$$\int_{-\pi}^{\pi} \left[ f(x) - S_k(x)^2 \right] dx = \int_{-\pi}^{\pi} \left[ f(x) \right]^2 dx - 2 \int_{-\pi}^{\pi} f(x) S_k(x) dx + \int_{-\pi}^{\pi} \left[ S_k(x) \right]^2 dx$$

which reduces to

$$\int_{-\pi}^{\pi} f(x) S_k(x) dx = \int_{-\pi}^{\pi} f(x) \left[ \frac{A_0}{2} + \sum_{n=1}^{k} \left( A_n \cos nx + B_n \sin nx \right) \right] dx$$

$$= \frac{\pi A_0^2}{2} + \pi \sum_{n=1}^{k} \left( A_n^2 + B_n^2 \right) \tag{5.5.2}$$

also

$$\int_{-\pi}^{\pi} \left[ S_k(x) \right]^2 dx = \int_{-\pi}^{\pi} \left[ \frac{A_0}{2} + \sum_{n=1}^{k} \left( A_n \cos nx + B_n \sin nx \right) \right]^2 dx$$

$$= \frac{\pi A_0^2}{2} + \pi \sum_{n=1}^{k} \left( A_n^2 + B_n^2 \right) \tag{5.5.3}$$

Therefore

$$\int_{-\pi}^{\pi} \left[ f(x) - S_k(x) \right]^2 dx = \int_{-\pi}^{\pi} \left[ f(x) \right]^2 dx - \left[ \frac{\pi A_0^2}{2} + \pi \sum_{n=1}^{k} \left( A_n^2 + B_n^2 \right) \right] \geq 0 \tag{5.5.4}$$

from which it follows that

$$\frac{A_0^2}{2} + \sum_{n=1}^{k} \left( A_n^2 + B_n^2 \right) \leq \int_{-\pi}^{\pi} \left[ f(x) \right]^2 dx \tag{5.5.5}$$

for all k. Further, since

$$\frac{1}{\pi}\int_{-\pi}^{\pi}[f(x)]^2\,dx$$

is independent of k, then

$$\frac{A_0^2}{2}+\sum_{n=1}^{\infty}\left(A_n^2+B_n^2\right)\le\frac{1}{\pi}\int_{-\pi}^{\pi}[f(x)]^2\,dx \qquad (5.5.6)$$

which is *Bessel's inequality*. Therefore, the series

$$\frac{A_0^2}{2}+\sum_{n=1}^{\infty}\left(A_n^2+B_n^2\right) \qquad (5.5.7)$$

converges, since it is nondecreasing and is bounded from above. That is

$$\lim_{n\to\infty}A_n=0=\lim_{n\to\infty}B_n \qquad (5.5.8)$$

is the necessary condition for convergence of the series, Equation 5.2.1.
Finally, the Fourier series is said to *converge in the mean* to f(x) when

$$\lim_{k\to\infty}\int_{-\pi}^{\pi}\left[f(x)-\left(\frac{A_0}{2}+\sum_{n=1}^{k}\left(A_n\cos nx+B_n\sin nx\right)\right)\right]^2\,dx=0 \qquad (5.5.9)$$

If Equation 5.5.9 holds, then one gets

$$\frac{A_0^2}{2}+\sum_{n=1}^{\infty}\left(A_n^2+B_n^2\right)=\frac{1}{\pi}\int_{-\pi}^{\pi}[f(x)]^2\,dx \qquad (5.5.10)$$

which is known as *Parseval's relation*, and the set of functions 1, cos x, sin x, cos 2x, sin 2x ... is said to be *complete*.

The following theorem can be used to establish point convergence of a given Fourier series (Myint-U and Debnath,[4] Churchill and Brown[1]).

*THEOREM 5.1*
*If f(x) is piecewise smooth and periodic with period* $2\pi$ *in* $-\pi\le x\le\pi$*, then for any x*

$$\frac{A_0}{2}+\sum_{n=1}^{\infty}\left(A_n\cos nx+B_n\sin nx\right)=\frac{1}{2}\left[f(x_+)+f(x_-)\right] \qquad (5.5.11)$$

where

$$A_n = \frac{1}{\pi} \int_{-\pi}^{\pi} f(x) \cos nx \, dx, \quad n \geq 0$$

and

$$B_n = \frac{1}{\pi} \int_{-\pi}^{\pi} f(x) \sin nx \, dx, \quad n \geq 1$$

Proof of Theorem 5.1 can be found elsewhere (Churchill and Brown[1]), however, an example of its usage is given below.

Previously (see Example 5.2.1), the Fourier series of $x + x^2$ on $[-\pi, \pi]$ was found to be

$$\frac{\pi^2}{3} + \sum_{n=1}^{\infty} \left[ \frac{4}{n^2} (-1)^n \cos nx - \frac{2}{n} (-1n) \sin nx \right]$$

since $f(x)$ is piecewise smooth, the series converges and

$$x + x^2 = \frac{\pi^2}{3} + \sum_{n=1}^{\infty} \left[ \frac{4}{n^2} (-1)^n \cos nx - \frac{2}{n} (-1n) \sin nx \right]$$

at points of continuity. However, at points of discontinuity, such as $x = \pi$, Theorem 5.1 gives

$$\frac{1}{2} \left[ (\pi + \pi^2) + (-\pi + \pi^2) \right] = \frac{\pi^2}{3} + \sum_{n=1}^{\infty} \frac{4}{n^2} (-1)^n \cos nx$$

where

$$f(\pi_-) = \pi + \pi^2 : \text{ approaching from the left}$$

and

$$f(\pi_+) = f(-\pi_+) = -\pi + \pi^2 : \text{ approaching from the right}$$

Therefore, the series converges to

$$\frac{\pi^2}{6} = \sum_{n=1}^{\infty} \frac{1}{n^2} \qquad (5.5.12)$$

at the points of discontinuity.

Another useful result regarding convergence is given below. This result is known as the *Riemann-Lebesgue Lemma*.

*LEMMA 5.1*
*If g(x) is piecewise continuous on* a ≤ x ≤ b, *then*

$$\lim_{\lambda \to \infty} \int_a^b g(x) \sin \lambda x \, dx = 0 \qquad (5.5.13)$$

*Proof*
Consider

$$I(\lambda) = \int_a^b g(x) \sin \lambda x \, dx \qquad (5.5.14)$$

then, if

$$x = t + \frac{\pi}{\lambda}$$

$$\sin \lambda x = \sin \lambda \left( t + \frac{\pi}{\lambda} \right) = -\sin \lambda t$$

and I(λ) becomes

$$I(\lambda) = - \int_{a-\frac{\pi}{\lambda}}^{b-\frac{\pi}{\lambda}} g\left( t + \frac{\pi}{\lambda} \right) \sin \lambda t \, dt \qquad (5.5.15)$$

But t is a dummy variable in Equation 5.5.15; therefore, I(λ) can be rewritten as

$$I(\lambda) = - \int_{a-\frac{\pi}{\lambda}}^{b-\frac{\pi}{\lambda}} g\left( x + \frac{\pi}{\lambda} \right) \sin \lambda x \, dx \qquad (5.5.16)$$

Then adding Equation 5.5.14 and Equation 5.5.16 gives

$$2I(\lambda) = \int_a^b g(x)\sin\lambda x\,dx - \int_{a-\pi/\lambda}^{b-\pi/\lambda} g\left(x+\pi/\lambda\right)\sin\lambda x\,dx$$

$$= \int_{a-\pi/\lambda}^a g\left(x+\pi/\lambda\right)\sin\lambda x\,dx + \int_{b-\pi/\lambda}^b g(x)\sin\lambda x\,dx + \int_a^{b-\pi/\lambda}\left[g(x)-g\left(x+\pi/\lambda\right)\right]\sin\lambda x\,dx$$

If $g(x)$ is continuous in $a \le x \le b$, then $g(x)$ is necessarily bounded; that is, there exists a number M such that $|g(x)| \le M$. Therefore

$$\left|\int_{a-\pi/\lambda}^a g\left(x+\pi/\lambda\right)\sin\lambda x\,dx\right| = \left|\int_a^{a-\pi/\lambda} g(x)\sin\lambda x\,dx\right| \le \frac{\pi M}{\lambda}$$

and

$$\left|\int_{b-\pi/\lambda}^b g(x)\sin\lambda x\,dx\right| \le \frac{\pi M}{\lambda}$$

giving

$$|I(\lambda)| \le \frac{M\pi}{\lambda} + \int_a^{b-\pi/\lambda}\left|g(x)-g\left(x+\pi/\lambda\right)\right|dx \tag{5.5.17}$$

as a consequence. Further, since $g(x)$ is a continuous function on a closed interval $a \le x \le b$, it is uniformly continuous on $a \le x \le b$ such that

$$\left|g(x)-g\left(x+\pi/\lambda\right)\right| < \frac{\varepsilon}{b-a}$$

for all $\lambda > \Lambda$ and all $x$ in $a \le x \le b$. By choosing $\lambda$ such that $\dfrac{\pi M}{\lambda} < \dfrac{\varepsilon}{2}$ whenever $\lambda > \Lambda$ gives

$$|I(\lambda)| < \frac{\varepsilon}{2} + \frac{\varepsilon}{2} = \varepsilon$$

The second result makes use of the Dirichlet kernel, $D_n(x)$ and is given here without proof.

*LEMMA 5.2*
*If a given function $g(x)$ is piecewise continuous on the interval $0 < x < \pi$ and the right-hand derivative $g'_R(0)$ exists, then*

$$\lim_{n\to\infty} \int_0^\pi g(x)D_n(x)\,dx = \frac{\pi}{2}g(0_+)$$

where $D_n(x)$ is given as

$$D_n(x) = \frac{1}{2} + \sum_{n=1}^N \cos nx$$

or

$$D_n(x) = \frac{\sin\left[(2n+1)\,x\!/\!2\right]}{2\sin\dfrac{x}{2}}$$

## Example 5.5.1

The least-square properties of Fourier series: quite often one is interested in approximating a periodic function, f, over a period. Therefore, the trigonometric partial sum $\bar{S}_K$ which best approximates such a periodic function over one full period, p, is the one for which the total squared error

$$\bar{E} = \int_0^p \left[f(x) - \bar{S}_K(x)\right]^2 dx \tag{5.5.18}$$

is a minimum. This example shows that among all trigonometric sums

$$\bar{S}_K(x) = \frac{\bar{A}_0}{2} + \sum_{n=1}^K \left(\bar{A}_n \cos\frac{n\pi x}{p} + \bar{B}_n \sin\frac{n\pi x}{p}\right)$$

the *K*th partial sum

$$S_K(x) = \frac{A_0}{2} + \sum_{n=1}^K \left(A_n \cos\frac{n\pi x}{p} + B_n \sin\frac{n\pi x}{p}\right)$$

of the Fourier series of a periodic function is for every value K the best least-square approximation to f over one period (and can be extended to any number of periods).

*Solution*
Let

$$E = \int_0^P \left[ f(x) - S_K(x) \right]^2 dx$$

be the error associated with the particular partial sum $S_k(x)$. Then it is important to compare $\overline{E}$ with E. Expansion of Equation 5.5.18 gives

$$\overline{E} = \int_0^P \left[ f^2(x) - 2f(x)\overline{S}_K(x) + \overline{S}_K^2(x) \right] dx$$

$$= \int_0^P f^2(x)dx - 2\int_0^P f(x)\left[ \frac{\overline{A}_0}{2} + \sum_{n=1}^K \left( \overline{A}_n \cos\frac{n\pi x}{p} + \overline{B}_n \sin\frac{n\pi x}{p} \right) \right] dx$$

$$+ \int_0^P \left[ \frac{\overline{A}_0}{2} + \sum_{n=1}^K \left( \overline{A}_n \cos\frac{n\pi x}{p} + \overline{B}_n \sin\frac{n\pi x}{p} \right) \right]^2 dx$$

The integral in the second term simplifies to

$$-2p\left[ \frac{\overline{A}_0 A_0}{2} + \sum_{n=1}^K \left( \overline{A}_n A_n + \overline{B}_n B_n \right) \right]$$

The third term reduces to

$$p\left[ \frac{\overline{A}_0^2}{2} + \sum_{n=1}^K \left( \overline{A}_n^2 + \overline{B}_n^2 \right) \right]$$

such that

$$\overline{E} = \int_0^P f^2(x)dx - 2p\left[ \frac{\overline{A}_0 A_0}{2} + \sum_{n=1}^K \left( \overline{A}_n A_n + \overline{B}_n B_n \right) \right] + p\left[ \frac{\overline{A}_0^2}{2} + \sum_{n=1}^K \left( \overline{A}_n^2 + \overline{B}_n^2 \right) \right]^2 dx$$

By making the following substitutions:

$$\overline{A}_0 = A_0, \overline{A}_n = A_n, \text{and} \overline{B}_n = B_n$$

the quantity E becomes

$$E = \int_0^p f^2(x)dx - p\left[\frac{A_0^2}{2} + \sum_{n=1}^K \left(A_n^2 + B_n^2\right)\right]$$

Upon subtracting E from $\overline{E}$ results in

$$\overline{E} - E = p\left\{\left(\frac{\overline{A}_0 - A_0}{2}\right)^2 + \sum_{n=1}^K \left[\left(\overline{A}_n - A_n\right)^2 + \left(\overline{B}_n - B_n\right)^2\right]\right\}$$

From this result, one can conclude that for every possible set of coefficients $(\overline{A}_n, \overline{B}_n)$, $\overline{E} \geq E$. Further, equality holds only if each squared difference is zero, that is, if and only if $\overline{A}_0 = A_0$, $\overline{A}_n = A_n$, and $\overline{B}_n = B_n$ which means that $\overline{S}_K(x) = S_k(x)$.

---

## 5.6   Fourier Integrals

In the previous sections, some theory and a few applications involving the expansion of a function $f(x)$ into a Fourier series were discussed. The function $f(x)$ had period 2L, with L assumed finite. An important question is what happens when L approaches infinity? Below, this question is explored.

Suppose $f(x)$ satisfies the following conditions:

$f(x)$ and $f'(x)$are piecewise continuous in every finite interval        (I)

$\int_{-\infty}^{\infty} |f(x)|\, dx$ converges, that is, is absolutely integrable in $(-\infty, \infty.)$   (II)

Then

$$f(x) = \int_0^\infty \left[A(\omega)\cos \omega x + B(\omega)\sin \omega x\right]d\omega \qquad (5.6.1)$$

where

$$A(\omega) = \frac{1}{\pi} \int_{-\infty}^{\infty} f(x) \cos \omega x \, dx$$

(5.6.2)

$$B(\omega) = \frac{1}{\pi} \int_{-\infty}^{\infty} f(x) \sin \omega x \, dx$$

The condition (I) is true if x is a point of continuity of f(x), and f(x) must be replaced by

$$\frac{f(x^+) + f(x^-)}{2}$$ if x is a point of discontinuity.

Equation 5.6.1 is referred to as a Fourier integral expansion of f(x).

## Example 5.6.1

Find the Fourier integral expansion of $f(x) = \begin{cases} 0 & x < 0 \\ e^{-ax} & x > 0 \end{cases}$  $a > 0$

*Solution*
Using Equation 5.6.1 with

$$A(\omega) = \frac{1}{\pi} \int_{-\infty}^{\infty} f(x) \cos \omega x \, dx$$

$$= \frac{1}{\pi} \int_{-\infty}^{\infty} e^{-ax} \cos \omega x \, dx$$

$$= \frac{1}{\pi} \left[ \frac{a}{a^2 + \omega^2} \right]$$

and

$$B(\omega) = \frac{1}{\pi} \int_{-\infty}^{\infty} e^{-ax} \sin \omega x \, dx$$

$$= \frac{1}{\pi} \left[ \frac{\omega}{a^2 + \omega^2} \right]$$

Therefore, the given function becomes

$$f(x) = \int_0^\infty \left[ A(\omega)\cos\omega x + B(\omega)\sin\omega x \right] d\omega$$

$$= \int_0^\infty \frac{1}{\pi} \left[ \frac{a}{a^2 + \omega^2}\cos\omega x + \frac{\omega}{a^2 + \omega^2}\sin\omega x \right] d\omega$$

Some other useful forms of Fourier's integral are

$$f(x) = \frac{1}{\pi} \int_0^\infty \int_{-\infty}^\infty f(u)\cos\omega(x - u)\, du\, d\omega \qquad (5.6.3)$$

and

$$f(x) = \frac{1}{2\pi} \int_{-\infty}^\infty \int_{-\infty}^\infty f(u)e^{i\omega(x-u)}\, du\, d\omega \qquad (5.6.4)$$

## Example 5.6.2

Use Equation 5.6.4 to find the Fourier representation of

$$f(x) = \begin{cases} 0 & x < 0 \\ e^{-ax} & x > 0 \end{cases} \qquad a > 0.$$

*Solution*

$$f(x) = \frac{1}{2\pi} \int_{-\infty}^\infty \int_{-\infty}^\infty f(u)e^{i\omega(x-u)}\, du\, d\omega$$

$$= \frac{1}{2\pi} \int_{-\infty}^\infty \int_{-\infty}^\infty e^{-ax}e^{-i\omega x}e^{i\omega x}\, dx\, d\omega$$

where the substitution

$$f(u)e^{-i\omega u}\, du \equiv e^{-ax}e^{-i\omega x}\, dx$$

has been made. However,

$$\int_{-\infty}^{\infty} e^{-(a+i\omega)x} dx = \int_{0}^{\infty} \overline{e}^{(a+i\omega)x} dx$$

$$= \frac{1}{a+i\omega}$$

Therefore,

$$f(x) = \frac{1}{2\pi} \int_{-\infty}^{\infty} \frac{1}{a+i\omega} e^{i\omega t} d\omega$$

If $f(x)$ is either an odd or an even function, the Fourier integral can be reduced to a sine and cosine representation. That is, for $f(x)$ odd

$$f(x) = \frac{2}{\pi} \int_{0}^{\infty} \sin \omega x \, d\omega \int_{0}^{\infty} f(u) \sin \omega u \, du \qquad (5.6.5)$$

and for $f(x)$ even

$$f(x) = \frac{2}{\pi} \int_{0}^{\infty} \cos \omega x \, d\omega \int_{0}^{\infty} f(u) \cos \omega u \, du \qquad (5.6.6)$$

Equations 5.6.5 and 5.6.6 are the Fourier sine and cosine integral representation of $f(x)$, respectively.

### Example 5.6.3
Find the Fourier cosine and sine representation of $f(t) = e^{-at}$, $0 < t < \infty$, $a > 0$.

*Solution*
Equation (5.6.6) suggests that

$$\int_{0}^{\infty} f(u) \cos \omega u \, du \equiv \int_{0}^{\infty} e^{-at} \cos \omega t \, dt$$

$$= \frac{a}{a^2 + \omega^2}$$

Therefore, the cosine representation of the given function is

$$f(x) = \frac{2}{\pi} \int_0^\infty \frac{a}{a^2 + \omega^2} \cos \omega t \, d\omega$$

Similarly, the sine representation of $e^{-at}$ is

$$f(x) = \frac{2}{\pi} \int_0^\infty \frac{\omega}{a^2 + \omega^2} \sin \omega t \, d\omega$$

## Example 5.6.4

Find the Fourier cosine integral representation of the function

$$f(x) = \begin{cases} 1, & 0 < x < 1 \\ 0, & x \ge 1 \end{cases}$$

*Solution*
From Equation 5.6.6:

$$f(x) = \frac{2}{\pi} \int_0^\infty \cos \omega x \, d\omega \int_0^\infty f(u) \cos \omega u \, du$$

therefore

$$1 = f(x) = \frac{2}{\pi} \int_0^\infty \cos \omega x \, d\omega \int_0^1 \cos \omega u \, du$$

$$= \frac{2}{\pi} \int_0^\infty \frac{\sin \omega}{\omega} \cos \omega x \, d\omega$$

The form given by Equation 5.6.4 is particularly useful, in that if

$$F(\omega) = \int_{-\infty}^\infty f(u) e^{-i\omega u} \, du \qquad\qquad (5.6.7)$$

then

$$f(x) = \frac{1}{2\pi} \int_{-\infty}^\infty F(\omega) e^{i\omega x} \, d\omega \qquad\qquad (5.6.8)$$

Equations 5.6.7 and 5.6.8 are the *Fourier transform pair*, where F($\omega$) is the Fourier transform of f(x) and Equation 5.6.8 is the inverse transform. The customary notation for the Fourier transform is $F\{f(x)\}$ and its inverse is denoted by $F^{-1}\{F(\omega)\}$. These notations will be used in this book.

Again, if f(x) is an odd function, then Equation 5.6.5 can be reinterpreted as

$$F_s(\omega) = \int_0^\infty f(u)\sin \omega u\, du \qquad (5.6.9)$$

with inverse

$$f(x) = \frac{2}{\pi}\int_0^\infty F_s(\omega)\sin \omega x\, d\omega \qquad (5.6.10)$$

here, $F_s(\omega)$ is the *Fourier sine transform* of f(x). If f(x) is an even function, Equation 5.6.6 can be reinterpreted as

$$F_c(\omega) = \int_0^\infty f(u)\cos \omega u\, du \qquad (5.6.11)$$

with inverse transform given by

$$f(x) = \frac{2}{\pi}\int_0^\infty F_c(\omega)\cos \omega x\, d\omega \qquad (5.6.12)$$

where $F_c(\omega)$ is the *Fourier cosine transform* of f(x).

Similar to Laplace transforms, there is a convolution theorem for Fourier transforms (Wylie and Barrett,[7] Greenberg,[8] Myint-U and Debnath,[4] Zauderer,[9] and O'Neil[2]) which states that the Fourier transform of the convolution of two functions f(x) and g(x) is equal to the product of their Fourier transforms. That is,

$$F\{f * g\} = F\{f(x)\}F\{g(x)\} \qquad (5.6.13)$$

and the convolution obeys the commutative, associative, and distributive laws of algebra. The convolution of the functions f(x) and g(x) is defined to be

$$f * g = \int_{-\infty}^\infty f(u)g(x-u)\, du \qquad (5.6.14)$$

**Example 5.6.5**

Solve for f(x) in the integral equation given by

$$\int_0^\infty f(x)\sin\alpha x\,dx = \begin{cases} 1-\alpha & 0 \le \alpha \le 1 \\ 0 & \alpha > 1 \end{cases}$$

*Solution*

According to Equation 5.6.9

$$F_s(\alpha) = \int_0^\infty f(x)\sin\alpha x\,dx = \begin{cases} 1-\alpha & 0 \le \alpha \le 1 \\ 0 & \alpha > 1 \end{cases}$$

is the Fourier sine transform of f(x). Then, by Equation 5.6.10,

$$f(x) = \frac{2}{\pi}\int_0^\infty F_s(\alpha)\sin\alpha x\,d\alpha = \frac{2}{\pi}\int_0^1 (1-\alpha)\sin\alpha x\,d\alpha$$

$$= \frac{2(x-\sin x)}{\pi x^2}$$

**Example 5.6.6**

If the Fourier transforms of the functions $f_1(x)$ and $f_2(x)$ exist, what is the Fourier transform of $a_1 f_1 + a_2 f_2$ for $a_1$, $a_2$ constants?

*Solution*

$$F\{a_1 f_1 + a_2 f_2\} = \int_{-\infty}^\infty \left[a_1 f_1(u) + a_2 f_2(u)\right]e^{-i\omega u}du$$

$$= a_1 \int_{-\infty}^\infty f_1(u)e^{-i\omega u}du + a_2 \int_{-\infty}^\infty f_2(u)e^{-i\omega u}du \qquad (5.6.15)$$

$$a_1 F\{f_1\} + a_2 F\{f_2\}$$

The linearity property displayed by Example 5.6.6 is generally true for Fourier transforms and their inverses.

**Example 5.6.7**

If f(x) is continuous and f'(x) is at least piecewise continuous on $(-\infty, \infty)$, and if

$$\int_{-\infty}^{\infty} |f(x)|\, dx$$

and

$$\int_{-\infty}^{\infty} |f'(x)|\, dx$$

exist, then show that $F\{f'(x)\} = i\omega F\{f(x)\}$.

*Solution*

$$F\{f'(u)\} = \int_{-\infty}^{\infty} f'(u)e^{-i\omega u}\, du = F(\omega)$$

then using integration by parts with

$$p = e^{-i\omega u} \quad \text{and} \quad dq = f'(u)\, du$$

results in

$$F\{f'(u)\} = \lim_{H \to \infty}\left[ e^{-i\omega u} f(u)\right]_{-H}^{H} + i\omega \int_{-\infty}^{\infty} f(u)e^{-i\omega u}\, du$$

However, $f(x)$ is given as absolutely integrable; therefore, it must vanish at both $-\infty$ and $\infty$. Therefore, the final result is

$$F\{f'(u)\} = i\omega \int_{-\infty}^{\infty} f(u)e^{-i\omega u}\, du \tag{5.6.16}$$

$$= i\omega F(\omega)$$

The result of this latter example can be extended to the general result:

$$F\{f^{(n)}(x)\} = (i\omega)^n F(\omega) \tag{5.6.17}$$

as long as successive derivatives of $f(x)$ are continuous and absolutely integrable on $(-\infty, \infty)$.

Additional properties of Fourier transforms can be found in most advanced engineering texts (O'Neil,[2] Wylie and Barrett,[7] and Greenberg[8]).

## Example 5.6.8

Obtain the solution of the initial-value problem of heat conduction in an infinite rod as described by

$$\frac{\partial T}{\partial t} = k\frac{\partial^2 T}{\partial x^2}, \quad -\infty < x < \infty, \quad t > 0$$

subject to

$$T(x,0) = f(x)$$

where T is the temperature distribution and is bounded, and k is a constant representing diffusivity.

*Solution*

Let $U(\omega, t) = F\{T(x, t)\} = \displaystyle\int_{-\infty}^{\infty} T(x, t)e^{-i\omega x} dx$

then the inverse transform is

$$T(x, t) = \frac{1}{2\pi}\int_{-\infty}^{\infty} U(\omega, t)e^{-i\omega x} d\omega$$

Therefore, the differential equation becomes

$$\frac{dU(\omega, t)}{dt} + k\omega^2 U = 0$$

and the given initial condition becomes

$$U(\omega, 0) = F(\omega)$$

This is now a first order linear ordinary differential equation with

$$U(\omega, t) = ce^{-k\omega^2 t}$$

as its general solution while treating $\omega$ as a parameter. The particular solution is

$$U(\omega, t) = F(\omega)e^{-k\omega^2 t}$$

Using Equation 5.6.8, the inverse transform is

$$T(x, t) = \frac{1}{2\pi} \int_{-\infty}^{\infty} U(\omega, t)e^{i\omega x} d\omega$$

$$= \frac{1}{2\pi} \int_{-\infty}^{\infty} F(\omega)e^{-k\omega^2 t}e^{i\omega x} d\omega$$

This is now in the appropriate form for which the convolution theorem for Fourier transform is useful to carry out the inversion. That is

$$f * g = \int_{-\infty}^{\infty} f(u)g(x - u) du$$

By comparison, $g(x)$ is the inverse of $G(\omega) = e^{-k\omega^2 t}$, which has the form

$$g(x) = \frac{1}{2\pi} \int_{-\infty}^{\infty} G(\omega)e^{i\omega x} d\omega$$

$$= \frac{1}{2\pi} \left[ \int_{-\infty}^{\infty} e^{-k\omega^2 t}e^{i\omega x} d\omega \right]$$

$$= \frac{1}{2\pi} \left[ \sqrt{\frac{\pi}{kt}} e^{-x^2/4kt} \right]$$

$$= \frac{1}{\sqrt{4\pi kt}} e^{-x^2/4kt}$$

Consequently, the final solution is

$$T(x, t) = \frac{1}{2\pi} \int_{-\infty}^{\infty} f(u) \frac{1}{\sqrt{4\pi kt}} \exp\left[ -\frac{(x - u)^2}{4kt} \right] du$$

## 5.7 Problems

1. Assuming that f(t), f′(t), f″(t) and f‴(t) are continuous and absolutely integrable on $(-\infty, \infty)$, show that

   a.   $F\{f''(t)\} = i\omega F\{f'(t)\} = (i\omega)^2 F(\omega)$

   b.   $F\{f'''(t)\} = i\omega F\{f''(t)\} = (i\omega)^3 F(\omega)$

2. If $F(\omega)$ is the Fourier transform of f(t) show that $F'(\omega) = F\{-it\ f(t)\}$. *Hint*: assume that f(t) is absolutely integrable on $(-\infty, \infty)$.

3. What error is made in approximating $f(t) = e^{-t}$ on $0 < t < 1$ by the sum of the first three nonzero terms of its Fourier series?

4. Determine the values of a and b, which make the line $y = a + bx$ the best least-square approximation to $e^x$ on $0 < x < 1$, using the least-square criterion given in Example 5.5.1 $E(a, b) = \int_0^1 \left[e^x - (a + bx)\right]^2 dx$ as the quantity to be minimized. Recall that for E to be a minimum, the quantities $\dfrac{\partial E}{\partial a}$ and $\dfrac{\partial E}{\partial b}$ must both be zero.

## References

1. Churchill, R.V. and Brown, J.W., *Fourier Series and Boundary Value Problems*, 4th ed., McGraw-Hill, New York, 1987.
2. O'Neil, P.V., *Advanced Engineering Mathematics*, 4th ed., PWS-Kent, Boston, 1995.
3. Spiegel, M.R., *Fourier Analysis with Applications to Boundary Value Problems*, McGraw-Hill, New York, 1974.
4. Myint-U.T. and Debnath, L., *Partial Differential Equations for Scientists and Engineers*, 3rd ed., Prentice-Hall, Englewood Cliffs, NJ, 1987.
5. Giordano, F.R. and Weir, M.D., *Differential Equations, a Modeling Approach*, Addison-Wesley, Reading, MA, 1991.
6. Boyce, W.E. and DiPrima, R.C., *Elementary Differential Equations and Boundary Value Problems*, 3rd ed., John Wiley, New York, 1977.
7. Wylie, C.R. and Barrett, L.C., *Advanced Engineering Mathematics*, 6th ed., McGraw-Hill, New York, 1995.
8. Greenberg, M.D., *Advanced Engineering Mathematics*, Prentice-Hall, Englewood Cliffs, NJ, 1988.
9. Zauderer, E., *Partial Differential Equations of Applied Mathematics*, John Wiley, New York, 1983.

# 6

# Partial Differential Equations

## 6.1  Introduction

In the previous five chapters, the essential tools necessary to tackle this chapter have been outlined. Herein we will follow the "standard mathematician's MO," that is, I will attempt to reduce these new problems (solution of partial differential equations) to those we already know how to solve (ordinary differential equations).

In the practice of chemical engineering science, there are many problems that must be described by two or more independent variables. For example, the equations of change for isothermal systems (Bird, Stewart, and Lightfoot[1]) are

$$\frac{\partial \rho}{\partial t} = -\left( \frac{\partial}{\partial x}\rho v_x + \frac{\partial}{\partial y}\rho v_y + \frac{\partial}{\partial z}\rho v_z \right) \tag{6.1.1}$$

$$\rho \frac{D\bar{v}}{Dt} = -\bar{\nabla}p - \left[ \bar{\nabla} \bullet \bar{\tau} \right] + \rho \bar{g} \tag{6.1.2}$$

$$\rho \frac{D\bar{v}}{Dt} = -\bar{\nabla}p + \mu \nabla^2 \bar{v} + \bar{g} \tag{6.1.3}$$

Equation 6.1.1 is the equation of continuity, and it describes the rate of change of density at a fixed point resulting from the changes in the mass velocity vector $\rho\bar{v}$. Equation 6.1.2 is the equation of motion, which states that a small volume element moving with the fluid is accelerated due to the forces acting on it. Equation 6.1.3 is the Navier-Stokes equation. These equations together with appropriate boundary and initial conditions make up a large portion of the research problems that are encountered in chemical engineering.

Equations 6.1.1 through 6.1.3 are partial differential equations, as opposed to ordinary differential equations, but the definitions of linearity and homogeneity remain the same as those given for second order ordinary differential equations.

As a reminder, a *linear operator* L (Giordano and Weir,[2] Myint-U and Debnath,[3] Haberman,[4] and Churchill and Brown[5]) must satisfy

$$L\left(c_1 u_1 + c_2 u_2\right) = c_1 L\left(u_1\right) + c_2 L\left(u_2\right) \tag{6.1.4}$$

for any two functions $u_1$ and $u_2$ where $c_1$ and $c_2$ are arbitrary constants. For example $\dfrac{\partial}{\partial t}$ and $\dfrac{\partial^2}{\partial x^2}$ are linear operators since they satisfy

$$\frac{\partial}{\partial t}\left(c_1 u_1 + c_2 u_2\right) = c_1 \frac{\partial u_1}{\partial t} + c_2 \frac{\partial u_2}{\partial t}$$

and

$$\frac{\partial^2}{\partial x^2}\left(c_1 u_1 + c_2 u_2\right) = c_1 \frac{\partial^2 u_1}{\partial x^2} + c_2 \frac{\partial^2 u_2}{\partial x^2}$$

The general linear second order partial differential equation in one dependent variable u and independent variables x and y is

$$A u_{xx} + B u_{xy} + C u_{yy} + D u_x + E u_y + F u = G \tag{6.1.5}$$

where the coefficients are functions of x and y and do not vanish simultaneously. Also, in Equation 6.1.5 it is assumed that u and the coefficients are twice continuously differentiable in some domain $\Re$. If $G = 0$ in Equation 6.1.5, then the equation is *homogeneous*.

A fundamental property of linear operators (Equation 6.1.4) allows solutions of linear equations to be added together. This property is the *Principle of Superposition* and can be stated as follows: *Given that $u_1, u_2, ..., u_k$ satisfy a linear homogeneous equation, then an arbitrary linear combination of these, $c_1 u_1 + c_2 u_2 + ... + c_k u_k$ also satisfies the same linear homogeneous equation.*

The concepts of linearity and homogeneity also apply to boundary conditions (see definitions in Chapter 4).

In the sections that follow, solution techniques for linear boundary value problems are developed. Specifically, the method of separation of variables in Section 6.2 is illustrated. In Section 6.3, the method of eigenfunction expansion is outlined. In Section 6.4, the method of Laplace transform is illustrated. The method of combination of variables is outlined in Section 6.5. In Section 6.6, Fourier integrals are introduced.

Each method is accompanied by at least two worked-out examples demonstrating the important steps in the construction of a solution to a given problem. Also, briefly introduced in this chapter is the method of regular perturbation, a technique that can be very helpful in estimating a solution to some nonlinear problems.

## 6.2  Separation of Variables

Separation of variables is one of the most widely used solution techniques for a system consisting of a second order partial differential equation (pde) with boundary and/or initial conditions. This solution technique requires that the pde be reduced to a system of ordinary differential equations. The spatial part of the pde forms the ordinary differential equation in the resulting boundary value problem. Typically, these boundary value problems are precisely those Sturm-Liouville problems discussed in Chapter 4. As a reminder, a boundary value problem consists of a differential equation together with appropriate boundary conditions. In separating the variables, a substitution is made that transforms the pde to an ordinary differential equation. This same substitution is used to transform the boundary conditions accompanying the pde into appropriate boundary conditions for the ordinary differential equation.

The time or time-like part of the pde usually results in an initial value type problem. The general solution of this initial value problem is then combined with the eigenfunctions resulting from solving the boundary value problem. Application of the initial condition usually results in a Fourier series.

In order for this technique to be successfully applied, both the partial differential equation and accompanying boundary conditions must be linear and homogeneous. To demonstrate the technique, we will consider the following model:

1. A one-dimensional rod of length, L, that is laterally insulated.
2. Heat is not internally generated.
3. The rod is uniform with constant density throughout its length.
4. The material of construction has constant specific heat and constant thermal conductivity.
5. The ends of the rod are kept at a fixed temperature, 0° in this case.
6. The rod has an initial temperature distribution that is a function of its length.

This model can be mathematically described for the temperature distribution, u, by

$$\frac{1}{k}\frac{\partial u}{\partial t} = \frac{\partial^2 u}{\partial x^2} \quad 0 < x < L, \quad t > 0 \tag{6.2.1}$$

$$\text{BC:} \quad \begin{aligned} u(0,t) &= 0 \\ &\qquad\qquad \text{for } t > 0 \\ u(L,t) &= 0, \end{aligned} \tag{6.2.2}$$

$$\text{IC: } u(x,0) = f(x), \quad 0 < x < L \tag{6.2.3}$$

where k is thermal diffusivity ($L^2$/time) and these boundary conditions (BC) are described as *fixed homogeneous*.

*Solution*
Assume that

$$u(x,t) = X(x)T(t) \tag{6.2.4}$$

where X(x) represents a function of x only and T(t) represents a function of t only. Then Equation 6.2.4 can be substituted into Equation 6.2.1 such that:

$$\frac{\partial u}{\partial t} = X(x)\frac{dT}{dt} = X(x)T'(t)$$

$$\frac{\partial u}{\partial X} = \frac{dX}{dx}T(t) = X'(x)T(t)$$

$$\frac{\partial^2 u}{\partial x^2} = \frac{d^2 X}{dx^2}T(t) = X''(x)T(t)$$

giving

$$\frac{1}{k}X(x)T'(t) = X''(x)T(t)$$

which can be rewritten as

$$\frac{1}{k}\frac{T'}{T} = \frac{X''}{X} \tag{6.2.5}$$

Now, observe that the left-hand side of Equation 6.2.5 is a function of t only while the right-hand side is a function of x only. This can be a true statement only if Equation 6.2.5 is equal to a constant. That is

$$\frac{1}{k}\frac{T'}{T} = \frac{X''}{X} = \lambda \qquad (6.2.6)$$

where the number $\lambda$ is called the *separation constant*. Equation 6.2.6 can now be recasted as a system of two ordinary differential equations

$$\frac{1}{k}\frac{T'}{T} = \lambda \qquad (6.2.7)$$

and

$$\frac{X''}{X} = \lambda \qquad (6.2.8)$$

In order to further define $\lambda$, we sequentially substitute the BC, Equation 6.2.2, into Equation 6.2.4 thus

$$u(0,t) = X(0)T(t) \implies X(0) = 0$$

where the argument that T(t) is an arbitrary function independent of x has been applied to conclude that $X(0) = 0$. The result

$$u(L,t) = X(L)T(t) \implies X(L) = 0$$

is arrived at using the same reasoning. Therefore, Equation 6.2.8 is to be subjected to the boundary conditions

$$X(0) = 0 \qquad (6.2.2A)$$

and

$$X(L) = 0 \qquad (6.2.2B)$$

Equation 6.2.8 together with Equations 6.2.2A and 6.2.2B constitute a *homogeneous boundary value* problem as defined in Chapter 4. Furthermore, Equation 6.2.8 together with Equations 6.2.2A and 6.2.2B is an example of a *regular Sturm-Liouville* problem, where $p(x) = 1$, $r(x) = 1$, and $q(x) = 0$. Also, $A_1 = B_1 = 1$ and $A_2 = B_2 = 0$ for the boundary conditions (Chapter 4). The three cases of $\lambda$ are examined in the way of Chapter 4, that is, $\lambda = 0$ and $\lambda > 0$ both produce the trivial solution. The third case, $\lambda < 0$, say $\lambda = -\beta^2$ gives

$$X'' + \beta^2 X = 0$$

$$X(0) = 0$$

$$X(L) = 0$$

which solves to

$$\lambda_n = -\frac{n^2\pi^2}{L^2}, \quad n = 1, 2, \ldots$$

and

$$X_n(x) = A_{1,n} \sin\frac{n\pi x}{L}$$

The numbers defined by $\lambda_n$ are the eigenvalues and the functions $X_n(x)$ are the corresponding eigenfunctions as discussed in Chapter 4.

Now, attention is given to Equation 6.2.7, which solves to

$$T_n(t) = ce^{\lambda_n kt}$$

where c is an arbitrary constant.

Therefore, Equations 6.2.1, 6.2.2, and 6.2.4 result in

$$u_n(x,t) = \left( A_{1,n} \sin\frac{n\pi x}{L} \right) ce^{\lambda_n kt}$$

for each n. However, by the Principle of Superposition the solution of Equations 6.2.1 and 6.2.2, using 6.2.4, can be represented as

$$u(x,t) = \sum_{n=1}^{\infty} B_n \sin\frac{n\pi x}{L} \exp\left( -\frac{n^2\pi^2}{L^2} kt \right) \tag{6.2.9}$$

We now use the given initial condition, Equation 6.2.3, to define $B_n$ in the following way:

$$u(x,0) = \sum_{n=1}^{\infty} B_n \sin\frac{n\pi x}{L} = f(x) \tag{6.2.10}$$

Equation 6.2.10 can be recognized as the Fourier sine series representation of f(x). That is

$$B_n = \frac{2}{L} \int_0^L f(x) \sin \frac{n\pi x}{L} dx, \quad \text{for } n \geq 1 \tag{6.2.11}$$

Therefore, the described model has Equations 6.2.9 and 6.2.11 as its solution.

## 6.2.1   Boundary Conditions

In this section, we will use examples to demonstrate the four standard types (classes) of boundary conditions that usually accompany a pde.

*Class 1.* Bar with zero temperature at both ends

$$\frac{\partial u}{\partial t} = \frac{\partial^2 u}{\partial x^2}, \quad 0 < x < 1, \quad t > 0 \tag{6.2.12}$$

$$\text{BC:} \quad u(0,t) = u(1,t) = 0 \tag{6.2.13}$$

$$\text{IC:} \quad u(x,0) = \begin{cases} 1, & 0 < x < 0.5 \\ 0, & 0.5 < x < 1 \end{cases} \tag{6.2.14}$$

*Solution*

**Step 1:** Substitute u = XT into Equation 6.2.12 to obtain

$$\frac{X''}{X} = \text{constant} \tag{5.2.15}$$

and

$$\frac{T'}{T} = \text{same constant} \tag{6.2.16}$$

Interpret the boundary conditions, Equation 6.2.13 in terms of the substitution u = XT:

$$u(0,t) = X(0)T(t) = 0 \quad \text{and} \quad u(1,t) = X(1)T(t) = 0$$

Then, for nontrivial solution to exist we get:

$$X(0) = 0 \quad \text{and} \quad X(1) = 0 \tag{6.2.17}$$

**Step 2:** Usually there are three cases of the constant to consider in solving Equation 6.2.15. However, in the general model considered in Equations 6.2.1, 6.2.2, and 6.2.3, we found out that only the trivial solution resulted if the constant was zero or positive. Nontrivial solution resulted only for a negative constant, say $-\lambda^2$, that is, designating the constant to be $-\lambda^2$, Equation 6.2.15 has general solution

$$X(x) = c_1 \cos \lambda x + c_2 \sin \lambda x$$

Then applying the conditions given by Equation 6.2.17, we get

$$\lambda = n\pi, \quad n = 1, 2, 3, \ldots$$

and

$$X_n(x) = c_2 \sin n\pi x$$

**Step 3:** Solve Equation 6.2.16

$$\frac{T'}{T} = -\lambda^2 = -n^2\pi^2$$

to get

$$T_n(t) = c_3 \exp\left(-n^2\pi^2 t\right)$$

Therefore, for each n we have the product solution

$$u_n(x, t) = X_n T_n = B_n \sin n\pi x \exp\left(-n^2\pi^2 t\right)$$

**Step 4:** By the superposition principle we get

$$u(x, t) = \sum_{n=1}^{\infty} B_n e^{-n^2\pi^2 t} \sin n\pi x$$

as the solution to Equations 6.2.12 and 6.2.13, with $B_n$ to be determined.

**Step 5:** Here we determine the constants $B_n$ with the aid of Equation 6.2.14. That is

$$\sum_{n=1}^{\infty} B_n \sin n\pi x = \begin{cases} 1, & 0 < x \, 0.5, \\ 0, & 0.5 < x < 1 \end{cases}$$

Then, following the discussion on Fourier series in Chapter 5, we get

$$B_n = \frac{2}{1} \int_0^{0.5} 1 \cdot \sin n\pi x \, dx$$

$$= \frac{2}{n\pi} \left( 1 - \cos \frac{n\pi}{2} \right)$$

Finally, the solution to the model described by Equations 6.2.12, 6.2.13, and 6.2.14 is

$$u(x,t) = \sum_{n=1}^{\infty} \frac{2}{n\pi} \left( 1 - \cos \frac{n\pi}{2} \right) e^{-n^2\pi^2 t} \sin n\pi x$$

It is important to note that the above models include the fixed homogeneous type boundary conditions and they both result in a Fourier sine series.

Class 2: *Bar with insulated boundary(ies)*

As a second model, consider the previous model with one modification. Suppose we have the situation in which the ends of the rod are insulated instead of being at a fixed temperature of 0°. Then in this scenario, no heat flows out of, or into, either end. This is the so-called *insulated boundary conditions*. This new model can be described as

$$\frac{1}{k} \frac{\partial u}{\partial t} = \frac{\partial^2 u}{\partial x^2} \quad 0 < x < L, \quad t > 0 \tag{6.2.1}$$

subject to

$$\frac{\partial u(0,t)}{\partial x} = 0$$

$$\text{BC: } \frac{\partial u(L,t)}{\partial x} = 0, \quad t > 0 \tag{6.2.18}$$

$$\text{IC: } u(x,0) = f(x), \quad 0 < x < L \tag{6.2.3}$$

In order to solve this new model by the separation of variables technique, the same procedure used above will be followed.

**Step 1:** Substitute Equation 6.2.4 into Equation 6.2.1 to get

$$\frac{X''}{X} = \text{constant} \tag{6.2.19}$$

and

$$\frac{T'}{kT} = \text{same constant} \tag{6.2.20}$$

Also,

$$\frac{\partial(XT)}{\partial x} = X'T$$

Therefore, the interpretation of the boundary conditions given by Equation 6.2.18 results in

$$u_x(0,t) = X'(0)T(t) = 0 \quad \text{and} \quad u_x(L,t) = X'(L)T(t) = 0$$

such that

$$X'(0) = 0 \quad \text{and} \quad X'(L) = 0 \tag{6.2.18A}$$

for nontrivial solutions.

**Step 2:** In finding the solution to Equation 6.2.19, there will be three cases of the constant to consider.

*Case I:* Suppose the *constant* is zero. Then Equation 6.2.19 becomes

$$X'' = 0$$

with the general solution

$$X(x) = c_1 x + c_2$$

The conditions given by Equation 6.2.18A when applied to this general solution result in

$$X(x) = c_2 \tag{6.2.21}$$

a constant. This implies that the separation constant, or eigenvalue, can be zero.

*Case II:* Suppose the *constant* is positive, say $\lambda^2$. Then Equation 6.2.19 becomes

$$X'' - \lambda^2 X = 0$$

with general solution

$$X(x) = c_3 e^{\lambda x} + c_4 e^{-\lambda x}$$

Applying Equation 6.2.18A to this general solution results in

$$c_3 = c_4 = 0$$

for $\lambda^2 \neq 0$, and this case yields only the trivial solution. Therefore, the separation constant, $\lambda$ is not positive.

*Case III:* Here we suppose the *constant* is negative, say $-\lambda^2$. Then Equation 6.2.19 results in

$$X'' + \lambda^2 X = 0$$

with general solution

$$X(x) = c_5 \cos \lambda x + c_6 \sin \lambda x$$

Application of Equation 6.2.18A to evaluate $c_5$ and $c_6$ results in $c_6 = 0$ and $c_5 \neq 0$ and in order to obtain a nontrivial solution. That is

$$X'(L) = 0 = -c_5 \lambda \sin \lambda L + c_6 \lambda \cos \lambda L$$

results in

$$\sin \lambda L = 0 \implies \lambda L = n\pi$$

$$\text{or} \quad \lambda = \frac{n\pi}{L}, \quad n = 1, 2, 3, \ldots$$

and

$$X_n(x) = c_5 \cos \frac{n\pi x}{L} \qquad (6.2.22)$$

**Step 3:** Solve Equation 6.2.20 to get

$$T(t) = c_7$$

for the case of the separation constant being zero (step 2, case I). Therefore, $u(x, t) = XT$ produces a constant in the case of a zero separation constant, that is

$$u(x,t) = XT = c_2 c_7 = a \ \text{constant} = \frac{A_0}{2}$$

where $\dfrac{A_0}{2}$ is used to reemphasize the relationship to a Fourier cosine series as discussed in Chapter 5. In the case for a negative separation constant, Equation 6.2.20 solves to

$$T_n(t) = c_8 e^{-k\lambda_n^2 t}$$

Then for each $n$,

$$u_n(x,t) = X_n T_n = A_n e^{-k(n\pi/L)^2 t} \cos \frac{n\pi x}{L}$$

**Step 4:** By applying the Superposition Principle, we get the result

$$u(x,t) = \frac{A_0}{2} + \sum_{n=1}^{\infty} A_n e^{-k(n\pi/L)^2 t} \cos \frac{n\pi x}{L}$$

which is the solution to Equations 6.2.1 and 6.2.18. In this case we note that a Fourier cosine series results when the boundary conditions are of the insulated type.

**Step 5:** Here we determine the constants $A_n$ ($n \geq 0$) with the aid of the given IC, in this case Equation 6.2.3. Thus,

$$u(x,0) = f(x) = \frac{A_0}{2} + \sum_{n=1}^{\infty} A_n \cos \frac{n\pi x}{L}$$

where the constants $A_0$ and $A_n$ are the coefficients of the Fourier cosine series for $f(x) = u(x, 0)$. For example, suppose $f(x)$ is given as

$$f(x) = u(x,0) = \begin{cases} 1, & 0 < x < 0.5 \\ 0, & 0.5 < x < 1 \end{cases}$$

Then

$$A_0 = \frac{2}{1} \int_0^{0.5} dx = 1$$

and for $n \geq 1$

$$A_n = \int_0^{0.5} 1 \cdot \cos n\pi x \, dx = \frac{2}{n\pi} \sin \frac{n\pi}{2}$$

Therefore,

$$u(x,t) = \frac{1}{2} + \sum_{n=1}^{\infty} \frac{2}{n} \sin \frac{n\pi}{2} e^{-k(n\pi)^2 t} \cos n\pi x$$

**Class 3:** *Bar with periodic boundary conditions*

A third model can be created from the first one by considering the *mixed* or *periodic boundary conditions* instead of the fixed temperature at the ends. Here the previously considered straight rod is formed into a circular ring with perfectly joined ends such that the temperature is continuous across the joint. That is

$$u(-L,t) = u(L,t).$$

Also, the derivative of the temperature function, $u$, with respect to length can be assumed continuous

$$u_x(-L,t) = u_x(L,t)$$

This new model can be described as

$$\frac{1}{k}\frac{\partial u}{\partial t} = \frac{\partial^2 u}{\partial x^2} \quad -L < x < L, \quad t > 0 \tag{6.2.23}$$

$$\text{BC:} \quad \begin{array}{c} u(-L,t) = u(L,t) \\ u_x(-L,t) = u_x(L,t), \quad t > 0 \end{array} \tag{6.2.24}$$

$$\text{IC:} \quad u(x,0) = f(x), \quad -L < x < L. \tag{6.2.25}$$

Again, the solution procedure is the same as above.

**Step 1:** Substitute u = XT into Equation 6.2.23 to get

$$\frac{X''}{X} = \text{cons tan t} \tag{6.2.19}$$

$$\frac{T'}{kT} = \text{same cons tan t} \tag{6.2.20}$$

The boundary conditions in Equation 6.2.24 are interpreted in terms of u = XT to be

$$X(-L) = X(L) \tag{6.2.24A}$$

$$X'(-L) = X'(L). \tag{6.2.24B}$$

It should be observed here that Equations 6.2.19, 6.2.24A, and 6.2.24B together form a periodic Sturm-Liouville problem on [–L, L] as discussed in Chapter 4.

**Step 2:** Again there are three cases of the separation constant to consider.

*Case I:* Separation constant is zero. This gives the solution to Equation 6.2.19 as

$$X(x) = c_1 x + c_2$$

which is subject to the boundary conditions, Equations 6.2.24A and 6.2.24B. That is, at –L

$$X(-L) = -c_1 L + c_2$$

and at L

$$X(L) = c_1 L + c_2$$

such that

$$X(-L) = -c_1 L + c_2 = X(L) = c_1 L + c_2 \implies -c_1 = c_1$$

Therefore, $c_1 = 0$. Further, since $X' = c_1$, no new information is obtained except that

$$X(x) = c_2 \text{ a constant}$$

This constant will be conveniently denoted as $\dfrac{A_0}{2}$ to correspond with the notation for the Fourier series coefficient discussed in Chapter 5. That is

$$X(x) = \frac{A_0}{2}$$

This is taken to mean that zero is an eigenvalue with the corresponding eigenfunction a constant.

*Case II*: Separation constant is positive, say $\lambda^2$. Then Equation 6.2.19 gives the general solution

$$X(x) = c_3 e^{\lambda x} + c_4 e^{-\lambda x}.$$

Application of Equation 6.2.24A results in

$$X(-L) = c_3 e^{-\lambda L} + c_4 e^{\lambda L} = c_3 e^{\lambda L} + c_4 e^{-\lambda L} = X(L)$$

or

$$(c_3 - c_4)\left[ e^{\lambda L} - e^{-\lambda L} \right] = 0 \implies c_3 = c_4$$

since $e^{\lambda L} \neq e^{-\lambda L}$

Substitution of Equation 6.2.25A into the general solution yields

$$X'(-L) = \lambda c_3 e^{-\lambda L} - \lambda c_4 e^{\lambda L} = \lambda c_3 e^{\lambda L} - \lambda c_4 e^{-\lambda L} = X'(L) \qquad (6.2.26)$$

But since $c_3 = c_4$, Equation 6.2.26 can be rearranged to give

$$c_3\left(e^{-\lambda L} - e^{\lambda L}\right) = -c_3\left(e^{-\lambda L} - e^{\lambda L}\right) \Rightarrow c_3 = -c_3$$

Therefore, $c_3 = 0 = c_4$. From this, we can conclude that only the trivial solution is obtained for a positive separation constant.

*Case III:* Separation constant is negative, say $-\lambda^2$. Equation 6.2.19 then solves to

$$X(x) = c_5 \cos \lambda x + c_0 \sin \lambda x \tag{6.2.27}$$

Then, applying Equation 6.2.24A to Equation 6.2.27 results in

$$X(-L) = c_5 \cos \lambda L - c_6 \sin \lambda L = c_5 \cos \lambda L + c_6 \sin \lambda L = X(L)$$

or

$$c_6 \sin \lambda L = 0$$

and applying Equation 6.2.25A to Equation 6.2.27 results in

$$X'(-L) = \lambda c_5 \sin \lambda L + \lambda c_6 \cos \lambda L = -c_5 \lambda \sin \lambda L + c_6 \lambda \cos \lambda L = X'(L)$$

which reduces to

$$c_5 \sin \lambda L = 0$$

Therefore, for nontrivial solution to exist

$$\sin \lambda L = 0 \Rightarrow \lambda L = n\pi, \quad n = 1, 2, 3, \ldots$$

giving the eigenvalues

$$\lambda = \frac{n\pi}{L}$$

and

$$X(x) = c_5 \cos \frac{n\pi x}{L} + c_6 \sin \frac{n\pi x}{L}$$

are the corresponding eigenfunctions.

**Step 3:** As before, Equation 6.2.20 solves to

$$T(t) = c_7$$

in the case of the zero eigenvalue and

$$T(t) = c_8 e^{-\lambda^2 kt} = c_8 e^{-k(n\pi/L)^2 t}$$

in the case of the negative eigenvalues $-\lambda_n^2$. In terms of the assumption $u(x, t) = XT$ we get

$$u_n(x,t) = \left[ A_n \cos\frac{n\pi x}{L} + \sin\frac{n\pi x}{L} \right] e^{-k(n\pi/L)^2 t}$$

for each n.

**Step 4:** By the principle of superposition, the temperature profile $u(x, t)$ is given by

$$u(x,t) = \frac{A_0}{2} + \sum_{n=1}^{\infty} \left[ A_n \cos\frac{n\pi x}{L} + B_n \sin\frac{n\pi x}{L} \right] e^{-k(n\pi/L)^2 t}$$

which is the solution to Equations 6.2.23 and 6.2.24 with the as-yet-undefined Fourier coefficients.

**Step 5:** Application of Equation 6.2.25 define the constants $A_0$, $A_n$, and $B_n$. That is,

$$u(x,0) = f(x) = \frac{A_0}{2} + \sum_{n=1}^{\infty} \left[ A_n \cos\frac{n\pi x}{L} + B_n \sin\frac{n\pi x}{L} \right]$$

where

$$A_n = \frac{1}{L} \int_{-L}^{L} f(x) \cos\frac{n\pi x}{L} dx, \quad n \geq 0$$

$$B_n = \frac{1}{L} \int_{-L}^{L} f(x) \sin\frac{n\pi x}{L} dx, \quad n \geq 1$$

**Class 4:** *Bar with fixed nonhomogeneous time-independent boundary(ies)*
So far we have dealt with models that are strictly homogeneous. However, there are cases of nonhomogeneous problems that can be adjusted in such a way that the method of separation of variables can still be applied. One such case is that in which the boundary conditions are nonzero constants (fixed nonhomogeneous). This case is illustrated below.

A model can be constructed by maintaining the ends of the rod at fixed nonzero temperatures, that is, at $x = 0$, $u = T_1$, and at $x = L$, $u = T_2$. This model is the so-called *fixed nonhomogeneous time-independent* boundary conditions case. However, the technique of separation of variables requires the boundary conditions and the differential equation to be homogeneous. Therefore, the model given by

$$\frac{1}{k}\frac{\partial u}{\partial t} = \frac{\partial^2 u}{\partial x^2} \quad 0 < x < L, \quad t > 0 \tag{6.2.1}$$

$$\text{BC:} \quad \begin{matrix} u(0,t) = T_1 \\ u(L,t) = T_2, \quad t > 0 \end{matrix} \tag{6.2.28}$$

$$\text{IC:} \quad u(x,0) = f(x), \quad 0 < x < L \tag{6.2.3}$$

is a nonhomogeneous problem that must be reduced to a homogeneous one. The substitution of

$$u(x,t) = w(x,t) + v(x) \tag{6.2.29}$$

where $v(x)$ and $w(x, t)$ represent the steady-state and transient solution, respectively, reduces Equations 6.2.1 and 6.2.28 to

$$\frac{1}{k}\frac{\partial w}{\partial t} = \frac{\partial^2 w}{\partial x^2} \quad 0 < x < L, \quad t > 0 \tag{6.2.1A}$$

and

$$\begin{matrix} w(0,t) = 0 \\ w(L,t) = 0, \quad t > 0 \end{matrix} \tag{6.2.30}$$

with $v(x)$ defined by

$$v''(x) = 0 \tag{6.2.31}$$

$$v(0) = T_1$$

$$v(L) = T_2 \tag{6.2.32}$$

Equations 6.2.1A and 6.2.30 now form a homogeneous boundary value problem, with the initial condition given by

$$w(x,0) = f(x) - v(x) \qquad 0 < x < L \tag{6.2.3A}$$

This model (Equation 6.2.1A and Equation 6.2.30) were solved previously with the fixed homogeneous boundary conditions.

Equation 6.2.31 together with the boundary conditions given by Equation 6.2.32 form the steady-state model, and its solution is given as

$$v(x) = T_1 + \left( \frac{T_2 - T_1}{L} \right) x \tag{6.2.33}$$

The final solution of this nonhomogeneous model is given in Table 6.1. Also, the other discussed models are summarized in Table 6.1, according to boundary condition type. Examples demonstrating the use of this table are also given.

Other models may be developed from different combinations of the boundary conditions. In those cases, the 5 steps outlined above should yield the appropriate solutions, provided the conditions of linearity and homogeneity are satisfied.

There are plenty of examples in this section to illustrate the method of separation of variables. It is hoped that the reader will explore the references for deeper discussions involving these methods.

### Example 6.2.1

Consider the conduction of heat in a copper rod 100 cm in length whose ends are maintained at 0°C for all t > 0. Derive an expression for the temperature u(x, t) if the initial temperature distribution in the rod is

$$u(x,0) = 50, \quad 0 \le x \le 100$$

and the thermal diffusivity of copper is 1.14 cm²/sec.

*Solution*

This is a case of the model described by Equations 6.2.1, 6.2.2, and 6.2.3. From Table 6.1, the fixed homogeneous boundary conditions correspond to eigenvalues

$$\lambda = \frac{n\pi}{L} = \frac{n\pi}{100}$$

**TABLE 6.1**

Typical Boundary Conditions Associated with the Heat Equation

| Boundary Conditions | Eigenvalues | Series Solution |
|---|---|---|
| Fixed, homogeneous:<br>$u(0, t) = 0$<br>$u(L, t) = 0$ | $\lambda = \dfrac{n\pi}{L}$<br><br>$n = 1, 2, 3, \ldots$ | $u(x,t) = \displaystyle\sum_{n=1}^{\infty} B_n e^{-k\lambda^2 t} \sin \lambda x$<br>where $B_n$ are the Fourier sine series coefficients of $f(x)$ |
| Insulated:<br>$u_x(0, t) = 0$<br>$u_x(L, t) = 0$ | $\lambda = \dfrac{n\pi}{L}$<br><br>$n = 0, 1, 2, \ldots$ | $u(x,t) = \dfrac{A_0}{2} + \displaystyle\sum_{n=1}^{\infty} A_n e^{-k\lambda^2 t} \cos \lambda x$<br>where $A_0$ and $A_n$ are the Fourier cosine series coefficients of $f(x)$ |
| Periodic:<br>$u(-L, t) = u(L, t)$<br>$u_x(-L, t) = u_x(L, t)$ | $\lambda = \dfrac{n\pi}{L}$<br><br>$n = 0, 1, 2, \ldots$ | $u(x,t) = \dfrac{A_0}{2} + \displaystyle\sum_{n=1}^{\infty} \left[ A_n \cos \lambda x + B_n \sin \lambda x \right] e^{-k\lambda^2 t}$<br>where $A_0, A_n,$ and $B_n$ are the Fourier coefficients of $f(x)$ |
| Fixed, nonhomogeneous:<br>$u(0, t) = T_1$<br>$u(L, t) = T_2$ | $\lambda = \dfrac{n\pi}{L}$<br><br>$n = 1, 2, 3, \ldots$ | $u(x, t) = w(x, t) + v(x)$<br><br>$v(x) = T_1 + \left( \dfrac{T_2 - T_1}{L} \right) x$<br><br>$w(x,t) = \displaystyle\sum_{n=1}^{\infty} B_n e^{-k\lambda^2 t} \sin \lambda x$<br>where $B_n$ are the Fourier sine series coefficients of $f(x) - v(x)$ |

*Source:* Adapted from Giordano, F.R. and Weir, M.D., *Differential Equations, A Modeling Approach*, Addison-Wesley, Reading MA, 1991. With permission.

and series solution

$$u\left(x,t\right) = \sum_{n=1}^{\infty} B_n e^{-k\left(n\pi/100\right)^2 t} \sin \frac{n\pi x}{100}$$

Then, the initial value

$$u\left(x,0\right) = 50 = \sum_{n=1}^{\infty} B_n \sin \frac{n\pi x}{100}$$

defines $B_n$, that is, for $n \geq 1$

$$B_n = \frac{\displaystyle\int_0^{100} 50 \sin \frac{n\pi x}{100}}{\displaystyle\int_0^{100} \sin^2 \frac{n\pi x}{100} dx} dx = \frac{1}{50} \int_0^{100} 50 \sin \frac{n\pi x}{100} dx = \frac{100}{n\pi}\left(1 - \cos n\pi\right) = \frac{200}{\left(2n-1\right)\pi}$$

Therefore,

$$u(x,t) = \sum_{n=1}^{\infty} \frac{200}{(2n-1)\pi} e^{-1.14(2n-1)^2 \pi^2 / 10^4 t} \sin \frac{(2n-1)\pi x}{100}$$

## Example 6.2.2

Consider an aluminum rod (k = 0.86 cm²/sec) of length $\ell$ to be previously at the uniform temperature of 25°C. Suppose that at time t = 0, the end x = 0 is cooled to 0°C while the end x = $\ell$ is heated to 60°C, and both thereafter are maintained at those temperatures.

(a) Find the temperature distribution in the rod at any time t.

(b) For $\ell$ = 20 cm., use the first term in the series for the temperature distribution to find the approximate temperature at x = 5 cm when t = 30 sec; when t = 60 sec.

(c) Use the first two terms in the series for the temperature distribution to find an approximate value of u(5, 30). What is the percentage difference between the one- and two-term approximations? Does the third term in the series have any appreciable effect for this value of t?

(d) Use the first term in the series for the temperature distribution to estimate the time interval that must elapse before the temperature at x = 5 cm comes within 1% of its steady-state value.

*Solution*
The model for this phenomenon is

$$\frac{1}{k} \frac{\partial u}{\partial t} = \frac{\partial^2 u}{\partial x^2} \quad 0 < x < l, \quad t > 0$$

subject to

$$u(0,t) = 0$$

$$\text{BC: } u(l,t) = 60$$

$$\text{IC: } (x,0) = 25$$

This is a problem with fixed nonhomogeneous boundary conditions and the eigenvalues and series solution are given in Table 6.1. That is,

$$u(x,t) = w(x,t) + v(x)$$

where

$$v(x) = T_1 + \left(\frac{T_2 - T_1}{l}\right)X = \frac{60x}{l}$$

and

$$w(x,t) = \sum_{n=1}^{\infty} B_n e^{-k\lambda^2 t} \sin \lambda x$$

where

$$\lambda = \frac{\lambda x}{l}$$

(a) In order to define $B_n$, the initial condition is applied

$$w(x,0) = u(x,0) - v(x) = 25 - \frac{60x}{l} = \sum_{n=1}^{\infty} B_n \sin \frac{n\pi x}{l}$$

Therefore, for $n \geq 1$

$$B_n = \frac{2}{l} \int_0^l \left(25 - \frac{60x}{l}\right) \sin \frac{n\pi x}{l} dx = \frac{10}{n\pi}(7 \cos n\pi + 5)$$

and the temperature distribution in the rod at any time t is

$$u(x,t) = \frac{60x}{l} + \sum_{n=1}^{\infty} \frac{10}{n\pi}(7 \cos n\pi + 5)e^{-0.86(n\pi/l)^2 t} \sin \frac{n\pi x}{l}$$

(b) For $\ell = 20$ cm, $x = 5$ cm, and $t = 30$ sec, the first term of the series gives

$$u(5,30) \cong 15 + \frac{10}{\pi}(-2)(0.53)\sin \frac{\pi}{4} = 12.6°C.$$

When t = 60 sec

$$u(5,60) = 15 + \frac{10}{\pi}(-2)(0.28)(0.707) = 13.7°C$$

(c) Using the first two terms of the series, we get

$$u_2(5,30) = 15 + \frac{10}{\pi}(-2)(0.530)(0.707) + \frac{10}{2\pi}(12)(0.078) = 14.1°C$$

Then the percentage difference between the one- and two-term approximations is

$$\left[\frac{u_2(5,30) - u(5,30)}{u_2(5,30)}\right] \times 100 = 11\%$$

For t = 30 sec, the third term of the series is −4.876E-03, which does not have any appreciable effect on either of the two approximations.

(d) Time interval that must elapse before the temperature at x = 5 cm comes within 1% of its steady-state value with the first term of the series can be estimated as follows:

$$u(5,\tau) = 15 + \frac{10}{\pi}(-2)(0.707)e^{-0.0212\tau}$$

where $\tau$ indicates the time interval. Then

$$\left|\frac{0.15}{-4.5}\right| = e^{-0.0212\tau} \Rightarrow \tau = 160\,\text{sec}$$

Therefore, a time interval of 160 sec would have to elapse before the temperature at x = 5 cm comes within 1% of the steady-state value.

## Example 6.2.3

Consider a uniform rod of length $\ell$ having an initial temperature distribution given by f(x), $0 < x < \ell$.

Assume that the temperature at the end x = 0 is held at 0°C, while the end x = l is insulated, so that no heat passes through it.

(a) Show that the fundamental solutions of the partial differential equation and boundary conditions are

$$u_n(x,t) = e^{-(2n-1)^2\pi^2 kt/4l^2}\sin\frac{(2n-1)\pi x}{2l}, n = 1,2,3,..$$

(b) Find a formal series expansion for the temperature u(x, t),

$$u_n(x,t) = \sum_{n=1}^{\infty} c_n u_n(x,t)$$

that also satisfies the initial condition u(x, 0) = f(x).

*Solution*
The model for this problem is

$$\frac{1}{k}\frac{\partial u}{\partial t} = \frac{\partial^2 u}{\partial x^2} \quad 0 < x < l, \quad t > 0$$

$$\begin{aligned} u(0,t) &= 0 \\ u_x(l,t) &= 0 \end{aligned} \quad t > 0$$

$$u(x,0) = f(x), \quad 0 \le x \le l$$

From Table 6.1, there is no model with this combination of boundary conditions. Therefore, the method must be directly applied; that is, let

$$u(x,t) = X(x)T(t)$$

Then, after substitution into the differential equation, one gets

$$\frac{1}{k}\frac{T'(t)}{T(t)} = \frac{X''(x)}{X(x)} = \lambda$$

or the boundary value problem

$$X'' - \lambda X = 0$$

$$X(0) = 0$$

$$X'(l) = 0$$

and

$$\frac{T'}{T} = \lambda k \Rightarrow T(t) = c_1 e^{\lambda kt}$$

We now consider the three cases of $\lambda$ that must satisfy the above boundary value problem.

That is, for $\lambda = 0$ we get the general solution:

$$X(x) = c_2 x + c_3$$

$$X(0) = 0 = c_3$$

$$X'(1) = 0 = c_2 \quad \therefore \lambda \neq 0$$

Suppose $\lambda > 0$, say $\lambda = \beta^2$, then

$$X(x) = c_4 e^{\beta x} + c_5 e^{-\beta x}$$

and

$$X(0) = c_4 + c_5 = 0 \quad \Rightarrow c_5 = -c_4$$

Therefore,

$$X(x) = c_4\left(e^{\beta x} - e^{-\beta x}\right) = 2c_4 \sinh \beta x$$

then

$$X'(1) = 2\beta c_4 \cosh \beta l = 0 \quad \Rightarrow c_4 = 0 \text{ and } c_5 = 0$$

since

$$\cosh \beta l \neq 0.$$

Therefore, $\lambda > 0$ yields the trivial solution.

As the third case, suppose $\lambda < 0$, $\lambda = -\alpha^2$, then the general solution

$$X(x) = c_6 \cos \alpha x + c_7 \sin \alpha x$$

results.

$$X(0) = 0 = c_6$$

$$X'(1) = \alpha c_7 - \cos \alpha l = 0$$

then for nontrivial solution to exist, $c_7 \neq 0$ and

$$\cos \alpha l = 0 \quad \Rightarrow \alpha l = (2n-1)\frac{\pi}{2}, \quad n = 1, 2, \ldots$$

$$\therefore \alpha = \frac{(2n-1)\pi}{2l}$$

are the eigenvalues and

$$X(x) = c_7 \sin \frac{(2n-1)\pi x}{2l}$$

are the eigenfunctions. Then, for each n

$$u_n(x,t) = T_n(t) X_n(x) = c_n e^{-(2n-1)^2 \pi^2 kt / 4l^2} \sin \frac{(2n-1)\pi x}{2l}$$

By the superposition principle,

$$u(x,t) = \sum_{n=1}^{\infty} c_n u_n(x,t) = \sum_{n=1}^{\infty} c_n e^{-(2n-1)^2 \pi^2 kt / 4l^2} \sin \frac{(2n-1)\pi x}{2l}$$

where

$$f(x) = u(x,0) = \sum_{n=1}^{\infty} c_n \sin \frac{(2n-1)\pi x}{2l}$$

That is, for $n \geq 1$

$$\int_0^l f(x) \sin \frac{(2n-1)\pi x}{2l} dx = c_n \int_0^l \sin^2 \frac{(2n-1)\pi x}{2l} dx$$

or

$$c_n = \frac{2}{l} \int_0^l f(x) \sin \frac{(2n-1)\pi x}{2l} dx$$

## Example 6.2.4
Consider the model described by

$$\frac{1}{k}u_t = u_{xx}, \quad 0 < xl, \quad t > 0$$

$$u(0,t) = 0, \quad u_x(l,t) + \gamma u(l,t) = 0, \quad t > 0 \qquad \text{(i)}$$

$$u(x,0) = f(x) \quad 0 \le x \le l$$

where $\gamma$ is a positive constant.

(a)  Let u(x, t) = X(x)T(t) and show that

$$X'' - \sigma X = 0, \quad X(0) = 0, \quad X'(l) + \gamma X(l) = 0 \qquad \text{(ii)}$$

and

$$T' - \sigma kT = 0$$

where $\sigma$ is the separation constant.

(b)  Assume that $\sigma$ is real, and show that problem (ii) has only trivial solutions if $\sigma \ge 0$.

(c)  If $\sigma < 0$, let $\sigma = -\lambda^2$, $\lambda > 0$. Show that problem (ii) has nontrivial solutions only if $\lambda$ is a solution of the equation

$$\lambda \cos \lambda l + \gamma \sin \lambda l = 0 \qquad \text{(iii)}$$

*Solution*

(a)  For u(x, t) = X(x)T(t), substitution into the partial differential equation results in

$$\frac{1}{k}\frac{T'}{T} = \frac{X''}{X} = \sigma$$

and

$$u(0,t) = X(0)T(t) \implies X(0) = 0$$

Also

$$u_x(x,t) = X'(x)T(t)$$

such that

$$u_x(l,t) = X'(l)T(t)$$

But,

$$u_x(l,t) + \gamma u(l,t) = 0 = X'(l)T(t) + \gamma X(l)T(t)$$
$$= X'(l) + \gamma X(l)$$

since T(t) is arbitrary. Therefore, we have

$$X'' - \sigma X = 0$$

subject to

$$X(0) = 0, \quad X'(l) + \gamma X(l) = 0$$

and

$$T' - \sigma k T = 0$$

(b) Given that

$$X'' - \sigma X = 0$$

and

$$X(0) = 0$$
$$X'(l) + \gamma X(l) = 0$$

then the general solution for the case $\sigma = 0$ is

$$X(x) = c_1 x + c_2$$

and

$$X(x) = c_3 e^{\sqrt{\sigma}x} + c_4 e^{-\sqrt{\sigma}x}, \quad \text{if } \sigma > 0$$

But,

$$X(0) = 0 = c_2 \quad \text{and} \quad X'(l) + \gamma X(l) = 0 = c_1 + c\gamma l \implies c_1 = 0$$

Therefore, the case $\sigma = 0$ gives only the trivial solution.
Also, for $\sigma > 0$, say $\sigma = \beta^2$, $\beta > 0$

$$X(0) = 0 = c_3 + c_4 \quad \Rightarrow c_4 = -c_3$$

and

$$X'(l) + \gamma X(l) = c_3 \beta \left( e^{\beta l} - e^{-\beta l} \right) + \gamma c_3 \left( e^{\beta l} + e^{-\beta l} \right) = 0$$

or

$$c_3 \left\{ 2\beta \sinh \beta l + 2\gamma \cosh \beta l \right\} = 0.$$

Thus, either

$$\frac{\beta}{\gamma} \tanh \beta l = -1 \Rightarrow \beta l < 0 \quad \text{or} \quad c_3 = 0 \quad \text{but} \quad \beta l > 0 \quad \therefore c_3 = 0$$

Therefore, the case $\sigma > 0$, also results in the trivial solution.

(c) For $\sigma < 0$, say $\sigma = -\lambda^2$, $\lambda > 0$ then

$$X'' + \lambda^2 X = 0$$

$$X(0) = 0$$

$$X'(l) + \gamma X(l) = 0$$

give

$$X(x) = c_5 \cos \lambda x + c_6 \sin \lambda x$$

$$X'(x) = \lambda c_5 \sin \lambda x + \lambda c_6 \cos \lambda x$$

then

$$X(0) = 0 = c_5, \quad \text{and}$$

$$X'(l) + \gamma X(l) = 0 = \lambda c_6 \cos \lambda l + \gamma c_6 \sin \lambda l$$

Therefore, for nontrivial solution to exist we must choose $c_6 \neq 0$ and

$$\lambda \cos \lambda l + \gamma \sin \lambda l = 0$$

This final equation can be solved to give

$$\lambda = -\gamma \tan \lambda l$$

### Example 6.2.5

In this fifth example for this section, we will consider the temperature distribution in a circular cylinder of finite length with insulated ends. Also, the lateral surface $\rho = c$ is kept at a temperature of $0°$ and the initial temperature distribution is a given function of $\rho$-only.

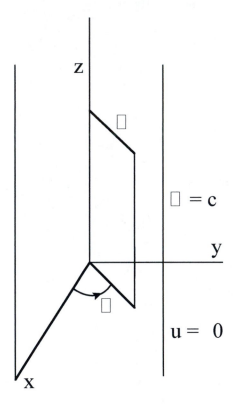

For the cylinder shown, the model under consideration for a homogeneous material is

$$\frac{\partial u}{\partial t} = k\left(\frac{\partial^2 u}{\partial \rho^2} + \frac{1}{\rho}\frac{\partial u}{\partial \rho}\right) \quad (0 < \rho < c,\ t > 0) \tag{6.2.34}$$

$$u(c,t) = 0 \quad (t > 0) \tag{6.2.35}$$

$$u(\rho, 0) = f(\rho) \quad (0 < \rho < c) \tag{6.2.36}$$

Observe that there is an unstated boundary condition at $\rho = 0$. However, one can argue that the temperature must be finite along the axis of the cylinder and so the temperature distribution must be bounded there. Alternatively, the boundary value problem that will result from reducing the partial differential equation is a singular Sturm-Liouville type. That is, the resulting second order ordinary differential equation, when put in Sturm-Liouville form, will satisfy the requirement of $\rho(0) = 0$. One boundary condition is given by Equation 6.2.35 while the other requirement is the solution must be bounded at $\rho = 0$. This type was discussed as Class 3, type 1 in Chapter 4.

*Solution*
Assume a solution of the form

$$u(\rho, t) = R(\rho)T(t) \tag{6.2.37}$$

then follow the five steps previously given.
Equation 6.2.34 gives

$$\frac{T'}{kT} = \frac{1}{R}\left(R'' + \frac{1}{\rho}R'\right) = -\lambda^2$$

That is

$$\rho R''(\rho) + R'(\rho) + \lambda^2 \rho R(\rho) = 0 \tag{6.2.34A}$$

subject to

$$R(c) = 0$$

and

$$T'(t) + \lambda^2 kT = 0 \tag{6.2.34B}$$

Equation 6.2.34A is Bessel's equation. Comparing Equation 6.2.34A with

$$y'' - \left(\frac{2a-1}{x}\right)y' + \left(b^2 c^2 x^{2c-2} + \frac{a^2 - \gamma^2 c^2}{x^2}\right)y = 0 \tag{3.6.4}$$

gives

$$2a - 1 = -1 \implies a = 0, \quad 2c - 2 = 0 \implies c = 1$$

$$b^2 c^2 = \lambda^2 \implies b = \lambda, \quad a^2 - \gamma^2 c^2 = 0 \implies \gamma = 0$$

Equation 3.6.4 was given in Chapter 3. Also, Ewuation 3.6.4 has fundamental solutions

$$y_1 = x^a J_\gamma \left( bx^c \right)$$

and

$$y_2 = x^a J_{-\gamma} \left( bx^c \right)$$

where $J_\gamma(\bullet)$ and $J_{-\gamma}(\bullet)$ are Bessel functions of the first kind of order $\gamma$ as discussed in Chapter 3. Since $\gamma = 0$ in this case, then we have the zero-order Bessel functions; that is, the general solution of Equation 6.2.34A is

$$R(\rho) = c_1 J_0(\lambda \rho) + c_2 Y_0(\lambda \rho) \tag{6.2.38}$$

where

$$Y_0(\lambda \rho) = \frac{2}{\pi} \left\{ \ln \left( \frac{\lambda \rho}{2} \right) + \gamma \right\} J_0(\lambda \rho)$$

$$+ \frac{2}{\pi} \left\{ \frac{\lambda^2 \rho^2}{2^2} - \frac{\lambda^4 \rho^4}{2^2 \cdot 4^2} \left( 1 + \frac{1}{2} \right) + \frac{\lambda^6 \rho 6}{2^2 \cdot 4^2 \cdot 6^2} \left( 1 + \frac{1}{2} + \frac{1}{3} \right) - \cdots \right\}$$

and

$$J_0(\lambda \rho) = 1 - \frac{\lambda^2 \rho^2}{2^2} + \frac{\lambda^4 \rho^4}{2^2 \cdot 4^2} - \frac{\lambda^6 \rho^6}{2^2 \cdot 4^2 \cdot 6^2} + \cdots$$

It is clear that the function $Y_0(\lambda \rho)$ is unbounded at $\rho = 0$. Therefore, in order to have a bounded solution, $c_2$ must be chosen as zero. Application of the other boundary condition gives

$$R(c) = 0 = c_1 J_0(\lambda c) \implies J_0(\lambda c) = 0$$

for a nontrivial solution to exist. This means that the eigenvalues $\lambda_i$ are defined by

$$J_0(\lambda_i c) = 0, \quad \text{for } i = 1, 2, \dots \tag{6.2.39}$$

and

$$R_k(\rho) = J_0(\lambda_i \rho)$$

are the corresponding eigenfunctions. Furthermore, Equation 6.2.34B solves to

$$T(t) = c_3 e^{-\lambda_i^2 kt}.$$

Therefore for each i,

$$u_i(\rho, t) = A_i J_0(\lambda_i \rho) e^{-\lambda_i^2 kt}$$

and by the Principle of Superposition

$$u(\rho, t) = \sum_{i=1}^{\infty} A_i J_0(\lambda_i \rho) e^{-\lambda_i^2 kt} \tag{6.2.40}$$

satisfies the given differential equation and boundary conditions. The initial condition is incorporated as follows

$$f(\rho) = \sum_{i=1}^{\infty} A_i J_0(\lambda_i \rho) \quad (0 < \rho < c)$$

Then, for $i \geq 1$,

$$\int_0^c \rho f(\rho) J_0(\lambda_i \rho) d\rho = A_i \int_0^c \rho [J_0(\lambda_i \rho)]^2 d\rho$$

That is

$$A_i = \frac{2 \int_0^c \rho f(\rho) J_0(\lambda_i \rho) d\rho}{c^2 [J_1(\lambda_i c)]^2} \tag{6.2.41}$$

Therefore, Equations 6.2.39, 6.2.40, and 6.2.41 define the solution to the model given by Equations 6.2.34 to 6.2.36.

## Example 6.2.6

In this example a model with a nonzero sink term is examined. That is, consider the problem described by

$$\frac{\partial u}{\partial t} = \frac{4\partial^2 u}{\partial x^2} - Au, \quad (0 < x < 9), \quad t > 0 \tag{6.2.42}$$

$$u(0,t) = u(9,t) = 0, \quad t > 0 \tag{6.2.43}$$

$$u(x,0) = 3x, \quad (0 < x < 9) \tag{6.2.44}$$

where A is a positive constant. Then except for the term $-Au$, this would be a homogeneous problem. One is therefore motivated to find a substitution that will recast the problem into a familiar homogeneous one. In order to accomplish this, consider the substitution

$$u(x,t) = w(x,t)e^{-At} \tag{6.2.45}$$

Then Equations 6.2.42 to 6.2.44 become

$$\frac{\partial w}{\partial t} = \frac{4\partial^2 w}{\partial x^2}, \quad (0 < x < 9), \quad t > 0 \tag{6.2.42A}$$

$$w(0,t) = w(9,t) = 0, \quad t > 0 \tag{6.2.43A}$$

$$w(x,0) = 3x, \quad (0 < x < 9) \tag{6.2.44A}$$

a problem that can be solved using Table 6.1. From Table 6.1, the case of fixed, homogeneous boundary condition gives

$$w(x,t) = \sum_{n=1}^{\infty} B_n e^{-4\lambda_n^2 t} \sin \lambda_n x$$

where $\lambda_n$ is defined by

$$\lambda_n = \frac{n\pi}{9}, \quad n = 1, 2, 3, \dots$$

and the $B_n$ are the Fourier sine series coefficients of $f(x) = 3x$. That is, for $n \geq 1$

$$B_n = \frac{2}{9}\int_0^9 3x\sin\frac{n\pi x}{9}\,dx = \frac{2}{3}\int_0^9 x\sin\frac{n\pi x}{9}\,dx$$

$$= \frac{2}{3}\left[-\frac{9x}{n\pi}\cos\frac{n\pi x}{9} + \left(\frac{9}{n\pi}\right)^2\sin\frac{n\pi x}{9}\right]_0^9 = -\frac{54}{n\pi}\cos n\pi$$

$$= \frac{54}{n\pi}(-1)^{n+1}$$

Therefore, the solution to Equations 6.2.42 to 6.2.44 is

$$u(x,t) = \sum_{n=1}^{\infty}\frac{54}{n\pi}(-1)^{n+1}e^{-\left(4\lambda_n^2+A\right)t}\sin\frac{n\pi x}{9}$$

## 6.3 The Nonhomogeneous Problem and Eigenfunction Expansion

In the previous section, the method of separation of variables was applied to some problems with special nonhomogeneous terms that could be recasted as homogeneous by using a suitable substitution. In this section, a method will be outlined that is applicable to those nonhomogeneous problems for which no simple substitution can be made to remove the nonhomogeneity. This method is called *Eigenfunction Expansion* (Myint-U and Debnath,[3] Haberman,[4] and Boyce and DiPrima[6]).

Consider the flow of heat in a rod of length L that is uniformly constructed. Further, the rod has temperature-independent heat sources distributed in some prescribed way throughout and is time-dependent. In addition, the temperature at the ends is allowed to be time-dependent. Then, for a prescribed initial temperature distribution, the following model is appropriate:

$$\frac{\partial u}{\partial t} = k\frac{\partial^2 u}{\partial x^2} + Q(x,t) \tag{6.3.1}$$

$$\text{BC: } u(0,t) = A(t) \tag{6.3.2}$$

$$u(L,t) = B(t) \tag{6.3.3}$$

$$\text{IC: } u(x,0) = f(x) \tag{6.3.4}$$

Equations 6.3.1 to 6.3.3 describe a nonhomogeneous partial differential equation with nonhomogeneous boundary conditions.

The associated homogeneous model is given by

$$\frac{\partial u}{\partial t} = k \frac{\partial^2 u}{\partial x^2} + Q(x,t) \tag{6.3.1A}$$

$$v(0,t) = 0 \tag{6.3.2A}$$

$$v(L,t) = 0 \tag{6.3.3A}$$

$$v(x,o) = f(x) \tag{6.3.4A}$$

and was solved in the previous section. Also recall from the previous section that the associated homogeneous model produced the regular Sturm-Liouville boundary value problem

$$\frac{d^2 \phi_n}{dx^2} + \lambda_n \phi_n = 0 \tag{6.3.5}$$

$$\phi_n(0) = 0 \tag{6.3.6}$$

$$\phi_n(L) = 0 \tag{6.3.7}$$

where $\lambda_n = (n\pi/L)^2$ with corresponding eigenfunctions $\phi_n(x) = \sin(n\pi x/L)$. Any *piecewise smooth* function can be expanded in terms of these eigenfunctions. Again, as in Chapter 5, by piecewise smooth we mean a function, $f(x)$ say, that is continuous on the closed interval $[a, b]$ and whose first derivative, $f'(x)$ is continuous on each of the subintervals $x_j < x < x_{j+1}$ and the limits $f(x_j^+)$ and $f(x_j^-)$ exist. Therefore, even though $u(x, t)$ satisfies nonhomogeneous boundary conditions, it is valid that

$$u(x,t) = \sum_{n=1}^{\infty} b_n(t)\phi_n(x) \tag{6.3.8}$$

except at $x = 0$ and $x = L$. This means that term-by-term differentiations with respect to x are not justified since $u(x, t)$ and $\phi_n(x)$ do not satisfy the same homogeneous boundary conditions. That is,

$$\frac{\partial^2 u}{\partial x^2} \neq \sum_{n=1}^{\infty} b_n(t) \frac{d^2 \phi_n}{dx^2}$$

However, term-by-term time derivatives are valid, such that

$$\frac{\partial u}{\partial t} = \sum_{n=1}^{\infty} \frac{db_n}{dt} \phi_n(x) \qquad (6.3.9)$$

Therefore, Equation 6.3.1 becomes

$$k \frac{\partial^2 u}{\partial x^2} + Q(x,t) = \sum_{n=1}^{\infty} \frac{db_n}{dt} \phi_n(x) \qquad (6.3.10)$$

a generalized Fourier series from which

$$\frac{db_n}{dt} = \frac{\int_0^L \left[ k \frac{\partial^2 u}{\partial x^2} + Q(x,t) \right] \phi_n(x) dx}{\int_0^L \phi_n^2(x) dx}, \quad n \geq 1$$

and can be rearranged to give

$$\frac{db_n}{dt} = \frac{\int_0^L k \frac{\partial^2 u}{\partial x^2} \phi_n(x) dx}{\int_0^L \phi_n^2(x) dx} + \frac{\int_0^L Q(x,t) \phi_n(x) dx}{\int_0^L \phi_n^2(x) dx} \qquad (6.3.11)$$

The quantity

$$\frac{\int_0^L Q(x,t) \phi_n(x) dx}{\int_0^L \phi_n^2(x) dx}$$

can be evaluated from known information; that is, we know the eigenfunctions $\phi_n(x)$ from solving Equations 6.3.1A to 6.3.3A and $Q(x, t)$ is given in

Equation 6.3.1. The resulting quantity is expected to be a function of t-only and is denoted as $q_n(t)$. That is,

$$q_n(t) = \frac{\displaystyle\int_0^L Q(x,t)\phi_n(x)dx}{\displaystyle\int_0^L \phi_n^2(x)dx} \qquad (6.3.12)$$

Equation 6.3.11 may now be simplified to

$$\frac{db_n}{dt} = q_n(t) + \frac{\displaystyle\int_0^L k\frac{\partial^2 u}{\partial x^2}\phi_n(x)dx}{\displaystyle\int_0^L \phi_n^2(x)dx} \qquad (6.3.13)$$

In order to evaluate the quantity

$$\int_0^L k\frac{\partial^2 u}{\partial x^2}\phi_n(x)dx$$

the Sturm-Liouville operator

$$L \equiv \frac{d}{dx}\left(p\frac{d}{dx}\right) + q$$

is reintroduced. In this problem,

$$p = 1, \quad q = 0$$

and

$$L = \frac{\partial}{\partial x^2}$$

Further, recall that $\dfrac{\partial}{\partial x} = \dfrac{d}{dx}$ with t held constant. Then the formula

$$\int_a^b [uL(v) - vL(u)]dx = p\left[u\frac{dv}{dx} - v\frac{du}{dx}\right]_a^b \qquad (6.3.14)$$

known as *Green's formula,* or the integral form of Lagrange's identity can be used instead of integration by parts. Therefore, for this problem, we have

$$\int_0^L \left( u \frac{\partial^2 v}{\partial x^2} - v \frac{\partial^2 u}{\partial x^2} \right) dx = \left[ u \frac{\partial v}{\partial x} - v \frac{\partial u}{\partial x} \right]_0^L \qquad (6.3.15)$$

The right-hand side of Equation 6.3.15 can be evaluated as

$$u(L,t) \left[ \frac{\partial v}{\partial x} \right]_{x=L} - v(L,t) \left[ \frac{\partial u}{\partial x} \right]_{x=L} - u(0,t) \left[ \frac{\partial v}{\partial x} \right]_{x=0} + v(0,t) \left[ \frac{\partial u}{\partial x} \right]_{x=}$$

$$= B(t) \frac{n\pi}{L} \cos n\pi - A(t) \frac{n\pi}{L} = \frac{n\pi}{L} \left[ B(t)(-1)^n - A(t) \right]$$

for

$$v = \phi_n(x) = \sin \frac{n\pi x}{L}$$

$$\frac{dv}{dx} = \frac{\partial v}{\partial x} = \frac{n\pi}{L} \cos \frac{n\pi x}{L}$$

$$v(0,t) = \phi_n(0) = 0$$

$$v(L,t) = \phi_n(L) = 0$$

Therefore, Equation 6.3.15 becomes

$$\int_0^L \left( u \frac{d^2 \phi_n(x)}{dx^2} - \phi_n(x) \frac{\partial^2 u}{\partial x^2} \right) dx = \frac{n\pi}{L} \left[ B(t)(-1)^n - A(t) \right] \qquad (6.3.16)$$

But Equation 6.3.5 can be rearranged to give

$$\frac{d^2 \phi_n}{dx^2} = -\lambda_n \phi_n$$

such that Equation 6.3.16 becomes

$$\int_0^L \phi_n(x) \frac{\partial^2 u}{\partial x^2} dx = -\lambda_n \int_0^L u \phi_n(x) dx - \frac{n\pi}{L} \left[ B(t)(-1)^n - A(t) \right] \qquad (6.3.17)$$

Then, substituting Equation 6.3.17 into Equation 6.3.13 gives

$$\frac{db_n}{dt} = q_n(t) + \frac{\int_0^L k\frac{\partial^2 u}{\partial x^2}\phi_n(x)\,dx}{\int_0^L \phi_n^2(x)\,dx}$$

$$\tag{6.3.18}$$

$$= q_n(t) - \frac{\lambda_n k\int_0^L u\phi_n(x)\,dx}{\int_0^L \phi_n^2(x)\,dx} + \frac{\frac{n\pi k}{L}\left[A(t)-(-1)^n B(t)\right]}{\int_0^L \phi_n^2(x)\,dx}$$

However, Equation 6.3.8 is also a generalized Fourier series, which means that for $n \geq 1$

$$b_n(t) = \frac{\int_0^L u(x,t)\phi_n(x)\,dx}{\int_0^L \phi_n^2(x)\,dx} \tag{6.3.19}$$

Therefore, Equation 6.3.19 can be used to simplify Equation 6.3.18 to

$$\frac{db_n}{dt} + k\lambda_n b_n = q_n(t) + \frac{\frac{n\pi k}{L}\left[A(t)-(-1)^n B(t)\right]}{\int_0^L \phi_n^2(x)\,dx} \tag{6.3.20}$$

a linear first order ordinary differential equation. The required initial condition for $b_n(t)$ comes from Equation 6.3.4. That is

$$u(x,0) = \sum_{n=1}^{\infty} b_n(0)\phi_n(x) = f(x)$$

such that for $n \geq 1$

$$b_n(0) = \frac{\int_0^L f(x)\phi_n(x)\,dx}{\int_0^L \phi_n^2(x)\,dx} \tag{6.3.21}$$

As an illustration on the use of this method, an elementary example follows.

## Example 6.3.1
Given the model

$$\frac{\partial u}{\partial t} = \frac{\partial^2 u}{\partial x^2} + \sin(5x)e^{-2t} \quad 0 < x < \pi, \quad t > 0$$

$$u(0,t) = 1 \quad t > 0$$

$$u(\pi,t) = 0$$

$$u(x,0) = 0, \qquad 0 \le x \le \pi$$

find the temperature distribution u(x, t).

*Solution*
The associated homogeneous model is given by Equations 6.3.1A to 6.3.3A and the boundary value problem satisfying the same homogeneous boundary conditions is given by Equations 6.3.5 to 6.3.7.
Then

$$\lambda_n = \left(\frac{n\pi}{L}\right)^2 = n^2$$

since $L = \pi$ and the corresponding eigenfunctions are

$$\phi_n(x) = \sin nx$$

Also, from the given differential equation

$$Q(x,t) = \sin(5x)e^{-2t}$$

Then the quantity

$$\int_0^L \phi_n^2(x)\,dx = \int_0^\pi \sin^2 nx \, dx = \frac{\pi}{2}$$

and

$$q_n(t) = \frac{\int\limits_0^L Q(x,t)\phi_n(x)\,dx}{\int\limits_0^L \phi_n^2(x)\,dx} = \frac{2}{\pi}e^{-2t}\int\limits_0^L \sin 5x \sin nx\,dx = 0, \text{ if } n \neq 5 \text{ (why ?)}$$

$$= \frac{2}{\pi}e^{-2t}\int\limits_0^L \sin^2 5x\,dx = e^{-2t} \text{ if } n = 5$$

Further, the other needed quantities are

$$u(0,t) = 1, \quad u(\pi,t) = 0, \quad v(0,t) = \phi_n(0) = 0$$

$$v(\pi,t) = \phi_n(\pi) = 0$$

since

$$\phi_n = \sin nx = v$$

$$\frac{dv}{dx} = n \cos nx$$

Therefore, the RHS of Equation 6.3.15 evaluates to $-n$, and Equation 6.3.20 becomes

$$\frac{db_n}{dt} + n^2 b_n = e^{-2t} - \frac{2n}{\pi}$$

subject to

$$b_n(0) = 0$$

since $f(x) = 0$. The first order linear differential equation solves to

$$b_5(t) = -\frac{5\pi + 46}{115\pi}e^{-25t} + \frac{1}{23}e^{-2t} + \frac{2}{5\pi}$$

using the integrating factor method described in Chapter 2. Finally, the temperature distribution is given by Equation 6.3.8 as

$$u(x,t) = \sum_{n=1}^{\infty} b_n(t)\phi_n(x) = \left[-\frac{5\pi + 46}{115\pi}e^{-25t} + \frac{1}{23}e^{-2t} + \frac{2}{5\pi}\right]\sin 5x$$

This method is suitable for nonhomogeneous problems when the non-homogeneity occurs as a time-dependent source term or as time-dependent boundary conditions.

If either the differential equation or the boundary conditions is nonhomogeneous, then direct substitution into Equation 6.3.8 may be applicable. For example, consider a plane wall of thickness L that is initially at a uniform temperature $T_i$. The ambient temperature on one side of the wall is suddenly changed to a new value of $T_\infty$ while the temperature on the other side of the wall is held at $T_i$. This can be mathematically modeled as

$$\frac{\partial^2 u}{\partial x^2} = \frac{1}{\alpha}\frac{\partial u}{\partial t}, \quad 0 < x < L, \quad t > 0 \tag{6.3.22}$$

$$u(0,t) = T_i \tag{6.3.23}$$

$$-\left[k\frac{\partial u}{\partial x}\right]_{x=L} = h\left[u(L,t) - T_\infty\right] \tag{6.3.24}$$

$$u(x,0) = T_i \tag{6.3.25}$$

The associated homogeneous problem was previously discussed as Example 6.2.4. Then since both u(x, t) and $\phi_n(x)$ satisfy the same homogeneous boundary conditions,

$$\frac{\partial^2 u}{\partial x^2} = \sum_{n=1}^{\infty} b_n(t)\frac{d^2\phi_n(x)}{dx^2} \tag{6.3.26}$$

is justifiable, and direct substitution can reduce the computational effort. Also,

$$\frac{\partial u}{\partial t} = \sum_{n=1}^{\infty} \frac{db_n}{dt}\phi_n(x) \tag{6.3.27}$$

Then from Example 6.2.4,

$$\phi_n(x) = \sin\lambda_n x, \quad \lambda_n = -\frac{h}{k}\tan\lambda_n L$$

$$\frac{d^2\phi_n}{dx^2} = -\lambda_n^2\phi_n$$

Substitution of these quantities along with Equations 6.3.26 and 6.3.27 into Equation 6.3.22 give

$$\sum_{n=1}^{\infty} \left( \frac{db_n}{dt} + \alpha \lambda_2^n b_n \right) \phi_n(x) = 0 \tag{6.3.28}$$

where Equation 6.3.8 has been used. Also, the required initial condition on the first order linear differential equation given by Equation 6.3.28 is

$$u(x,0) = T_i = \sum_{n=1}^{\infty} b_n(0) \phi_n(x)$$

which is a generalized Fourier series. Then,

$$b_n(0) = \frac{\displaystyle\int_0^L T_i \phi_n(x) dx}{\displaystyle\int_0^L \phi_n^2(x) dx}, \qquad \text{for} \qquad n \geq 1$$

In this case,

$$\int_0^L \phi_n^2(x) dx = \int_0^L \sin^2 \lambda_n x \, dx = \frac{L}{2} - \frac{1}{4\lambda_n} \sin 2\lambda_n L$$

and

$$\int_0^L T_i \phi_n(x) dx = \int_0^L T_i \sin \lambda_n x \, dx = T_i \left( \frac{1}{\lambda_n} - \frac{1}{\lambda_n} \cos \lambda_n L \right)$$

Therefore, for $n \geq 1$

$$b_n(0) = \frac{2T_i (1 - \cos \lambda_n L)}{\lambda_n \left[ L - \dfrac{\sin 2\lambda_n L}{2\lambda_n} \right]}$$

and

$$\frac{db_n}{dt} + \alpha \lambda_n^2 b_n = 0$$

which solves to

$$b_n(t) = \frac{2T_i\left(1 - \cos\lambda_n L\right)}{\lambda_n\left[L - \dfrac{\sin 2\lambda_n L}{2\lambda_n}\right]} e^{-\alpha\lambda_n^2 t}$$

Therefore,

$$u(x,t) = \sum_{n=1}^{\infty} \frac{2T_i\left(1 - \cos\lambda_n L\right)}{\lambda_n\left[L - \dfrac{\sin 2\lambda_n L}{2\lambda_n}\right]} e^{-\alpha\lambda_n^2 t} \sin\lambda_n x$$

So far, the methods that have been outlined are primarily applicable to problems involving finite space dimensions. In the next section, problems with semi-infinite domains will be discussed.

## 6.4   Laplace Transform Methods

In this section, the method of Laplace transform will be used. The properties of Laplace transforms, and especially Theorem 3.7 (Section 3.6.1), will be applicable. The Laplace transform was introduced earlier for use in solving ordinary differential equations. Now we will emphasize its use in solving partial differential equations.

Assuming that the Laplace transform of the dependent variable exists, the usual procedure to solve a pde is

1. Transform the pde to an ordinary differential equation.
2. Transform the accompanying boundary conditions to those suitable for use with the ordinary differential equation.
3. Solve the resulting problem using the techniques discussed (Chapters 1 to 3).
4. Invert the results to recover the solution to the pde.

The inversion step can be relatively easy if the terms of step 3 can be located in a table of Laplace transforms. Without such a convenient table a more difficult technique involving the Residue theorem has to be employed (see Example 6.4.4).

Here, a semi-infinite rod with one end at x = 0 and extending to infinity along the positive x-axis will be a typical physical model. For example, consider the following model:

## Example 6.4.1

$$\frac{\partial u}{\partial t} = \alpha^2 \frac{\partial^2 u}{\partial x^2} \quad (x > 0, \, t > 0) \tag{6.4.1}$$

$$u(x,0) = A \tag{6.4.2}$$

$$u(0,t) = \begin{cases} B, & 0 < t < t_0 \\ 0, & t > t_0 \end{cases} \tag{6.4.3}$$

Then applying the Laplace transform to Equation 6.4.1 means

$$L\left\{\frac{\partial u}{\partial t}\right\} = \alpha^2 L\left\{\frac{\partial^2 u}{\partial x^2}\right\}$$

giving

$$L\left\{\frac{\partial u}{\partial t}\right\} = sv(x,s) - u(x,0)$$

where

$$v(x,s) = L\{u(x,t)\} = \int_0^\infty e^{-st} u(x,t) \, dt$$

and,

$$L\left\{\frac{\partial^2 u}{\partial x^2}\right\} = \frac{\partial^2}{\partial x^2} \int_0^\infty e^{-st} u(x,t) \, dt = \frac{\partial^2}{\partial x^2}\left[L\{u(x,t)\}\right]$$

where differentiation is done treating s as a parameter, such that Equation 6.4.1 becomes

$$sv(x,s) - u(x,0) = \alpha^2 \frac{d^2 v(x,s)}{dx^2}$$

or

$$\frac{d^2 v}{dx^2} - \frac{s}{\alpha^2} v = -\frac{A}{\alpha^2} \tag{6.4.4}$$

using Equation 6.4.2. It should also be noted that the Laplace transform is carried out on the variable t while x is treated as a parameter. The general solution of Equation 6.4.4 is

$$v(x,s) = c_1(s)e^{\sqrt{s}x/\alpha} + c_2(s)e^{-\sqrt{s}x/\alpha} + \frac{A}{s} \tag{6.4.5}$$

where the arbitrary constants $c_1(s)$ and $c_2(s)$ may depend on s. For a bounded solution, the constant $c_1(s)$ must be chosen as zero. Also, to determine $c_2(s)$, it is necessary to take the Laplace transform of the given boundary condition, Equation 6.4.3. That is, in terms of the unit step function

$$u(0,t) = B\left[1 - U(t - t_0)\right]$$

where

$$U(t - t_0) = \begin{cases} 0 & \text{if } t < t_0 \\ 1 & \text{if } t \geq t_0 \end{cases}$$

is the unit step function. Then

$$L\{u(0,t)\} = B L\left\{\left[1 - U(t - t_0)\right]\right\} = \frac{B}{s} - \frac{B}{s}e^{-st_0} = v(0,s)$$

Therefore,

$$v(0,s) = c_2(s) + \frac{A}{s} = \frac{B}{s} - \frac{B}{s}e^{-st_0}$$

or

$$c_2(s) = \frac{B - A}{s} - \frac{B}{s}e^{-st_0}$$

such that Equation 6.4.5 becomes

$$v(x,s) = \left[\frac{B - A}{s} - \frac{B}{s}e^{-st_0}\right]e^{-\sqrt{s}\frac{x}{\alpha}} + \frac{A}{s} \tag{6.4.6}$$

Then taking the inverse transform of Equation 6.4.6) results in

$$u(x,t) = L^1\{v(x,s)\} = (B - A)\,\text{erfc}\left(\frac{x}{2\alpha\sqrt{t}}\right) - B\,\text{erf}\left(\frac{x}{2\alpha\sqrt{t - t_0}}\right) + A$$

where the quantities

$$\text{erfc}(x) = 1 - \text{erf}(x) = \frac{2}{\sqrt{\pi}} \int_{x}^{\infty} e^{-\xi^2} d\xi$$

and

$$\text{erf}(x) = \frac{2}{\sqrt{\pi}} \int_{0}^{x} e^{-\xi^2} d\xi$$

are the *complementary error function* and *error function*, respectively.
As a second example illustrating the method, consider

**Example 6.4.2**

$$\frac{\partial u}{\partial t} = \alpha^2 \frac{\partial^2 u}{\partial x^2} \quad (x > 0, \quad t > 0)$$

subject to

$$u(0,t) = 1 \quad (t > 0)$$
$$u(x,0) = 0 \quad (x > 0)$$

Then, as a procedure, the *first step* is to transform the partial differential equation. Thus

$$L\left\{\frac{\partial u}{\partial t}\right\} = sv(x,s) - u(x,0) = \alpha^2 L\left\{\frac{\partial^2 u}{\partial x^2}\right\} = \alpha^2 \frac{\partial^2 v}{\partial x^2}$$

resulting in an ordinary differential equation

$$\alpha^2 \frac{d^2 v}{dx^2} = sv(x,s)$$

since, in this case, the initial value is zero. The *second step* is to find the general solution of the ordinary differential equation, resulting in

$$v(x,s) = c_1(s) e^{-\frac{\sqrt{s}}{\alpha}x} + c_2 e^{\frac{\sqrt{s}}{\alpha}x}$$

and observe that a bounded solution is expected, in which case the arbitrary constant, $c_2(s)$ in this case, must be chosen as zero. The remaining constant in the equation can be determined by taking the Laplace transform of the left boundary condition and comparing that result to

$$v(0,s) = c_1(s)$$

That is,

$$L\{u(0,t)\} = L\{1\} = \frac{1}{s}$$

so that

$$v(0,s) = c_1(s) = \frac{1}{s}$$

Therefore,

$$v(x,s) = \frac{1}{s} e^{-\frac{\sqrt{s}}{\alpha}x}$$

The *third step* is to invert the Laplace transform or find the inverse transform of v(x,s) with the aid of a table (Spiegel[7]) to get

$$u(x,t) = L^{-1}\left\{\frac{1}{s} e^{-\frac{\sqrt{s}}{\alpha}x}\right\} = \text{erfc}\left[\frac{x}{2\alpha\sqrt{t}}\right]$$

where erfc($\bullet$) was defined in the previous example.

This third example demonstrates the flexibility of the Laplace transform method over the separation of variables method to efficiently solve some partial differential equations problems when the domain is finite.

**Example 6.4.3**

Consider

$$\frac{\partial u}{\partial t} = \frac{\partial^2 u}{\partial x^2}, \quad 0 < x < 1, \ t > 0$$

subject to

$$u(0,t) = 1, \qquad t > 0$$

$$u(1,t) = 0, \qquad t > 0$$

$$u(x,0) = 0, \qquad 0 < x < 1$$

First, take Laplace transform of the given differential equation to get

$$\frac{d^2 v(x,s)}{dx^2} = sv(x,s) - u(x,0)$$

which reduces to

$$\frac{d^2 v(x,s)}{dx^2} = sv(x,s) \tag{6.4.7}$$

where use of the initial condition is made and

$$v(x,s) = L\big[\{u(x,t)\}\big] = \int_0^\infty e^{-st} u(x,t) \, dt$$

In order to find the constants of integration for the general solution to Equation 6.4.7, the Laplace transform of the boundary conditions are needed. That is,

$$L\{u(0,t)\} = v(0,s) = \frac{1}{s}$$

$$L\{u(1,t)\} = v(1,s) = 0$$

Then, the general solution of Equation 6.4.7 can be represented as

$$v(x,s) = c_1(s)\sinh x\sqrt{s} + c_2(s)\cosh x\sqrt{s}$$

Employing the transformed boundary conditions give

$$v(1,s) = 0 = c_1(s)\sinh\sqrt{s} + c_2(s)\cosh\sqrt{s}$$

which can be rearranged to give

$$c_1(s) = -\frac{c_2(s)\cosh\sqrt{s}}{\sinh\sqrt{s}}$$

Then, use of the second boundary condition gives

$$v(0,s) = \frac{1}{s} = c_1(s)\sinh(0) + c_2(s)\cosh(0)$$

which results in

$$c_2(s) = \frac{1}{s}$$

Therefore,

$$c_1(s) = -\frac{1}{s}\frac{\cosh\sqrt{s}}{\sinh\sqrt{s}}$$

The solution may now be represented as

$$v(x,s) = -\frac{1}{s}\frac{\cosh\sqrt{s}}{\sinh\sqrt{s}}\sinh x\sqrt{s} + \frac{1}{s}\cosh x\sqrt{s}$$

$$= \frac{1}{s}\frac{\sinh\sqrt{s}\cosh x\sqrt{s} - \cosh\sqrt{s}\sinh x\sqrt{s}}{\sinh\sqrt{s}}$$

$$= \frac{1}{s}\frac{\sinh\left[(1-x)\sqrt{s}\right]}{\sinh\sqrt{s}}$$

Then

$$u(x,t) = L^{-1}\{v(x,s)\} = L^{-1}\left\{\frac{\sinh\left[(1-x)\sqrt{s}\right]}{s\sinh\sqrt{s}}\right\}$$

$$= 1 - x + \frac{2}{\pi}\sum_{n=1}^{\infty}\frac{(-1)^n}{n}e^{-n^2\pi^2 t}\sin n\pi(1-x)$$

As a note, the quantity $1 - x$ can be represented in its Fourier sine series form to be

$$1 - x = \frac{2}{\pi}\sum_{n=1}^{\infty}\frac{1}{n}\sin n\pi x$$

Another example follows.

## Example 6.4.4

Solve: $u_{xx} = u_y$ in $y > 0$,    $0 < x < a$,
subject to: $u(x, 0) = 1$, $u(0, y) = u(a, y) = 0$

*Solution*
Consider the Laplace transform of u with respect to y, that is,

$$\bar{u}(x,s) = \int_0^\infty e^{-sy} u(x,y) \, dy$$

Then the differential equation and time-like condition transforms to

$$\bar{u}_{xx} = s\bar{u} - 1 \Leftrightarrow \frac{d^2\bar{u}}{dx^2} = s\bar{u} - 1$$

The general solution to the transformed ordinary differential equation is

$$\bar{u}(x,s) = c_1 e^{\sqrt{s}x} + c_2 e^{-\sqrt{s}x} + c_3 \Leftrightarrow k_1 \cosh \sqrt{s}x + k_2 \sinh \sqrt{s}x + c_3$$

The particular constant $c_3$ can be evaluated by the method of undetermined coefficient, while the arbitrary constants $k_1$ and $k_2$ must be evaluated from the transformed boundary conditions on x. Following this procedure, $\bar{u}(x, s)$ is given by

$$\bar{u}(x,s) = \frac{1}{s} - \frac{\cosh \sqrt{s}\left(x - \frac{a}{2}\right)}{s \cosh \frac{\sqrt{s}a}{2}}.$$

The first term can be located in a table of Laplace transforms. The second term requires a little more effort, say, the residue theorem. That is, let

$$P(x,s) = -\cosh \sqrt{s}\left(x - \frac{a}{2}\right), Q(s) = s \cosh \frac{\sqrt{s}a}{2}$$

Then the residue at $s = 0$ is $\lim_{s \to 0} s \frac{P(x,s)}{Q(s)} = -1$. For $s \neq 0$, the residue is given by

$$p_n(y) = \frac{P(x,s_n)}{Q'(s_n)} e^{s_n y}$$

$$Q'(s) = \cosh \frac{\sqrt{s}a}{2} + \frac{a\sqrt{s}}{4} \sinh \frac{\sqrt{s}a}{2}$$

To determine the $s_n$, $Q(s)$ is set equal to zero, that is

$$\cosh\frac{\sqrt{sa}}{2} = \frac{e^{\sqrt{sa}/2}+e^{-\sqrt{sa}/2}}{2} = 0$$

simplifies to

$$e^{\sqrt{sa}} = -1$$

But the natural logarithm of a negative real number represents a multiple-valued function (Churchill and Brown,[8] Spiegel,[9] and Williams[10]), That is, if

$$z = e^w, \ w = \ln z = \ln r + i(\theta + 2n\pi), \ n = 0, \pm 1, \pm 2, \dots$$

where $z = re^{i\theta} = re^{i(\theta+2n\pi)}$. However, in our case, $\theta = \pi$, $r = 1$ and

$$\sqrt{sa} = i(\pi + 2n\pi), \quad n = 0, 1, 2, \cdots$$

Therefore,

$$s_n = -(1+2n)^2 \pi^2/a^2$$

Finally,

$$p_n(y) = \frac{4\sin\left[(1+2n)\pi x/a\right]}{(1+2n)\pi}e^{\left[-(1+2n)^2 y\pi^2/a^2\right]}$$

Therefore, the final result is

$$u(x,y) = \sum_{n=0}^{\infty}\frac{4\sin\left[(1+2n)\pi x/a\right]\exp\left[-(1+2n)^2 y\pi^2/a^2\right]}{(1+2n)\pi}$$

Additional examples involving the use of the Residue theorem are discussed in Chapter 7.

---

## 6.5   Combination of Variables

Another technique that is sometimes employed to reduce partial differential equations to ordinary differential equations is combination of variables or a similarity transformation.

The process of normalization can be used to establish the applicability of combining the independent variables of the given partial differential equation. For example, consider the flow of a fluid near a wall suddenly set in motion. Following Bird, Stewart, and Lightfoot,[1] the problem statement is as follows:

A semi-infinite body of liquid with constant density ($\rho$) and viscosity ($\mu$) is bounded on one side by a flat surface (the x z-plane). Initially, the fluid and the solid surface are at rest, but at time $t = 0$, the solid surface is set in motion in the positive x-direction with a velocity V ( shown below).

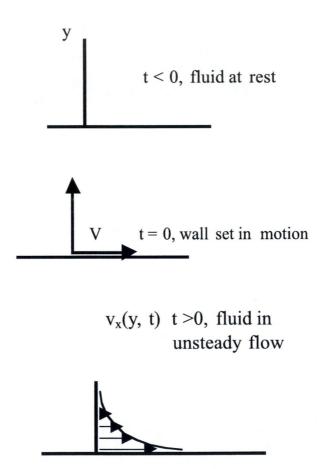

$t < 0$, fluid at rest

V       $t = 0$, wall set in motion

$v_x(y, t)$   $t > 0$, fluid in
unsteady flow

One would like to know the velocity profile as a function of y and t. If there is no pressure gradient or gravity force in the x-direction and the flow is laminar, then this simplified problem can be solved in the following way.

*Solution*
In rectangular coordinates (x, y, z):

$$\frac{\partial \rho}{\partial t} + \frac{\partial}{\partial x}(\rho v_x) + \frac{\partial}{\partial y}(\rho v_y) + \frac{\partial}{\partial z}(\rho v_z) = 0 \tag{6.5.1}$$

is the equation of continuity (Bird, Stewart, and Lightfoot[1]). The x-component equation of motion in terms of velocity gradients for a Newtonian fluid with constant $\rho$ and $\mu$ is (Bird, Stewart, and Lightfoot[1])

$$\rho\left(\frac{\partial v_x}{\partial t} + v_x \frac{\partial v_x}{\partial x} + v_y \frac{\partial v_x}{\partial y} + v_z \frac{\partial v_x}{\partial z}\right)$$

$$= -\frac{\partial P}{\partial x} + \mu\left(\frac{\partial^2 v_x}{\partial x^2} + \frac{\partial^2 v_x}{\partial y^2} + \frac{\partial^2 v_x}{\partial z^2}\right) + \rho g_x \tag{6.5.2}$$

However,

$$v_y = v_z = 0 \quad \text{and} \quad v_x = v_x(y,t)$$

such that Equation 6.5.1 reduces to

$$\frac{\partial v_x}{\partial x} = 0$$

and Equation 6.5.2 reduces to

$$\frac{\partial v_x}{\partial t} = \gamma \frac{\partial^2 v_x}{\partial y^2} \tag{6.5.3}$$

where $\gamma = \mu/\rho$. The initial and boundary conditions are

$$\text{at } t \le 0, \quad v_x = 0 \quad \text{for all } y \tag{6.5.4}$$

$$\text{at } y = 0, \quad v_x = V \quad \text{for all } t > 0 \tag{6.5.5}$$

$$\text{at } y = \infty, \quad v_x = 0 \quad \text{for all } t > 0 \tag{6.5.6}$$

Now suppose

$$\frac{v_x}{V} = \phi(\eta) \tag{6.5.7}$$

where

$$\eta = \frac{y}{\sqrt{4\gamma t}} \qquad (6.5.8)$$

such that

$$\frac{\partial\left(\frac{v_x}{V}\right)}{\partial t} = -\frac{1}{2}\frac{\eta}{t}\phi'; \qquad \frac{\partial^2\left(\frac{v_x}{V}\right)}{\partial y^2} = \frac{\eta^2}{y^2}\phi''$$

then Equation 6.5.3 becomes

$$\phi'' + 2\eta\phi' = 0 \qquad (6.5.9)$$

subject to

$$\phi = 1 \quad \text{at} \quad \eta = 0 \qquad (6.5.10)$$

$$\phi = 0 \quad \text{at} \quad \eta = \infty \qquad (6.5.11)$$

Note that the initial and boundary conditions are combined to form Equation 6.5.11. Equation 6.5.9 solves to give the general solution

$$\phi(\eta) = c_1 \int_0^\eta e^{-\eta^2} d\eta + c_2$$

where the lower limit of the integral is a convenient selection. However, notice that an alternate choice of lower limit would only affect $c_2$, but $c_2$ is an arbitrary constant.

With the aid of Equations 6.5.10 and 6.5.11, the solution to Equation 6.5.9 is

$$\phi(\eta) = 1 - \frac{2}{\sqrt{\pi}} \int_0^\eta e^{-\eta^2} d\eta = 1 - \text{erf}\left(\frac{y}{4\gamma t}\right) \qquad (6.5.12)$$

where the error function

$$\text{erf}(x) = \frac{2}{\sqrt{\pi}} \int_0^x e^{-\xi^2} d\xi$$

was defined in the previous section.

As a second example, consider a semi-infinite solid which is subjected to a step change in its surface temperature T (Myers[11]). Suppose the model can be described by

$$\frac{1}{\alpha}\frac{\partial T}{\partial t} = \frac{\partial^2 T}{\partial x^2} \tag{6.5.13}$$

with

$$T(0,t) = T_c \tag{6.5.14}$$

and

$$\lim_{x \to \infty} T(x,t) = T_i \tag{6.5.15}$$

as boundary conditions, and the initial condition given by

$$T(x,0) = T_i \tag{6.5.16}$$

Further, suppose that the normalized variables are

$$\overline{T} = \frac{T - T_i}{T_0}, \quad \overline{x} = \frac{x}{x_0} \quad \text{and} \quad \overline{t} = \frac{t}{t_0} \tag{6.5.17}$$

Substitution into Equation 6.5.13 gives

$$\frac{x_0^2}{\alpha t_0}\frac{\partial \overline{T}}{\partial \overline{t}} = \frac{\partial^2 \overline{T}}{\partial \overline{x}^2} \tag{6.5.18}$$

Equation 6.5.14 can be restated as

$$\overline{T}(0,\overline{t}) = \frac{T_c - T_i}{T_0}$$

then the choice of $T_0 = T_c - T_i$ gives

$$\overline{T}(0,\overline{t}) = 1 \tag{6.5.19}$$

Equations 6.5.15 and 6.5.16 become

$$\lim_{\bar{x}\to\infty} \overline{T}(\bar{x},\bar{t}) = 0 \qquad (6.5.20)$$

and

$$\overline{T}(\bar{x},0) = 0 \qquad (6.5.21)$$

respectively.

Equation 6.5.18 can be further simplified if the quantity

$$\frac{x_0^2}{\alpha t_0}$$

is taken as 1, since neither $x_0$ nor $t_0$ has been assigned any physical significance (see Chapter 8 on scaling). That is, if

$$\frac{x_0^2}{\alpha t_0} = 1$$

then Equation 6.5.18 becomes

$$\frac{\partial \overline{T}}{\partial \bar{t}} = \frac{\partial^2 \bar{r}}{\partial \bar{x}} \qquad (6.5.22)$$

and a possible choice of a combined variable is

$$\xi = \frac{\bar{x}^2}{\bar{t}} = \frac{\left(x/x_0\right)^2}{t/t_0} = \frac{x^2}{t}\left(\frac{t_0}{x_0^2}\right) = \frac{x^2}{\alpha t} \qquad (6.5.23)$$

Using Equation 6.5.23, the conditions described by Equations 6.5.20 and 6.5.21 can now be collapsed into one condition similar to Equation 6.5.11, namely

$$\overline{T}(\infty) = 0 \qquad (6.5.24)$$

since $\xi \to \infty$ as $\bar{x} \to \infty$ or $\bar{t} \to 0$.

Therefore, Equation 6.5.13 or 6.5.22 can be reduced to a second order ordinary differential equation following substitution of Equation 6.5.23.

Given that a combined variable is available, its substitution into the partial differential equation can be a delicate process and is demonstrated with the following example.

Consider the model described by Equation 6.5.13 subject to

$$T = T_0 \quad \text{at} \quad t = 0, \text{ for all } x \tag{6.5.25}$$

$$T = T_s \quad \text{at} \quad x = 0, \text{ for all } t \tag{6.5.26}$$

$$T \to T_s, t \to \infty, x > 0 \tag{6.5.27}$$

$$T \to T_0, x \to \infty, t > 0 \tag{6.5.28}$$

and further suppose that the combined variable is

$$\eta = \frac{x}{\sqrt{4\alpha t}} \tag{6.5.29}$$

Then

$$T(x,t) = f(\eta) \tag{6.5.30}$$

which implies that

$$dT(x,t) = df(\eta) \tag{6.5.31}$$

The chain rule applied to Equation 6.5.30 gives

$$\frac{\partial T}{\partial x} dx + \frac{\partial T}{\partial t} dt = \frac{df}{d\eta} d\eta \tag{6.5.32}$$

also, the total derivative of $\eta(x, t)$ gives

$$d\eta = \frac{\partial \eta}{\partial x} dx + \frac{\partial \eta}{\partial t} dt \tag{6.5.33}$$

such that Equation 6.5.32 becomes

$$\frac{\partial T}{\partial x} dx + \frac{\partial T}{\partial t} dt = f'(\eta) \left[ \frac{\partial \eta}{\partial x} dx + \frac{\partial \eta}{\partial t} dt \right]$$

Then equating like coefficients of both sides gives

$$dx : \frac{\partial T}{\partial x} = f'(\eta)\frac{\partial \eta}{\partial x} \tag{6.5.34}$$

$$dt : \frac{\partial T}{\partial t} = f'(\eta)\frac{\partial \eta}{\partial t} \tag{6.5.35}$$

In order to find the second derivative, the following device is useful:

$$\text{let } H(x,t) = \frac{\partial T}{\partial x} \tag{6.5.36}$$

$$\text{and } \phi(\eta,t) = f'(\eta)\frac{\partial \eta}{\partial x} \tag{6.5.37}$$

then

$$dH(x,t) = d\phi(\rho,t) \tag{6.5.38}$$

such that

$$\frac{\partial H}{\partial x}dx + \frac{\partial H}{\partial t}dt = \frac{\partial \phi}{\partial \eta}d\eta + \frac{\partial \phi}{\partial t}dt$$

that is

$$\frac{\partial H}{\partial x}dx + \frac{\partial H}{\partial t}dt = \frac{\partial \phi}{\partial \eta}\left[\frac{\partial \eta}{\partial x}dx + \frac{\partial \eta}{\partial t}dt\right] + \frac{\partial \phi}{\partial t}dt$$

and equating like coefficients of both sides results in

$$dx: \qquad \frac{\partial H}{\partial x} = \frac{\partial \phi}{\partial \eta}\frac{\partial \eta}{\partial x}$$

and

$$dt: \qquad \frac{\partial H}{\partial t} = \frac{\partial \phi}{\partial \eta}\frac{\partial \eta}{\partial t} + \frac{\partial \phi}{\partial t}$$

Therefore,

$$\frac{\partial^2 T}{\partial x^2} = \frac{\partial \phi}{\partial \eta}\frac{\partial \eta}{\partial x} = f''(\eta)\left(\frac{\partial \eta}{\partial x}\right)^2$$

since $\dfrac{\partial \eta}{\partial x}$ is independent of x. Finally, Equation 6.5.13 becomes

$$f'(\eta)\frac{\partial \eta}{\partial t} = \alpha f''(\eta)\left(\frac{\partial \eta}{\partial x}\right)^2$$

or

$$f''(\eta) + 2\eta f'(\eta) = 0 \qquad (6.5.40)$$

using Equation 6.5.29.

## 6.6   Fourier Integral Methods

In chemical engineering, the use of Fourier integrals to solve problems is not as popular as separation of variables or Laplace transform. This is due to the fact that the incorporation of the boundary conditions associated with a particular application can usually be very challenging. For instance, consider a slab of finite thickness undergoing some heat transfer phenomena. Suppose the phenomena can be described by

$$V\frac{\partial T}{\partial x} = \frac{\partial^2 T}{\partial x^2} + \frac{\partial^2 T}{\partial y^2} \qquad (6.6.1)$$

$$\left.\frac{\partial T}{\partial x}\right|_{y=\pm 1} = \pm q e^{x/\delta}, \ x < 0$$

$$\qquad (6.6.2)$$

$$\left.\frac{\partial T}{\partial y}\right|_{y=\pm 1} = \pm q e^{-x/\Delta}$$

$$T(-\infty, y) = 0 \qquad (6.6.3)$$

and

$$T(0, \pm 1) = 1 \tag{6.6.4}$$

Then, using Fourier transforms, Equation 6.6.1 becomes

$$\frac{\partial^2 \theta}{\partial y^2} - \omega^2 \theta - i\omega V\theta = 0 \tag{6.6.5}$$

where

$$\theta(\omega, y) = \int_{-\infty}^{\infty} T(x, y) e^{-i\omega x} dx$$

is defined by Equation 5.6.7. The general solution of Equation 6.6.5 includes two arbitrary constants, one of which can be eliminated based on the fact that the temperature field is symmetric. That is, the solution to Equation 6.6.5 reduces to

$$\theta(\omega, y) = c_1 \cosh \sqrt{(\omega^2 + i\omega V)}\, y \tag{6.6.6}$$

Transforming the conditions given by Equation 6.6.2 results in

$$\left. \frac{\partial \theta}{\partial y} \right|_{y=1} = \int_{-\infty}^{0} q \exp\left( \frac{x}{\delta} - i\omega x \right) dx + \int_{0}^{\infty} q \exp\left( -(x/\Delta) - i\omega x \right) dx$$

which integrates to

$$\left. \frac{\partial \theta}{\partial y} \right|_{y=1} = q \left[ \frac{1}{(1/\delta) - i\omega} + \frac{1}{i\omega + (1/\Delta)} \right]$$

Therefore, Equation 6.6.6 becomes

$$\theta(\omega, y) = q \left[ \frac{1}{(1/\delta) - i\omega} + \frac{1}{i\omega + (1/\Delta)} \right] \left\{ \frac{1}{\sqrt{(\omega^2 + i\omega V)} \sinh \sqrt{(\omega^2 + i\omega V)}} \right\}$$

$$\times \cosh \sqrt{(\omega^2 + i\omega V)}\, y$$

Application of Equation 5.6.8 gives the inverse transform

$$T(x,y) = \frac{q}{2\pi} \int_{-\infty}^{\infty} \left[ \frac{\cosh\sqrt{(\omega^2 + i\omega V)}\, y \exp(i\omega x) d\omega}{[(1/\delta) - i\omega]\sqrt{(\omega^2 + i\omega V)} \sinh\sqrt{(\omega^2 + i\omega V)}} \right]$$

$$+ \frac{q}{2\pi} \int_{-\infty}^{\infty} \frac{\cosh\sqrt{(\omega^2 + i\omega V)}\, y \exp(i\omega x) d\omega}{[i\omega + (1/\Delta)]\sqrt{(\omega^2 + i\omega V)} \sinh\sqrt{(\omega^2 + i\omega V)}}$$

(6.6.7)

However, at this point it is not obvious how the conditions given by Equations 6.6.3 and 6.6.4 are to be used.

There are some cases of practical value for which Fourier integrals can be helpful. For example, consider a semi-infinite thin slab whose surface is insulated. Suppose that the surface temperature of the bar is initially $f(x)$, and a temperature of zero degrees is suddenly applied to the end $x = 0$ and is maintained.

(a) Show that the solution to this boundary value problem can be represented as

$$u(x,t)\frac{1}{\pi} = \int_0^{\infty} \int_0^{\infty} f(v)e^{-k\lambda^2 t} \sin\lambda v \sin\lambda x \, d\lambda \, dv$$

(b) If $f(x)$ is a constant, say $u_0$, show that

$$u(x,t) = u_0 \text{erf}\left(\frac{x}{2}\sqrt{kt}\right)$$

*Solution*
The boundary value problem can be described by

$$\frac{\partial u}{\partial t} = k\frac{\partial^2 u}{\partial x^2} \quad x > 0, t > 0$$

$u(x,0) = f(x)$, $u(0,t) = 0$, $|u(x,t)| < M$ (due to physical reasons).

By separation of variables, the differential equation has a solution of the form

$$u(x,t) = e^{-k\lambda^2 t}\left(A \cos\lambda x + B \sin\lambda x\right)$$

Using the condition at $x = 0$ reduces the solution to

$$u(x,t) = e^{-k\lambda^2 t}(B \sin \lambda x)$$

Since there are no restrictions on $\lambda$, B can be replaced by a function $B(\lambda)$. Further, integration over $\lambda$ from 0 to $\infty$ can be carried out analogous to the superposition principle that was applied to discrete values of $\lambda$. Therefore, a possible solution is

$$u(x,t)\int_0^\infty B(\lambda)e^{-k\lambda^2 t} \sin \lambda x \, d\lambda$$

Application of the condition at $t = 0$ results in

$$u(x,0) = f(x) = \int_0^\infty B(\lambda)\sin \lambda x \, d\lambda$$

which is an integral equation for the determination of $B(\lambda)$. This integral equation can be interpreted with the aid of Equation 5.6.9, suggesting that $f(x)$ is an odd function. In which case

$$B(\lambda) = \frac{2}{\pi}\int_0^\infty f(x)\sin \lambda x \, dx = \frac{2}{\pi}\int_0^\infty f(v)\sin \lambda v \, dv$$

Therefore,

$$u(x,t) = \int_0^\infty B(\lambda)e^{-k\lambda^2 t} \sin \lambda x \, d\lambda = \frac{2}{\pi}\int_0^\infty \int_0^\infty f(v)e^{-k\lambda^2 t} \sin \lambda v \sin \lambda x \, d\lambda \, dv$$

Since

$$\sin \lambda v \sin \lambda x = \frac{1}{2}\left[\cos \lambda(v-x) - \cos \lambda(v+x)\right]$$

then

$$u(x,t) = \frac{1}{\pi}\int_0^\infty \int_0^\infty f(v)e^{-k\lambda^2 t}\left[\cos \lambda(v-x) - \cos \lambda(v+x)\right]d\lambda \, dv$$

$$= \frac{1}{\pi}\int_0^\infty f(v)\left[\int_0^\infty e^{-k\lambda^2 t}\cos \lambda(v-x)d\lambda - \int_0^\infty e^{-k\lambda^2 t}\cos \lambda(v+x)d\lambda\right]dv$$

But

$$\int_0^\infty e^{-\alpha\lambda^2}\cos\beta\lambda\,d\lambda = \frac{1}{2}\sqrt{\frac{\pi}{\alpha}}e^{-\beta^2/4\alpha}$$

Therefore,

$$u(x,t)=\frac{1}{2\sqrt{\pi kt}}\left[\int_0^\infty f(v)e^{-(v-x)^2/4kt}dv - \int_0^\infty f(v)e^{-(v+x)^2/4kt}dv\right]$$

By letting $(v-x)/2\sqrt{kt} = p$ in the first integral and $(v+x)/2\sqrt{kt} = p$ in the second integral reduces $u(x,t)$ to

$$u(x,t)=\frac{1}{\sqrt{\pi}}\left[\int_{-x/2\sqrt{kt}}^\infty f\left(2p\sqrt{kt}+x\right)e^{-p^2}dp - \int_{-x/2\sqrt{kt}}^\infty f\left(2p\sqrt{kt}-x\right)e^{-p^2}dp\right]$$

but $f(x)$ is a constant and can be taken outside of the integrals, thus giving

$$u(x,t)=\frac{u_0}{\sqrt{\pi}}\left[\int_{-x/2\sqrt{kt}}^\infty e^{-p^2}dp - \int_{x/2\sqrt{kt}}^\infty e^{-p^2}dp\right]$$

$$=\frac{u_0}{\sqrt{\pi}}\int_{-x/2\sqrt{kt}}^{x/2\sqrt{kt}} e^{-p^2}dp = \frac{2u_0}{\sqrt{\pi}}\int_0^{x/2\sqrt{kt}} e^{-p^2}dp = u_0\,\mathrm{erf}\left(x/2\sqrt{kt}\right)$$

From the very brief introduction of Fourier integrals in Chapter 5 together with the above examples, the engineer should get the sense that these integrals are more useful for general problems than specific practical problems. When the various transforms are to be employed, one has to be careful in making sure that the boundary conditions can be utilized *a priori*.

---

## 6.7 Regular Perturbation Approaches

Sometimes a problem is encountered in which the governing equation is almost identical to a simpler problem one already knows how to solve. Hopefully, some modification of the new problem can resolve the difference and lead to useful results. For example, consider the following model for a species A in a reacting system (Loney[12]):

$$\frac{1}{r}\frac{\partial}{\partial r}\left(\frac{r\partial C_A}{\partial r}\right) + \frac{\partial}{\partial z}\left(\frac{\partial C_A}{\partial z}\right) = 0 \tag{6.7.1}$$

where r and z are the radial and axial coordinates, respectively, as shown in Figure 6.1.

Suppose Equation 6.7.1 is subject to the following boundary conditions:

$$\frac{\partial C_A}{\partial z} = 0 \quad \text{at} \quad z = 0 \tag{6.7.2}$$

$$\frac{\partial C_A}{\partial r} = 0 \quad \text{at} \quad r = 0 \tag{6.7.3}$$

$$C_A\left(R_W, z\right) = C_{Abi}; \quad 0 < z < \delta \tag{6.7.4}$$

$$-D_{AB}\frac{\partial}{\partial z}C_A = \frac{k_1 C_A C_{tot}}{K + K' C_A} \quad \text{at} \quad z = \delta \tag{6.7.5}$$

Then, except for Equation 6.7.5, the method of separation of variables is applicable. Observe that the terms of the denominator of the right-hand side of Equation 6.7.5 are all positive physical quantities. That is, the right-hand side of Equation 6.7.5 is expected to have a power series expansion. First, the variables are redefined in terms of dimensionless quantities by

$$\frac{C_A\left(r, z\right)}{C_{Abi}} = F\left(\xi, \zeta\right)$$

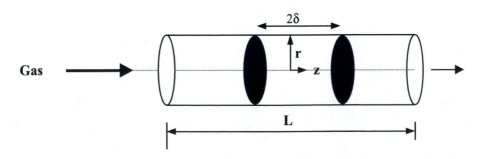

**FIGURE 6.1**
CVD reactor.

where

$$\xi = \frac{r}{R_w}; \quad \zeta = \frac{z}{\delta}; \quad a = \frac{\delta}{R_w}$$

Then Equations 6.7.1 to 6.7.5 become

$$\frac{-1}{a^2} \frac{\partial^2 F}{\partial \zeta^2} = \frac{\partial^2 F}{\partial \xi^2} + \frac{1}{\xi} \frac{\partial F}{\partial \xi} \tag{6.7.6}$$

$$\frac{\partial F(\xi, 0)}{\partial \zeta} = 0 \tag{6.7.7}$$

$$\frac{\partial F(0, \zeta)}{\partial \xi} = 0 \tag{6.7.8}$$

$$F(1, \zeta) = 1 \tag{6.7.9}$$

$$\frac{-D_{AB} C_{ABi}}{\delta} \frac{\partial F(\xi, 1)}{\partial \zeta} = \text{Rxn.rate (heterogeneous)} \tag{6.7.10}$$

By considering the heterogeneous reaction rate expression to be

$$\text{rate} = k_0 C_{Abi} \left( 1 - \varepsilon F + \varepsilon^2 F^2 - \varepsilon^3 F^3 + \cdots \right) \tag{6.7.11}$$

where

$$F = F_0 + \varepsilon F_1 + \varepsilon^2 F^2 + \cdots \tag{6.7.12}$$

and

$$\varepsilon = \frac{K' C_{A0}}{K} < 1 \tag{6.7.13}$$

The otherwise nonlinear problem can be reduced to a set of linear problems by substituting Equations 6.7.11 and 6.7.12 into Equations 6.7.6 to 6.7.10 to get

$$\frac{-1}{a^2} \left( \frac{\partial^2 F_0}{\partial \zeta^2} + \varepsilon \frac{\partial^2 F_1}{\partial \zeta^2} + \varepsilon^2 \frac{\partial^2 F_2}{\partial \zeta^2} + \cdots \right) = \frac{\partial^2 F_0}{\partial \xi^2} + \frac{\varepsilon \partial^2 F_1}{\partial \xi^2} + \varepsilon^2 \frac{\partial^2 F_2}{\partial \xi^2} + \cdots$$

$$+ \frac{1}{\xi} \left( \frac{\partial F_0}{\partial \xi} + \varepsilon \frac{\partial F_1}{\partial \xi} + \varepsilon^2 \frac{\partial F_2}{\partial \xi} + \cdots \right) \tag{6.7.14}$$

$$\frac{\partial F_0}{\partial \zeta} + \varepsilon \frac{\partial F_1}{\partial \zeta} + \varepsilon^2 \frac{\partial F_2}{\partial \zeta} + \cdots = 0 \quad \text{at} \quad \zeta = 0 \tag{6.7.15}$$

$$\frac{\partial F_0}{\partial \xi} + \frac{\varepsilon \partial F_1}{\partial \xi} + \varepsilon^2 \frac{\partial F_2}{\partial \xi} + \cdots = 0 \quad \text{at} \quad \xi = 0 \tag{6.7.16}$$

$$F_0 + \varepsilon F_1 + \varepsilon^2 F_2 + \cdots = 1 \quad \text{at} \quad \xi = 1 \tag{6.7.17}$$

$$-\frac{D_{AB}C_{ABi}}{\delta}\left(\frac{\partial F_0}{\partial \zeta} + \varepsilon \frac{\partial F_1}{\partial \zeta} + \varepsilon^2 \frac{\partial F_2}{\partial \zeta} + \cdots\right)$$
$$\tag{6.7.18}$$
$$= k_0 C_{ABi}\left[1 - \varepsilon\left(F_0 + \varepsilon F_1 + \varepsilon^2 F_2 + \cdots\right) + \varepsilon^2\left(F_0 + \varepsilon F_1 + \varepsilon^2 F_2 + \cdots\right)^2 + O\left(\varepsilon^3\right)\right]$$

Then equate like powers of $\varepsilon$ in both the differential equation and the given conditions as follows

$$\varepsilon^0: \qquad -\frac{1}{a^2}\frac{\partial^2 F_0}{\partial \zeta^2} = \frac{\partial^2 F_0}{\partial \xi^2} + \frac{1}{\xi}\frac{\partial F_0}{\partial \xi}$$

$$\frac{\partial F_0\left(\xi, 0\right)}{\partial \zeta} = 0$$

$$\frac{\partial F_0\left(0, \zeta\right)}{\partial \xi} = 0$$

$$F_0\left(1, \zeta\right) = 1$$

$$-\frac{D_{AB}C_{Abi}}{\delta}\frac{\partial F_0\left(\xi, 1\right)}{\partial \zeta} = k_0 C_{Abi}$$

$$\varepsilon: \qquad -\frac{1}{a^2}\frac{\partial^2 F_1}{\partial \zeta^2} = \frac{\partial^2 F_1}{\partial \xi^2} + \frac{1}{\xi}\frac{\partial F_1}{\partial \xi}$$

$$\frac{\partial F_1\left(\xi, 0\right)}{\partial \zeta} = 0$$

$$\frac{\partial F_1\left(0, \zeta\right)}{\partial \xi} = 0$$

$$F_1\left(1, \zeta\right) = 0$$

$$-\frac{D_{AB}C_{Abi}}{\delta}\frac{\partial F_1\left(\xi, 1\right)}{\partial \zeta} = k_0 C_{Abi} F_0$$

$\varepsilon^2$:
$$-\frac{1}{a^2}\frac{\partial^2 F_2}{\partial \zeta^2} = \frac{\partial^2 F_2}{\partial \xi^2} + \frac{1}{\xi}\frac{\partial F_2}{\partial \xi}$$

$$\frac{\partial F_2(\xi,0)}{\partial \zeta} = 0$$

$$\frac{\partial F_2(0,\zeta)}{\partial \xi} = 0$$

$$F_2(1,\zeta) = 0$$

$$-\frac{D_{AB}C_{Abi}}{\delta}\frac{\partial F_2(\xi,1)}{\partial \zeta} = k_0 C_{Abi}\left(F_0^2 + F_1\right)$$

Each set of linear problems (respective coefficient of $\varepsilon$) can now be solved by the method of separation of variables, and the above process can be continued up to any desired power of $\varepsilon$.

As a second example, consider the cooling of a given system (Aziz and Na[13]) in which the specific heat is not a constant. Further, suppose that the system has volume V, surface area A, density $\rho$, initial temperature $T_i$ and specific heat C. At an initial time, the system is exposed to a convective environment with heat transfer coefficient h and a temperature $T_a$. If the specific heat C is given by

$$C = C_a\left[1 + \beta\left(T - T_a\right)\right] \tag{6.7.19}$$

where $C_a$ is the specific heat at $T = T_a$ and $\beta$ is a given constant, what is the temperature profile for this cooling system?

*Solution*
The energy balance for this system results in

$$\rho V C \frac{dT}{dt} + hA\left(T - T_a\right) = 0 \tag{6.7.20}$$

subject to

$$t = 0, \quad T = T_1 \tag{6.7.21}$$

Then, for

$$\theta = \frac{T - T_i}{T_i - T_a}, \tau = \frac{t}{\rho V C_a/(hA)}, \varepsilon = \beta\left(T_i - T_a\right)$$

Equations 6.7.20 and 6.7.21 become

$$\left(1 + \varepsilon\theta\right)\frac{d\theta}{d\tau} + \theta = 0 \tag{6.7.22}$$

$$\tau = 0, \quad \theta = 1 \tag{6.7.23}$$

If $\varepsilon$ is much smaller than one, then the following infinite series solution can be adopted:

$$\theta = \theta_0 + \varepsilon\theta_1 + \varepsilon^2\theta_2 + \cdots = \sum_{n=0}^{\infty} \varepsilon^n\theta_n \tag{6.7.24}$$

Substitution into Equations 6.7.22 to 6.7.23 results in

$$\left[1 + \varepsilon\left(\theta_0 + \varepsilon\theta_1 + \varepsilon^2\theta_2 + \cdots\right)\right]\left(\frac{d\theta_0}{d\tau} + \varepsilon\frac{d\theta_1}{d\tau} + \varepsilon^2\frac{d\theta_2}{d\tau} + \cdots\right)$$

$$+\theta_0 + \varepsilon\theta_1 + \varepsilon^2\theta_2 + \cdots = 0$$

or

$$\frac{d\theta_0}{d\tau} + \theta_0 + \varepsilon\left(\frac{d\theta_1}{d\tau} + \theta_1 + \theta_0\frac{d\theta_0}{d\tau}\right) + \varepsilon^2\left(\frac{d\theta_2}{d\tau} + \theta_2 + \theta_0\frac{d\theta_1}{d\tau} + \theta_1\frac{d\theta_0}{d\tau}\right) + \cdots = 0$$

and

$$\theta_0 + \varepsilon\theta_1 + \varepsilon^2\theta_2 + \cdots = 1 \quad \text{at} \quad \tau = 0$$

Then equating coefficients of like powers of $\varepsilon$ gives

$$\varepsilon^0: \quad \frac{d\theta_0}{d\tau} + \theta_0 = 0$$

$$\theta_0 = 1 \quad \text{at} \quad \tau = 0$$

$$\varepsilon: \quad \frac{d\theta_1}{d\tau} + \theta_1 + \theta_0\frac{d\theta_0}{d\tau} = 0$$

$$\theta_1 = 0 \quad \text{at} \quad \tau = 0$$

and

$$\varepsilon^2: \quad \frac{d\theta_2}{d\tau} + \theta_2 + \theta_0 \frac{d\theta_1}{d\tau} + \theta_1 \frac{d\theta_0}{d\tau} = 0$$

$$\theta_2 = 0 \quad \text{at} \quad \tau = 0$$

for coefficients of $\varepsilon$ up to the second power. The resulting solutions for the respective set of new problems are

$$\theta_0 = e^{-\tau}; \quad \theta_1 = e^{-\tau} - e^{-2\tau}; \quad \theta_2 = e^{-\tau} - 2e^{-2\tau} + \frac{3}{2}e^{-3\tau}$$

such that a three-term solution is

$$\theta_0 = e^{-\tau} + \varepsilon\left(e^{-\tau} - e^{-2\tau}\right) + \varepsilon^2\left(e^{-\tau} - 2e^{-2\tau} + \frac{3}{2}e^{-3\tau}\right) \qquad (6.7.25)$$

Notice that the direct solution of Equations 6.7.22 and 6.7.23 results in

$$\ln(\theta) + \varepsilon(\theta - 1) = -\tau$$

which compares well with the three-term result (Aziz and Na[13]) for $\varepsilon = 0, 0.2$ and $-0.2$.

Another example is plane Couette flow with variable viscosity (Aziz and Na[13] and Turian and Bird[14]). Consider the steady flow of an incompressible Newtonian fluid between two infinite parallel plates separated by a distance, a, as shown in Figure 6.2. Each plate is maintained at a temperature $T_0$. The upper plate is allowed to move with a uniform velocity V. The thermal conductivity, k, of the fluid is constant while its viscosity, $\mu$, is allowed to vary according to

$$\mu = \mu_0 e^{-\alpha(T-T_0)} \qquad (6.7.26)$$

where $\mu_0$ is the viscosity at $T_0$ and $\alpha$ is a given constant.

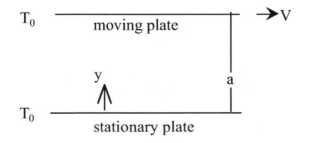

**FIGURE 6.2**
Plane couette flow.

The respective momentum and energy equations (Bird, Stewart, and Lightfoot[1]) are

$$\frac{d}{d\mu}\left(\mu\frac{du}{dy}\right)=0 \tag{6.7.27}$$

and

$$\frac{d^2T}{dy^2}+\frac{\mu}{k}\left(\frac{du}{dy}\right)^2=0 \tag{6.7.28}$$

The boundary conditions are

$$y=0, \quad u=0, \quad T=T_0$$

$$y=a, \quad u=V, \quad T=T_0$$

where u is axial velocity and T is the fluid temperature. By introducing the dimensionless quantities

$$\theta=\frac{T-T_0}{T_0}, Y=\frac{y}{a}, U=\frac{u}{V}, \beta=\alpha T_0 \quad \text{and} \quad \varepsilon=\frac{\mu_0 V^2}{kT_0}$$

Equations 6.7.27 and 6.7.28 together with their boundary conditions become

$$\frac{d}{dY}\left(e^{-\beta\theta}\frac{dU}{dY}\right)=0 \tag{6.7.29}$$

$$\frac{d^2\theta}{dY^2}+\varepsilon e^{-\beta\theta}\left(\frac{dU}{dY}\right)^2=0$$

$$Y=0, \quad U=0, \quad \theta=0 \tag{6.7.30}$$

$$Y=1, \quad U=1, \quad \theta=0$$

Notice that $\varepsilon$ is the Brinkman number, which is the ratio of viscous to conduction heating. For small $\varepsilon$ or negligible viscous heating effects, one may assume that

$$U=U_0+\varepsilon U_1+\varepsilon^2 U_2+\cdots \tag{6.7.31}$$

$$\theta = \theta_0 + \varepsilon\theta_1 + \varepsilon^2\theta_2 + \cdots \tag{6.7.32}$$

Upon substituting Equations 6.7.31 and 6.7.32 into Equations 6.7.29 and 6.7.30, the term $\exp[-\beta(\theta_0 + \varepsilon\theta_1 + \varepsilon^2\theta_2 + \ldots)]$ arises and needs to be expanded. That is, if we take three terms of the series given by Equation 6.7.32, then

$$\exp\left[-\beta\left(\theta_0 + \varepsilon\theta_1 + \varepsilon^2\theta_2 + \cdots\right)\right] = \exp\left(-\beta\theta_0\right)\exp\left(-\varepsilon\beta\theta_1\right)\exp\left(-\varepsilon^2\beta\theta_2\right)$$

$$= e^{-\beta\theta_0}\left[1 - \varepsilon\left(\beta\theta_1\right) + \varepsilon^2\frac{\left(\beta\theta_1\right)^2}{2}\right]\left(1 - \varepsilon^2\beta\theta_2\right)$$

$$= e^{-\beta\theta_0}\left\{1 - \varepsilon\left(\beta\theta_1\right) + \varepsilon^2\left[\frac{\left(\beta\theta_1\right)^2}{2} - \beta\theta_2\right]\right\}$$

$$= e^{-\beta\theta_0} - \varepsilon e^{-\beta\theta_0}\beta\theta_1 + \varepsilon^2 e^{-\beta\theta_0}\left[\frac{\left(\beta\theta_1\right)^2}{2} - \beta\theta_2\right]$$

where only terms up to $\varepsilon^2$ have been retained.

Similarly, the term $(dU/dY)^2$ is expanded as

$$\left(\frac{dU}{dY}\right)^2 = \left(\frac{dU_0}{dY} + \varepsilon\frac{dU_1}{dY} + \varepsilon^2\frac{dU_2}{dY}\right)^2$$

$$= \left(\frac{dU_0}{dY}\right)^2 + \varepsilon\left(2\frac{dU_0}{dY}\frac{dU_1}{dY}\right) + \varepsilon^2\left[2\frac{dU_0}{dY}\frac{dU_1}{dY} + \left(\frac{dU_1}{dY}\right)^2\right]$$

where only the first three terms of Equation 6.7.31 are retained. Employing these expansions together with Equations 6.7.29 and 6.7.30 and equating coefficients of like powers of $\varepsilon$ result in the sets

$$\varepsilon^0: \quad \frac{d}{dY}\left(e^{-\beta\theta_0}\frac{dU_0}{dY}\right) = 0$$

$$\frac{d^2\theta_0}{dY^2} = 0$$

$$Y = 0, \quad U_0 = 0, \quad \theta_0 = 0$$

$$Y = 1, \quad U_0 = 1, \quad \theta_0 = 0$$

$$\epsilon: \quad \frac{d}{dY}\left[e^{-\beta\theta_0}\left(\frac{dU_1}{dY} - \beta\theta_1\frac{dU_0}{dY}\right)\right] = 0$$

$$\frac{d^2\theta_1}{dY^2} + e^{-\beta\theta_0}\left(\frac{dU_0}{dY}\right)^2 = 0$$

$$Y = 0, \quad U_1 = 0, \quad \theta_1 = 0$$

$$Y = 1, \quad U_1 = 0, \quad \theta_1 = 0$$

$$\epsilon^2: \quad \frac{d}{dY}\left(e^{-\beta\theta_0}\left\{\frac{dU_2}{dY} - \beta\theta_1\frac{dU_1}{dY} + \left[\frac{(\beta\theta_1)^2}{2} - \beta\theta_2\right]\frac{dU_0}{dY}\right\}\right) = 0$$

$$\frac{d^2\theta_2}{dY^2} + e^{-\beta\theta_0}\left[2\frac{dU_0}{dY}\frac{dU_1}{dY} - \beta\theta_1\left(\frac{dU_0}{dY}\right)^2\right] = 0$$

$$Y = 0, \quad U_2 = 0, \quad \theta_2 = 0$$

$$Y = 1, \quad U_2 = 0, \quad \theta_2 = 0$$

for terms up to $\epsilon^2$. The solution sequence should be $\theta_0$, $U_0$, $\theta_1$, $U_1$, $\theta_2$, and $U_2$ to get the following:

$$\theta_0 = 0$$

$$U_0 = Y$$

$$\theta_1 = \frac{1}{2}Y(1-Y)$$

$$U_1 = -\frac{1}{12}\beta\left(Y - 3Y^2 + 2Y^3\right)$$

$$\theta_2 = -\frac{1}{24}\beta\left(Y - 2Y^2 + 2Y^3 - Y^4\right)$$

$$U_2 = \frac{1}{120}\beta^2\left(Y - 5Y^2 + 10Y^3 - 10Y^4 + 4Y^5\right)$$

These results for three terms may be substituted into Equations 6.7.31 and 6.7.32 for comparison to the exact solution given by Turian and Bird[14]:

$$e^{\beta\theta} = \left(1 + \frac{\epsilon\beta}{8}\right)\sec h^2\left[(2Y-1)\sinh^{-1}\left(\frac{\epsilon\beta}{8}\right)^{1/2}\right]$$

$$U = \frac{1}{2}\left\{\left(1 + \frac{8}{\varepsilon\beta}\right)^{1/2} \tanh\left[(2Y-1)\sinh^{-1}\left(\frac{\varepsilon\beta}{8}\right)^{1/2}\right] + 1\right\}$$

The above are three examples of *regular perturbation.* The method is applicable to both partial and ordinary differential equations. The next example is one in which the meaning of the term *regular* is emphasized.

Consider now the first order problem

$$y' + \varepsilon y = 0; \quad y(0) = 1, \quad \varepsilon < 1$$

Then, as discussed in Chapter 2, the solution is

$$y = e^{-\varepsilon x}$$

which can be expanded in a Taylor series about $\varepsilon = 0$, to be

$$y = 1 - \varepsilon x + \frac{\varepsilon^2 x}{2!} - \frac{\varepsilon^3 x^3}{3!} + \cdots$$

Suppose we did not know the analytical solution, but suspected that we could try an approximation of the form

$$y = y_0(x) + \varepsilon y_1(x) + \varepsilon^2 y_2(x) + \cdots \sum_{n=0}^{\infty} \varepsilon^n y_n(x)$$

Then substitution into the differential equation results in

$$y_0'(x) + \varepsilon y_1'(x) + \varepsilon^2 y_2'(x) + \cdots + \varepsilon\left[y_0'(x) + \varepsilon y_1'(x) + \varepsilon^2 y_2'(x) + \cdots\right] = 0.$$

By equating coefficients of like powers of $\varepsilon$, the following differential equations result:

$$\varepsilon^0: \quad y_0'(x) = 0$$

$$\varepsilon: \quad y_1'(x) + y_0(x) = 0$$

$$\varepsilon^2: \quad y_2'(x) + y_1(x) = 0$$

$$\vdots$$

Also, the condition $y(0) = 1$, becomes

$$y(0)=1=y_0(0)+\varepsilon y_1(0)+\varepsilon^2 y_2(0)+\cdots$$

which, on equating coefficients of like powers of $\varepsilon$ gives

$$\varepsilon^0: \quad y_0(0)=1$$

$$\varepsilon: \quad y_1(0)=0$$

$$\varepsilon^2: \quad y_2(0)=0$$

$$\vdots$$

such that

$$y_0(x)=1$$

$$y_1(x)=x$$

$$y_2(x)=\frac{x^2}{2}$$

Therefore,

$$y=1+\varepsilon x+\varepsilon^2 x^2/2+\cdots$$

One can observe that this final solution is identical to the Taylor series expansion of the analytical solution. Therefore, whenever the perturbed solution has a Taylor series expansion such as demonstrated, the perturbation is considered as regular.

In general, given a linear or nonlinear differential equation

$$L(u,\varepsilon)=0$$

that depends on the small positive parameter $\varepsilon$ and with appropriate boundary or initial conditions (data) that may also depend on $\varepsilon$, we say that such a problem is a regular perturbation problem if the reduced problem consists of

$$L(v,0)=0$$

and the reduced data has a unique solution (Zauderer[15]). As a further demonstration of the regular perturbation method, consider the two-dimensional Helmholtz equation:

$$L(u,\varepsilon) = u_{xx} + u_{yy} + \varepsilon^2 u = 0$$

in the unit disk $x^2 + y^2 < 1$ subject to the boundary condition

$$u(x,y) = 1; \quad x^2 + y^2 = 1$$

Assume that

$$u(x,y) = \sum_{n=0}^{\infty} u_n(x,y)\varepsilon^{2n}$$

is a solution. Then, by definition of a solution to a differential equation

$$u_{xx}u_{yy} + \varepsilon^2 u = \nabla^2 u + \varepsilon^2 u = \nabla^2\left[\sum_{n=0}^{\infty} u_n\varepsilon^{2n}\right] + \sum_{n=0}^{\infty} u_n\varepsilon^{2n+2}$$

$$= \nabla^2 u_0 + \sum_{n=1}^{\infty}\left[\nabla^2 u_n + u_{n-1}\right]\varepsilon^{2n} = 0$$

where summation and differentiation are assumed interchangeable. The boundary condition becomes

$$u(x,y) = u_0(x,y) + \sum_{n=1}^{\infty} u_n(x,y)\varepsilon^{2n} = 1; \quad x^2 + y^2 = 1$$

Therefore, equating like powers of $\varepsilon$ gives

$$\varepsilon^0: \quad \nabla^2 u = 0$$

$$u_0(x,y) = 1$$

$$\varepsilon^{2n}: \quad \nabla^2 u_n = -u_{n-1}; \quad n \geq 1$$

$$u_n(x,y) = 0; \quad x^2 + y^2 = 1. \quad n \geq 1$$

Notice that the perturbation method has replaced the Helmholtz equation with a system of Laplace and Poisson equations. By observing that polar coordinates can be used and that there is no $\theta$ dependence, that is $u = u(r)$, we get

$$\nabla^2 u_0 = 0 = \frac{\partial^2 u_0}{\partial r^2} + \frac{1}{r}\frac{\partial u_0}{\partial r} = 0; \quad u_0(1) = 1$$

whose bounded solution is $u_0 = 1$. Also, the equation for $u_1$ is

$$\frac{\partial^2 u_1}{\partial r^2} + \frac{1}{r}\frac{\partial u_1}{\partial r} = -u_0; \quad u_1(1) = 0$$

and results in

$$u_1(r) = \frac{1-r^2}{4}$$

Therefore, to leading orders the reduced problem solution is

$$u = 1 + \varepsilon^2 \frac{\left(1-r^2\right)}{4} + O\left(\varepsilon^4\right)$$

where the symbol $O(\varepsilon^4)$ means terms of order $\varepsilon^4$.

It can be shown that the polar coordinate form of the Helmholtz equation

$$\frac{\partial^2 u}{\partial r^2} + \frac{1}{r}\frac{\partial u}{\partial r} + \varepsilon^2 u = 0; \quad u_0(1) = 1$$

which is bounded at the origin solves to

$$u = \frac{J_0(\varepsilon r)}{J_0(\varepsilon)}$$

and $J_0(z)$ has the series expansion

$$J_0(z) = 1 - \frac{z^2}{4} + O\left(z^4\right)$$

Therefore,

$$u = \frac{J_0(\varepsilon r)}{J_0(\varepsilon)} = \frac{1 - (\varepsilon r)^2/4 + O\left(\varepsilon^4\right)}{1 - (\varepsilon)^2/4 + O\left(\varepsilon^4\right)}$$

$$= 1 + \varepsilon^2 \left(1 - r^2\right)/4 + O\left(\varepsilon^4\right)$$

It is interesting to note that $\varepsilon$ must be smaller than the first zero of the Bessel function in order for the reduced problem to yield a useful result.

As a final example of regular perturbation in this chapter, consider the freezing of a saturated liquid in a semi-infinite region (Aziz and Na[13]). A saturated liquid is initially at its freezing temperature $T_f$. At some time, the face is maintained at a constant subfreezing temperature $T_0(T_0 < T_f)$. As heat is removed from the liquid, it freezes. If the freezing front at any time t is at $x_f$ and if it can be assumed that the uniform liquid remains at $T_f$ throughout the process, what is the temperature profile $T(x, t)$ in the solid phase?

*Solution*
A model for this problem is

$$\frac{1}{\alpha}\frac{\partial T}{\partial t} = \frac{\partial^2 T}{\partial x^2}$$

$$T(0,t) = T_0, \quad T(x_f,t) = T_f$$

$$k\frac{\partial T}{\partial x}\bigg|_{x=x_f} = \rho\lambda\frac{dx_f}{dt}$$

where $\alpha$ is thermal diffusivity, k and $\rho$ are thermal conductivity and density, respectively of the solid phase and $\lambda$ is the latent heat.
Introducing the dimensionless quantities

$$\theta = \frac{T_f - T}{T_f - T_0}, \quad X = \frac{x}{x_s}, \quad X_f = \frac{x_f}{x_s}, \quad \tau = \frac{kt}{\rho Cx_s^2}, \varepsilon = \frac{C(T_f - T_0)}{\lambda}$$

where $x_s$ is a reference distance and C is the solid phase specific heat. Further, by changing the variables from $(X, \tau)$ to $(X, X_f)$, the dimensionless model becomes

$$\frac{\partial^2\theta}{\partial X^2} = -\varepsilon\frac{\partial\theta}{\partial X_f}\frac{\partial\theta}{\partial X}\bigg|_{X=X_f}$$

$$\theta(X=0, X_f) = 1, \quad \theta(X=X_f, X_f) = 0$$

$$\frac{dX_f}{d\tau} = -\varepsilon\frac{\partial\theta}{\partial X}\bigg|_{X=X_f}$$

Here, the quantity $\varepsilon$ is the *Stefan number*, which represents the ratio of sensible to latent heat during a phase change. For a process with small $\varepsilon$, or a comparatively large latent heat such as in this case, let

$$\theta = \theta_0 + \varepsilon\theta_1 + \varepsilon^2\theta_2 + \cdots$$

Then following substitution into the dimensionless model and equating coefficients of like powers of $\varepsilon$ results in

$$\varepsilon^0: \quad \frac{\partial^2 \theta_0}{\partial X^2} = 0$$

$$\theta_0\left(X = 0, X_f\right) = 1, \theta_0\left(X = X_x, X_f\right) = 0$$

$$\varepsilon: \quad \frac{\partial^2 \theta_1}{\partial X^2} = -\frac{\partial \theta_0}{\partial X_f} \frac{\partial \theta_0}{\partial X}\bigg|_{X=X_f}$$

$$\theta_f\left(X = 0, X_f\right) = 0, \quad \theta_1\left(X = X_f, X_f\right) = 0$$

and

$$\varepsilon^2: \quad \frac{\partial^2 \theta_2}{\partial X^2} = -\left(\frac{\partial \theta_0}{\partial X_f} \frac{\partial \theta_1}{\partial X}\bigg|_{X=X_f} + \frac{\partial \theta_1}{\partial X_f} \frac{\partial \theta_0}{\partial X}\bigg|_{X=X_f}\right)$$

$$\theta_2\left(X = 0, X_f\right) = 0, \quad \theta_2\left(X = X_f, X_f\right) = 0$$

for coefficients of terms up to $\varepsilon^2$. The three-term expansion becomes

$$\theta = 1 - \frac{X}{X_f} - \frac{1}{6} \varepsilon \frac{X}{X_f}\left[1 - \left(\frac{X}{X_f}\right)^2\right] + \frac{1}{360} \varepsilon^2 \frac{X}{X_f}\left[19 - 10\left(\frac{X}{X_f}\right)^2 - 9\left(\frac{X}{X_f}\right)^4\right]$$

In order to derive an expression for the progress of the freezing front with respect to time, recall that

$$\frac{dX_f}{d\tau} = -\varepsilon \frac{\partial \theta}{\partial X}\bigg|_{X=X_f} = \frac{1}{X_f}\left(\varepsilon - \frac{1}{3} \varepsilon^2 + \frac{7}{45} \varepsilon^3\right) + \text{ higher order terms,}$$

subject to the condition that

$$X_f = 0 \text{ at } \tau = 0$$

Solution of this latter equation results in

$$X_f^2 = 2\tau\left(\varepsilon - \frac{1}{3} \varepsilon^2 + \frac{7}{45} \varepsilon^3\right) + \text{ higher order terms,}$$

or

$$\tau = \frac{1}{2} X_f^2 \varepsilon^{-1} \left( 1 - \frac{1}{3} \varepsilon + \frac{7}{45} \varepsilon^2 \right)^{-1}$$

$$= \frac{1}{2} \varepsilon^{-1} X_f^2 + \frac{1}{6} X_f^2 - \frac{1}{45} \varepsilon X_f^2 + \text{ terms of order } \varepsilon^2.$$

Hopefully, these examples have demonstrated the importance of the regular perturbation method as a technique that can be used to reduce some nonlinear problems to sets of linear problems. And, as was stated earlier, we like to reduce new problems to old ones that we know how to solve.

## 6.8  Problems

1. In each of the following problems, use the method of separation of variables to solve the one-dimensional heat equation. Also, define the steady-state temperature at the midpoint of the region.

   (a) $\dfrac{\partial T}{\partial t} = \dfrac{1}{2} \dfrac{\partial^2 T}{\partial x^2}$, $T(0,t) = T(3,t) = 0$, $T(x,0) = \sin \pi x$, $0 < x < 3$

   (b) $\dfrac{\partial T}{\partial t} = \dfrac{\partial^2 T}{\partial x^2}$, $\dfrac{\partial T(0,t)}{\partial x} = \dfrac{\partial T(\pi,t)}{\partial x} = 0$, $T(x,0) = \cos x$, $0 < x < \pi$

   (c) $\dfrac{\partial T}{\partial t} = \dfrac{\partial^2 T}{\partial x^2}$, $T(-\pi,t) = T(\pi,t)$, $\dfrac{\partial T(-\pi,t)}{\partial x} = \dfrac{\partial T(\pi,t)}{\partial x}$

   $T(x,0) = x + \pi$, $-\pi < x < \pi$

   (d) $\dfrac{\partial T}{\partial t} = \dfrac{1}{2} \dfrac{\partial^2 T}{\partial x^2}$, $T(0,t) = 100$, $T(2,t) = 50$, $T(x,0) = 100 - 13x$,
   $0 < x < 2$

2. Solve the heat equation

$$\frac{\partial T}{\partial t} = \frac{\partial^2 T}{\partial x^2} - T$$

subject to

$$T(0,t) = T(1,t) = 0, \quad T(x,0) = \begin{cases} 1, & 0 < x < 0.5 \\ 0, & 0.5 < x < 1 \end{cases}$$

3. Use the method of eigenfunction expansions to solve, without reducing to homogeneous boundary conditions

$$\frac{\partial w}{\partial t} = k \frac{\partial^2 w}{\partial x^2}$$

$$w(x,0) = h(x), \quad \left. \begin{array}{l} w(0,t) = A \\ w(L,t) = B \end{array} \right\} \text{constants}$$

4. Use the method of Laplace transform to solve the following problem:

$$\frac{\partial u}{\partial t} = v \frac{\partial^2 u}{\partial y^2}$$

$$\text{at} \quad t = 0, u = 0$$

$$y = 0, u = V \;\; (\text{const.})$$

$$y = \infty, u = 0$$

5. A fluid of constant density, $\rho$, and viscosity, $\mu$, is contained in a very long horizontal pipe of length L and radius R. Initially, the fluid is at rest. At $t = 0$, a pressure gradient $(p_0 - p_L)/L$ is impressed on the system. Can the method of Laplace transform be used to determine how the velocity profiles change with time?

*Hint:* Use cylindrical coordinates and assume that the z-component of velocity, $v_z$, is a function of r and t, while the r- and $\theta$-components of velocity are both zero. Then combine the equations of continuity and motion to get (Bird, Stewart, and Lightfoot[1])

$$\rho \frac{\partial v_z}{\partial t} = \frac{p_0 - p_L}{L} + \mu \frac{1}{r} \frac{\partial}{\partial r} \left( r \frac{\partial v_z}{\partial r} \right)$$

subject to the conditions:

$$v_z(r,0) = 0$$

$$v_z(0,t) \text{ is finite}$$

$$v_z(R,t) = 0$$

6. Find a bounded solution to Laplace's equation $\nabla^2 w = 0$ for the half plane $y > 0$ if $w$ takes on the value $f(x)$ on the x-axis. *Hint: (i) Separate the variables and set each side equal to $-\lambda^2$, (ii) Argue that the lack of restrictions on $\lambda$ allow the replacement of the arbitrary constants with arbitrary functions of $\lambda$, thus leading to the Fourier integral defined by Equation 5.6.1.*

Answer: $w(x,y) = \dfrac{1}{\pi} \displaystyle\int\limits_{0}^{\infty} \int\limits_{u=-\infty}^{\infty} e^{-\lambda y} f(u) \cos \lambda(u-x) du d\lambda$

7. The surface of a cylinder of radius R is maintained at a constant temperature $T_0$, while the initial temperature was zero throughout (Walas[16]). Use Laplace transform to derive the temperature distribution.

Model: $\dfrac{\partial T}{\partial t} = k\left(\dfrac{\partial^2 T}{\partial r^2} + \dfrac{1}{r}\dfrac{\partial T}{\partial r}\right)$

$$T(r,0) = 0, \quad T(R,t) = T_0$$

$$\lim_{r \to 0} T(r,t) \text{ is finite.}$$

Answer: $T(r,t) = T_0\left\{1 + \dfrac{2}{R}\displaystyle\sum_{n=1}^{\infty} \dfrac{J_0(\alpha_n r)}{\alpha_n J_0'(\alpha_n R)}\exp(-k\alpha_n^2 t)\right\}$

where $J_0(\alpha_n R) = 0$ defines the $\alpha_n$.

8. Given the model: $\dfrac{\partial u}{\partial t} = k\dfrac{\partial^2 u}{\partial x^2} + q$

subject to: $u(0,t) = 0$, $u(x,0) = 0$; $u(\infty,t)$ is finite

derive $u(x,t) = \dfrac{x}{\sqrt{4k\pi t^3}}\exp\left(\dfrac{-x^2}{4kt}\right) - q\displaystyle\int_0^t \mathrm{erfc}\left(\dfrac{x}{\sqrt{4kt}}\right)dt + qt$

where q is a constant. *Hint: Use the formula*

$$L^{-1}\left\{\frac{\overline{u}(s)}{s}\right\} = \int_0^t u(t)\,dt$$

*where* $\overline{u}(s)$ *is a known (tabulated) transform to carry out the inversion* (Walas[16]).

9.  Show that the solution to the Helmholtz equation

$$\frac{\partial^2 u}{\partial r^2} + \frac{1}{r}\frac{\partial u}{\partial r} + \varepsilon^2 u = 0; \quad u_0(1) = 1$$

which is bounded at the origin solves to

$$u = \frac{J_0(\varepsilon r)}{J_0(\varepsilon)}.$$

## References

1.  Bird, R.B., Stewart, W.E., and Lightfoot, E.N., *Transport Phenomena*, John Wiley & Sons, New York, 1960.
2.  Giordano, F.R. and Weir, M.D., *Differential Equations, a Modeling Approach*, Addison-Wesley, Reading, MA, 1991.
3.  Myint-U, T. and Debnath, L., *Partial Differential Equations for Scientists and Engineers*, 3rd ed., Prentice-Hall, Englewood Cliffs, NJ, 1987.
4.  Haberman, R., *Elementary Applied Partial Differential Equations with Fourier Series and Boundary Value Problems*, Prentice-Hall, Englewood Cliffs, NJ, 1983.
5.  Churchill, R.V. and Brown, J.W., *Fourier Series and Boundary Value Problems*, 4th ed., McGraw-Hill, New York, 1987.
6.  Boyce, W.E. and DiPrima, R.C., *Elementary Differential Equations and Boundary Value Problems*, 3rd ed., John Wiley & Sons, New York, 1977.
7.  Spiegel, M.R., *Mathematical Handbook*, McGraw-Hill, New York, 1968.
8.  Churchill, R.V. and Brown, J.W., *Complex Variables and Applications*, McGraw-Hill, New York, 1984.
9.  Spiegel, M.R., *Complex Variables*, McGraw-Hill, New York, 1964.
10. Williams, W.E., *Partial Differential Equations*, Oxford, 1980.
11. Myers, G.E., *Analytical Methods in Conduction Heat Transfer*, McGraw-Hill, New York, 1971.
12. Loney, N.W., On using a boundary perturbation to linearize a system of non-linear pdes, *Chemical Eng. Educ.*, Winter, 58, 1996.
13. Aziz, A. and Na, T.Y., *Perturbation Methods in Heat Transfer*, Hemisphere, Washington, 1984.
14. Turian, R.M. and Bird, R.B., Viscous heating in the cone-and-plate viscometer II, *Chemical Eng. Sci.*, 18, 689, 1963.

15. Zauderer, E., *Partial Differential Equations of Applied Mathematics*, John Wiley & Sons, New York, 1983.
16. Walas, S.M., *Modeling with Differential Equations in Chemical Engineering*, Butterworth-Heinemann, Boston, 1991.

# 7

# Applications of Partial Differential Equations in Chemical Engineering

## 7.0 Introduction

The primary objective of this chapter is to present several worked-out examples. These examples reflect typical chemical engineering science research results, and it is recommended that the reader dedicate some time to work through these examples. The examples also demonstrate the application of much of the mathematics discussed in the previous chapters.

Examples are provided from heat transfer; mass transfer; simultaneous diffusion and convection; simultaneous diffusion and chemical reaction; simultaneous diffusion, convection, and chemical reaction; and viscous flow.

Some examples may be worked out in more detail than others. Those worked-out examples with missing details provide an opportunity for the reader to check his or her knowledge of the mathematical technique being employed.

## 7.1 Heat Transfer

### Example 7.1.1
A sheet of polymer 1/2-in. thick is to be cured at 292°F for 50 min (Bennett and Myers[1]). If the sheet is initially at 70°F, and heat is applied from both surfaces, find the time required for the temperature at the center of the sheet to reach 290°F. It can be assumed that the surfaces are brought to 292°F as soon as curing is begun and held at that temperature throughout the process. The thermal diffusivity $k/\rho C_p$ of this polymer can be taken as 0.0028 sq ft/hr.

The differential equation for unsteady heat conduction in one direction is

$$\frac{\partial u}{\partial t} = \frac{k}{\rho C_p}\frac{\partial^2 u}{\partial x^2} \qquad 0 < x < 1/48 \qquad (7.1.1)$$

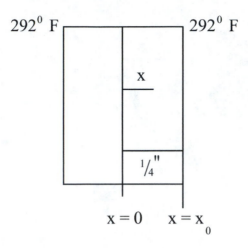

**FIGURE 7.1**
Unsteady heat conduction in polymer sheet.

The origin is chosen at the center of the sheet, so that $x$ is the distance from the center, as shown in the figure, and is in feet. Also, $x_0$ is the half thickness. The boundary conditions are

$$\frac{\partial u(0,t)}{\partial x} = 0$$

$$t > 0 \qquad\qquad (7.1.1A)$$

$$u(x_0, t) = 292°F$$

and the initial condition is

$$u(x,0) = 70°F \quad 0 < x < \frac{1}{48} \qquad\qquad (7.1.1B)$$

*Remark: It is good practice whenever possible to express the differential equation in dimensionless form so that the result can be portable.*
Let

$$Y(\xi,\tau) = \frac{292 - u(x,0)}{292 - 70}, \quad \xi = \frac{x}{x_0}, \quad \tau \frac{kt}{C_p \rho x_0^2} \qquad\qquad (7.1.2)$$

Then Equation 7.1.1 becomes

$$\frac{\partial Y}{\partial \tau} = \frac{\partial^2 Y}{\partial \xi^2} \qquad\qquad (7.1.3)$$

subject to

$$\frac{\partial Y(0,\tau)}{\partial \xi} = 0 \tag{7.1.4}$$

$$Y(1,\tau) = 0 \tag{7.1.5}$$

$$Y(\xi,0) = 1 \tag{7.1.6}$$

From our previous experience, we will try a solution of the form

$$Y(\xi,\tau) = f(\xi)g(\tau) \tag{7.1.7}$$

Then Equation 7.1.3 becomes

$$\frac{g'}{g} = \frac{f''}{f} = -\lambda^2$$

where the negative sign is chosen based on our experience using separation of variables up to now (*the reader should check this*). Therefore,

$$f'' + \lambda^2 f = 0 \tag{7.1.8}$$

subject to

$$f'(0) = 0$$

and

$$f(1) = 0$$

Also,

$$g(\tau) = c_1 e^{-\lambda^2 \tau} \tag{7.1.9}$$

The general solution of Equation 7.1.8 is

$$f(\xi) = c_2 \cos \lambda \xi + c_3 \sin \lambda \xi$$

and

$$f'(0) = 0 = \lambda c_3 \Rightarrow c_3 = 0$$

then for nontrivial solution to exist, $c_2 \neq 0$, so that

$$f(1) = 0 = c_2 \cos \lambda \quad \Rightarrow \lambda = \frac{(2n-1)\pi}{2}, n = 1, 2, \cdots$$

Therefore, for each n

$$Y_n(\xi, \tau) = A_n e^{-\lambda_n^2 t} \cos \lambda_n \xi$$

and by the principle of superposition

$$Y(\xi, \tau) = \sum_{n=1}^{\infty} A_n e^{-\lambda_n^2 t} \cos \lambda_n \xi$$

solves Equations 7.1.3, 7.1.4, and 7.1.5.
　　Then Equation 7.1.6 gives

$$Y(\xi, 0) = 1 = \sum_{n=1}^{\infty} A_n \cos \lambda_n \xi$$

such that for $n \geq 1$

$$\int_0^1 \cos \lambda_n \xi \, d\xi = A_n \int_0^1 \cos^2 \lambda_n \xi \, d\xi$$

or

$$A_n = \frac{4}{(2n-1)\pi} \sin \frac{(2n-1)\pi}{2}$$

$$= \frac{4}{(2n-1)\pi} (-1)^n$$

That is,

$$Y(\xi, \tau) = \sum_{n=1}^{\infty} (-1)^n \frac{4}{(2n-1)\pi} \exp\left[-\left(\frac{2n-1}{2}\pi\right)^2 \tau\right] \cos\left(\frac{2n-1}{2}\pi\xi\right)$$

Then the time required for the temperature at the center of the sheet to reach 290°F is estimated as follows:

$$Y(0,\tau) = \frac{292 - u(0,t)}{292 - 70} = \frac{292 - 290}{292 - 70} = 0.009$$

Therefore,

$$0.009 = \frac{4}{\pi} e^{\frac{-\pi^2}{4}\tau}$$

using only the first term of the series. That is, $\tau \cong 2.01$. Then from Equation 7.1.2

$$\tau = \frac{kt}{C_p \rho x_0^2}, \text{ or } t = \frac{C_p \rho}{k} x_0^2 \tau = \frac{2.01}{(0.0028)(48)^2} = 0.313 \text{ hr} = 18.7 \text{ min}$$

*Remark: The reader should check the validity of the approximation, which employed only the first term of the series.*

## Example 7.1.2

Consider a plane wall of thickness L (Myers[2]) initially at temperature $T_0$ when suddenly at t = 0, the surface temperature at x = 0 is changed to $T_\infty$, and the surface at x = L is suddenly exposed to a bath with an ambient temperature also of $T_\infty$.

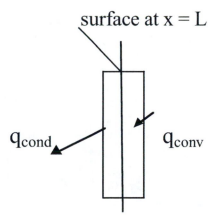

surface at x = L

$q_{cond}$     $q_{conv}$

**FIGURE 7.2**
Plane wall.

At the surface $x = L$

$$q_{conducted} = q_{convected}$$

such that the boundary condition becomes

$$-k\frac{\partial T}{\partial x} = h(T - T_\infty)$$

where h is the convective heat transfer coefficient and k is the thermal conductivity.

Determine the temperature distribution in this wall.

*Solution*
The mathematical description of the problem is

$$\frac{1}{\alpha}\frac{\partial T}{\partial t} = \frac{\partial^2 T}{\partial x^2}, \quad 0 < x < L, \quad t > 0$$

subject to the boundary conditions

$$T(0,t) = T_\infty$$

$$-k\frac{\partial T(L,t)}{\partial x} = h(T - T_\infty)$$

and the initial condition

$$T(x,0) = T_0$$

In order to employ the method of separation of variables, the differential equation and boundary conditions are required to be linear and homogeneous. Presently, the linearity requirement is satisfied, but the boundary conditions are not homogeneous. In this problem, recasting the variables into dimensionless quantities reduces the problem to a homogeneous one. That is, let

$$u(z,\tau) = \frac{T - T_\infty}{T_0 - T_\infty}, \quad z = \frac{x}{L} \text{ and } \tau = \frac{\alpha t}{L^2}$$

then the dimensionless problem becomes

$$\frac{\partial u}{\partial \tau} = \frac{\partial^2 u}{\partial z^2}$$

subject to

$$u(z,0) = 1$$

$$u(0,\tau) = 0$$

$$u_z(1,\tau) = -N_B u(1,\tau)$$

where $N_B$ is the *Biot number* defined by

$$N_B = \frac{hL}{k} = \frac{L/kA}{1/hA}$$

Notice that the Biot number compares the steady-state conduction resistance of a plane wall L-units thick to the convection resistance at the surface. Small or large Biot numbers can aid in deciding on the specific boundary condition to use in setting up a physical model. For example, a small Biot number implies that convection resistance is significant while a large Biot number implies that conduction resistance is significant.

Using the substitution

$$u(z,\tau) = y(z)f(\tau)$$

the dimensionless problem becomes

$$\frac{f'}{f} = \frac{y''}{y} = -\lambda^2$$

where the sign was chosen based on previous experience (Chapter 6). As a result, we get the two ordinary differential equations

$$f' + \lambda^2 f = 0, \quad \text{or} \quad f(\tau) = c_1 e^{-\lambda^2 \tau}$$

and

$$y'' + \lambda^2 y = 0$$

The conditions on the y-equation are derived as follows:

$$u(0,\tau) = 0 = y(0)f(\tau), \Rightarrow y(0) = 0$$

since $f(\tau)$ is arbitrary, and

$$u_z(1,\tau) = y'(1)f(\tau) = -N_B u(1,\tau) = y(1)f(\tau), \Rightarrow y'(1) = -N_B y(1)$$

Then the general solution of the y-equation is

$$y(z) = c_2 \cos \lambda z + c_3 \sin \lambda z$$

and the condition at $z = 0$ gives

$$y(0) = 0 = c_2$$

while the condition at $z = 1$ results in

$$y'(1) = \lambda c_3 \cos \lambda = -N_B c_3 \sin \lambda$$

Therefore, for a nontrivial solution, $c_3 \neq 0$ and the eigenvalues are given by

$$\lambda \cot \lambda = -N_B$$

The corresponding eigenfunctions are given by

$$y_n = c_{3,n} \sin \lambda_n z, \quad n = 1, 2, \ldots$$

since there are infinitely many eigenvalues. Then, for each n

$$u_n(z,\tau) = y_n(z)f_n(\tau) = B_n e^{-\lambda_n^2 \tau} \sin \lambda_n z$$

Further, by the principle of superposition

$$u(z,\tau) = \sum_{n=1}^{\infty} B_n e^{-\lambda_n^2 \tau} \sin \lambda_n z$$

is the solution to the dimensionless differential equation and the given boundary conditions. The initial condition will be satisfied if

$$u(z,0) = 1 = \sum_{n=1}^{\infty} B_n \sin \lambda_n z$$

such that for $n \geq 1$,

$$B_n = \frac{\int\limits_0^1 \sin \lambda_n z \, dz}{\left[ \dfrac{1}{2} - \dfrac{1}{4\lambda_n} \sin 2\lambda_n \right]} = \frac{1 - \cos \lambda_n}{\dfrac{\lambda_n}{2} - \dfrac{\sin 2\lambda_n}{4}}$$

$$= \frac{4(1 - \cos \lambda_n)}{2\lambda_n - \sin 2\lambda_n}$$

Finally, the dimensionless temperature distribution is

$$u(z, \tau) = \sum_{n=1}^\infty \frac{4(1 - \cos \lambda_n)}{2\lambda_n - \sin 2\lambda_n} e^{-\lambda_n^2 \tau} \sin \lambda_n z$$

where

$$\lambda_n \cot \lambda_n = -N_B, \quad n = 1, 2 \ldots$$

defines the $\lambda_n$. The results can now be reported in terms of the dimensioned variables.

## Example 7.1.3
Consider an infinitely long, solid cylinder of radius $r_0$ whose initial temperature is $T_0$ (Myers[2]). If the temperature of its outside surface is suddenly changed to $T_\infty$, determine the temperature distribution in the cylinder.

*Solution*
The governing equation is

$$\frac{1}{\alpha} \frac{\partial T}{\partial t} = \frac{\partial^2 T}{\partial r^2} + \frac{1}{r} \frac{\partial T}{\partial r}$$

and the boundary and initial conditions are

$$T(r_0, t) = T_\infty$$

and

$$T(r, 0) = T_0$$

respectively. It is important to point out that since we have a second order differential equation in r, there should be two boundary conditions involving r. The second boundary condition is that the temperature is expected to be finite at r = 0. This conclusion is based on our "engineering understanding" of the problem.

An alternative boundary condition at r = 0 is

$$\frac{\partial T}{\partial r} = 0$$

due to the symmetry of the temperature distribution in this problem. If, however, the temperature varies with angular position, then symmetry would not be valid. Further, to solve this problem using separation of variables, it is sufficient to express the temperature distribution as

$$w(r,t) = T - T_\infty$$

such that

$$\frac{1}{\alpha}\frac{\partial w}{\partial t} = \frac{\partial^2 w}{\partial r^2} + \frac{1}{r}\frac{\partial w}{\partial r}$$

subject to

$$w(0,t) < \infty$$

$$w(r_0,t) = 0$$

$$w(r,0) = T_0 - T_\infty$$

is homogeneous. Then letting

$$w(r,t) = R(r)g(t)$$

the pde reduces to

$$\frac{1}{\alpha}\frac{g'}{g} = -\lambda^2, \Rightarrow g(t) = c_1 e^{-\alpha\lambda^2 t}$$

and

$$R'' + \frac{1}{r}R' + \lambda^2 R = 0$$

subject to

$$R(0) < \infty \quad \text{and} \quad R(r_0, t) = 0$$

The R-equation and the accompanying boundary conditions constitute a singular Sturm-Liouville boundary value problem. Also the R-equation is Bessel's equation of order zero.

Bessel's equation can be solved indirectly by using the following procedure: Recall Equation 3.6.4:

$$y'' - \frac{2a-1}{x} y' + \left( b^2 c^2 x^{2c-2} + \frac{a^2 - v^2 c^2}{x^2} \right) y = 0$$

with general solution: $y = c_1 x^a J_v(bx^c) + c_2 x^a J_{-v}(bx^c)$

By comparing the coefficients of the first and zeroth derivatives in both equations we get

Coefficient of first derivative:
$$2a - 1 = -1, \quad \Rightarrow a = 0$$
$$2c - 2 = 0, \quad \Rightarrow c = 1$$

Coefficient of the zeroth derivative:
$$b^2 c^2 = \lambda^2, \quad \Rightarrow b = \lambda$$
$$a^2 - v^2 c^2 = 0, \quad \Rightarrow v = 0$$

resulting in a general solution

$$R(r) = c_2 J_0(\lambda r) + c_3 J_{-0}(\lambda r)$$

In customary notation, $Y_0(.)$ is used instead of $J_{-0}(.)$
Therefore, the general solution is

$$R(r) = c_2 J_0(\lambda r) + c_3 Y_0(\lambda r)$$

where $J_0(.)$ is the zero-order Bessel function of the first kind and $Y_0(.)$ is the zero-order Bessel function of the second kind. Further, since $R(r)$ must be finite at $r = 0$, then $c_3$ must be chosen as zero due to the fact that $Y_0$ is unbounded at $r = 0$. Also, for nontrivial solution $c_2 \neq 0$ such that

$$R(r_0) = 0 = c_2 J_0(r_0 \lambda), \quad \Rightarrow J_0(r_0 \lambda) = 0$$

The function $J_0(r_0\lambda)$ oscillates about the $r_0\lambda$ axis with infinitely many intersections, that is

$$r_0\lambda_1 = 2.4048$$

$$r_0\lambda_2 = 5.5201$$

$$r_0\lambda_3 = 8.6537$$

and more. These points of intersection can be read from a table of zeros of the Bessel function (Spiegel[3]). Therefore, the eigenvalues are defined by

$$J_0\left(r_0\lambda_n\right) = 0, \quad n = 1, 2, \dots$$

and

$$R_n = c_{2,n} J_0\left(\lambda_n r\right)$$

are the corresponding eigenfunctions. Then for each $n$

$$w_n\left(r, t\right) = R_n g_n = A_n e^{-\alpha\lambda_n^2 t} J_0\left(\lambda_n r\right)$$

By the principle of superposition

$$w\left(r, t\right) = \sum_{n=1}^{\infty} A_n e^{-\alpha\lambda_n^2 t} J_0\left(\lambda_n r\right)$$

satisfies the differential equation and the boundary conditions. The initial condition gives

$$w\left(r, 0\right) = \sum_{n=1}^{\infty} A_n J_0\left(\lambda_n r\right) = T_0 - T_\infty$$

which is recognizable as a generalized Fourier series. The Fourier coefficient can be evaluated with the aid of the orthogonality property discussed in the context of Sturm-Liouville problems (Chapter 4). That is,

$$\int_0^{r_0} r J_0\left(\lambda_m r\right) J_0\left(\lambda_n r\right) dr = 0 \quad \text{for} \quad m \neq n$$

where r is the weighting function determined from the Sturm-Liouville form of the differential equation. Therefore,

$$A_n = \frac{\displaystyle\int_0^{r_0} (T_0 - T_\infty) r J_0(\lambda_n r) dr}{\displaystyle\int_0^{r_0} r \left[ J_0(\lambda_n r) \right]^2 dr}, \quad n \geq 1$$

The integrals can be evaluated from standard tables containing integrals of Bessel functions (Spiegel[3]). The overall solution to this problem is

$$w(r,t) = \sum_{n=1}^{\infty} A_n e^{-\alpha \lambda_n^2 t} J_0(\lambda_n r)$$

with the eigenvalues defined by

$$J_0(r_0 \lambda_n) = 0, \quad n = 1, 2, \ldots$$

and the $A_n$ as given above.

## Example 7.1.4

Unsteady state heat conduction in one dimension (Jenson and Jeffreys[4]).

Consider a section of a flat wall of thickness L ft, whose height and length are both large in comparison to L. Suppose the temperature distribution is uniform throughout the wall initially, and heat is supplied at a fixed rate per unit area to one surface. Determine the temperature profile as a function of position and time.

*Solution*

Temperature is constant in any plane parallel to the surface (since initial temperature is uniform and every part of each wall surface sees the same conditions). Then one space coordinate is sufficient (see sketch). Further, for a section of unit area through the wall, the thermal equilibrium of a slice of the wall between a plane at distance x from the heated surface and parallel to a plane at x + Δx from the same surface gives the following:

Rate of heat input at distance x and time t is $-k \dfrac{\partial T}{\partial x}$

Rate of heat input at distance x and time t + Δt is

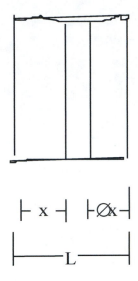

**FIGURE 7.3**
One-dimensional heat transfer.

$$-k\frac{\partial T}{\partial x} + \frac{\partial}{\partial t}\left(-k\frac{\partial T}{\partial x}\right)\Delta t$$

which comes from the first two terms of the Taylor series expansion of $T(x, t)$ while treating $x$ as a constant. Similarly,
   Rate of heat output at distance $x + \Delta x$ and time $t$ is

$$-k\frac{\partial T}{\partial x} + \frac{\partial}{\partial x}\left(-k\frac{\partial T}{\partial x}\right)\Delta x$$

which is the two-term Taylor series expansion of $T(x, t)$ holding $t$ constant. Also,
   Rate of heat output at distance $x + \Delta x$ and $t + \Delta t$ is

$$-k\frac{\partial T}{\partial x} + \frac{\partial}{\partial t}\left(-k\frac{\partial T}{\partial x}\right)\Delta t + \frac{\partial}{\partial x}\left[-k\frac{\partial T}{\partial x} + \frac{\partial}{\partial t}\left(-k\frac{\partial T}{\partial x}\right)\Delta t\right]\Delta x$$

The heat content of the element at time $t$ is $\rho C_p T \Delta x$, while that at $t + \Delta t$ is

$$\rho C_p\left(T + \frac{\partial T}{\partial t}\Delta t\right)\Delta x$$

such that the accumulation of heat in time $\Delta t$ is

$$\rho C_p \frac{\partial T}{\partial t}\Delta t \Delta x$$

Then by the conservation law: Input − Output = Accumulation
  we get

$$\left[-k\frac{\partial T}{\partial x} + \frac{1}{2}\frac{\partial}{\partial t}\left(-k\frac{\partial T}{\partial x}\right)\Delta t\right]\Delta t -$$

$$\left\{-k\frac{\partial T}{\partial x} + \frac{\partial}{\partial x}\left(-k\frac{\partial T}{\partial x}\right)\Delta x + \frac{1}{2}\frac{\partial}{\partial t}\left[-k\frac{\partial T}{\partial x} + \frac{\partial}{\partial x}\left(-k\frac{\partial T}{\partial x}\right)\Delta x\right]\Delta t\right\}\Delta t$$

$$= \rho C_p \frac{\partial T}{\partial x}\Delta t\Delta x$$

where we use the arithmetic average input and output rates during the time interval $\Delta t$. Following simplifications and taking the appropriate limits we get

$$\frac{\partial T}{\partial t} = \alpha\frac{\partial^2 T}{\partial x^2} \text{ for } \alpha = \frac{k}{\rho C_p}$$

where the physical properties are assumed constant. Further, notice that even though the differential equation was derived based on a finite thickness L, there is no thickness dependency, and the equation can be used to model semi-infinite domain ($x > 0$) problems. Therefore, if at

$$t = 0, \quad T = T_0$$

and at

$$x = 0, \quad T = T_1$$

$$L\left\{\frac{\partial T}{\partial t} = \alpha\frac{\partial^2 T}{\partial x^2}\right\}$$

gives

$$\alpha\frac{d^2 y(x,s)}{dx^2} = sy(x,s) - T_0, \quad \text{where } y(x,s) = L\{T(x,t)\}$$

subject to the boundednes of y(x, s) and

$$L\{T(0,t)\} = y(0,s) = \frac{T_1}{s}$$

Therefore,

$$y(x,s) = \frac{T_1 - T_0}{s} e^{-\sqrt{\frac{s}{\alpha}}x} + \frac{T_0}{s}$$

which inverts to

$$T(x,t) = (T_1 - T_0) \operatorname{erfc}\left(\frac{x}{2\sqrt{\alpha t}}\right) + T_0$$

## Example 7.1. 5

A solid rectangular slab at a uniform temperature $T_0$ has its four edges thermally insulated. The temperature of one exposed face is raised to and maintained at $T_1$, while the temperature of the other exposed face is held fixed at $T_0$. Determine how the temperature distribution varies with time.

*Solution*
Since the edges are insulated, heat is expected to flow in one direction, and the differential equation

$$\frac{\partial T}{\partial t} = \alpha \frac{\partial^2 T}{\partial x^2}$$

can be used to model the temperature distribution subject to the conditions

$$T(x,0) = T_0$$

$$T(0,t) = T_0$$

$$T(L,t) = T_1$$

Then using the method of Laplace transform we get

$$sU(x,s) - T(x,0) = \alpha \frac{d^2 U(x,s)}{dx^2}, \quad U(x,s) = L\{T(x,t)\}$$

or

$$sU(x,s) - T_0 = \alpha \frac{d^2 U(x,s)}{dx^2}$$

with general solution

$$U(x,s) = c_1(s)e^{-\sqrt{\frac{s}{\alpha}}x} + c_2(s)e^{-\sqrt{\frac{s}{\alpha}}x} + T_0\big/s \,.$$

The transformed boundary conditions become

$$L\{T(0,t)\} = U(0,s) = \frac{T_0}{s}$$

$$L\{T(L,t)\} = U(L,s) = \frac{T_1}{s}$$

Upon using the transformed boundary conditions, the constants of integration are determined such that

$$U(x,s) = \frac{T_1 - T_0}{s} \frac{e^{\sqrt{\frac{s}{\alpha}}x} - e^{-\sqrt{\frac{s}{\alpha}}x}}{e^{\sqrt{\frac{s}{\alpha}}L} - e^{-\sqrt{\frac{s}{\alpha}}L}} + T_0\big/s$$

$$= \frac{(T_1 - T_0)}{s} \frac{\sinh\sqrt{\frac{s}{\alpha}}x}{\sinh\sqrt{\frac{s}{\alpha}}L} + \frac{T_0}{s}$$

Two methods will be used to find $T(x, t) = L^{-1}\{U(x, s)\}$:

*Method I:*
In a suitable table of Laplace transform (Spiegel[3]), we can locate

$$L^{-1}\left\{\frac{\sinh x\sqrt{s}}{s\,\sinh a\sqrt{s}}\right\} = \frac{x}{a} + \frac{2}{\pi}\sum_{n=1}^{\infty}\frac{(-1)^n}{n}e^{-n^2\pi^2\alpha t/L^2}\sin\frac{n\pi x}{a}$$

then, if a is replaced by $\dfrac{L}{\sqrt{\alpha}}$ and $\dfrac{x}{\sqrt{\alpha}}$ replaces x,

$$L^{-1}\left\{\frac{\sinh x\sqrt{\frac{s}{\alpha}}}{s\,\sinh L\sqrt{\frac{s}{\alpha}}}\right\} = \frac{x}{L} + \frac{2}{\pi}\sum_{n=1}^{\infty}\frac{(-1)^n}{n}e^{-n^2\pi^2\alpha t/L^2}\sin\frac{n\pi x}{L}$$

Therefore,

$$T(x,t) = (T_1 - T_0)\left\{\frac{x}{L} + \frac{2}{\pi}\sum_{n=1}^{\infty}\frac{(-1)^n}{n}e^{-n^2\pi^2\alpha\,t/L^2}\sin\frac{n\pi x}{L}\right\} + T_0$$

*Method II:*

Consider the quantity

$$\frac{T_1 - T_0}{s}\frac{\sinh x\sqrt{\dfrac{s}{\alpha}}}{\sinh L\sqrt{\dfrac{s}{\alpha}}} = \frac{P(s)}{Q(s)}$$

Then

$$Q(s) = s\sinh L\sqrt{\frac{s}{\alpha}}$$

and Q(s) = 0 at either s = 0 or

$$\sqrt{\frac{s}{\alpha}}L = in\pi,\ n = 1,\ 2,\ \ldots$$

because sinh $in\pi$ = sin $n\pi$. The singular values

$$s = 0,\ s_n = \frac{-n^2\pi^2\alpha}{L^2}$$

are poles and in particular, simple poles as opposed to multiple poles. Further,

$$Q'(s) = \sinh L\sqrt{\frac{s}{\alpha}} + \frac{L}{2}\sqrt{\frac{s}{\alpha}}\cosh L\sqrt{\frac{s}{\alpha}}$$

such that

$$p_n(t) = \frac{P(s)}{Q'(s)}e^{s_n t} = \frac{(T_1 - T_0)\sinh\dfrac{in\pi x}{L}}{\sinh in\pi + \dfrac{in\pi}{2}\cosh in\pi}e^{-n^2\pi^2\alpha/L^2}$$

$$= \frac{(T_1 - T_0)\sin\dfrac{n\pi x}{L}e^{-n^2\pi^2\alpha t/L^2}}{\dfrac{n\pi}{2}\cos n\pi}$$

or

$$\rho_n(t) = (T_1 - T_0)\frac{2}{n\pi}(-1)^n e^{-n^2\pi^2\alpha t/L^2} \sin\frac{n\pi x}{L}$$

For the simple pole s = 0:

$$\rho_0(t) = (T_1 - T_0)\lim_{s \to 0}\frac{\sinh x\sqrt{\frac{s}{\alpha}}}{\sinh L\sqrt{\frac{s}{\alpha}}} = (T_1 - T_0)\frac{x}{L}$$

where the series expansion of the hyperbolic functions were used in the limiting process. Therefore,

$$f(t) = L^{-1}\left\{\frac{P(s)}{Q(s)}\right\} = (T_1 - T_0)\left\{\frac{x}{L} + \frac{2}{\pi}\sum_{n=1}^{\infty}\frac{(-1)^n}{n}e^{-n^2\pi^2\alpha t/L^2}\sin\frac{n\pi x}{L}\right\}$$

such that

$$T(x,t) = (T_1 - T_0)\left\{\frac{x}{L} + \frac{2}{\pi}\sum_{n=1}^{\infty}\frac{(-1)^n}{n}e^{-n^2\pi^2\alpha\, t/L^2}\sin\frac{n\pi x}{L}\right\} + T_0$$

Recall Table 6.1. One could certainly reduce the labor by employing that table where appropriate.

## 7.2　Mass Transfer

### Example 7.2.1

A slab of porous solid 1/2-in. thick is soaked in pure ethanol (Bennett and Myers[1]). The void space in the solid occupies 50% of its volume. The pores are fine, so that molecular diffusion can take place through the liquid in the passages: there is no convective mixing. The effective diffusivity of the system ethanol-water in the pores is one-tenth that in the free liquid.

If the slab is placed in a large well-agitated reservoir of pure water at 77°F, how long will it take for the mass fraction of ethanol at the center of the slab to fall to 0.009? Assume that there is no resistance to mass transfer in the water phase and that the concentration of ethanol in the water, and thus at the surface of the slab, is constant at zero.

*Solution*

Since the densities of alcohol and water differ by only 20%, then one may assume that total density remains constant such that

$$\frac{\partial C_A}{\partial t} = D_{AB} \frac{\partial^2 C_A}{\partial y^2}$$

is a reasonable mathematical description of the unsteady state process under discussion. If the following substitutions

$$\tau = \frac{D_{AB} t}{y_0^2}, \quad Y = \frac{C_A}{C_{tot}} = x_A, \quad \xi = \frac{y}{y_0}$$

are made, where the distance y is measured normal to the center of the slab, and $y_0$ is the half thickness of the slab, then we get

$$\frac{\partial Y}{\partial \tau} = \frac{\partial^2 Y}{\partial \xi^2}$$

subject to

$$\frac{\partial Y(0, \tau)}{\partial \xi} = 0$$

$$Y(1, \tau) = 0$$

$$Y(\xi, 0) = 1$$

This system of equations was solved in the context of heat transfer previously (example 7.1.1). There the solution was given as

$$Y(\xi, \tau) = \sum_{n=1}^{\infty} \frac{4}{(2n-1)\pi} \exp\left[-\left(\frac{2n-1}{2}\pi\right)^2 \tau\right] \cos\left(\frac{2n-1}{2}\pi\xi\right)$$

For $\xi = 0$ and $Y = 0.009$, and $D_{AB} = 1.0$ sq cm/sec it is determined that $\tau = 2.01$. Therefore,

$$t = \frac{\tau y_0}{D_{AB}} = \frac{(2.01)(1/48)^2}{(1/10)(1.0)(3.87)} = 0.0217 \text{ hr}, \quad \text{or} \quad 1.30 \text{ min}$$

Note that use is made of the solution from a previous problem that was solved in dimensionless form. Also, the analogy between heat conduction and molecular diffusion can best be exploited when dimensionless variables are utilized.

## Example 7.2.2

Diffusivities of Gases in Polymers (Felder, Spence and Ferrell[5]) is a model of diffusion through a membrane which separates two compartments of a continuous-flow permeation chamber. Essentially, at time t = 0, a penetrant is introduced into one compartment (the upstream compartment) and permeates through the membrane into a stream flowing through the other (downstream) compartment.

Assumptions:

1. Diffusion of the penetrant in the gas phase and absorption at the membrane surface are instantaneous processes.
2. Diffusion in the membrane is Fickian with a constant diffusivity D (cm²/sec).
3. The concentration of dissolved gas at the downstream surface of the membrane is always sufficiently low compared to that at the upstream surface, such that the downstream surface concentration may be set equal to zero.

Below are the models for a flat membrane of thickness h and a cylindrical membrane of inner radius a and outer radius b. The penetrant is introduced from the outside (cylindrical case). Determine the rate at which the gas permeates through the downstream membrane surface.

Flat Membrane

$$\frac{\partial C(t,x)}{\partial t} = D \frac{\partial^2 C(t,x)}{\partial x^2}$$

$$C(0,x) = 0$$

$$C(t,0) = C_1$$

$$C(t,h) = 0$$

<u>Cylinder</u>

$$\frac{\partial C(t,x)}{\partial t} = \frac{D}{r}\frac{\partial}{\partial r}\left(r\frac{\partial C}{\partial r}\right)$$

$$C(0,r) = 0$$

$$C(t,a) = 0$$

$$C(t,b) = C_1$$

*Solution*
For the flat membrane case:

$$\text{Let } L\{C(t,x)\} = y(s,x)$$

then the differential equation transforms to

$$\frac{d^2y}{dx^2} - \frac{s}{D}y = 0$$

whose general solution is

$$y(s,x) = k_1 \cosh\left(\sqrt{\frac{s}{D}}x\right) + k_2 \sinh\left(\sqrt{\frac{s}{D}}x\right)$$

subject to

$$L\{C(t,0)\} = \frac{C_1}{s}$$

and

$$L\{C(t,h)\} = 0$$

Therefore,

$$k_1 = \frac{C_1}{s}, \quad k_2 = \frac{-C_1\cosh\left(\sqrt{\frac{s}{D}}h\right)}{s\sinh\left(\sqrt{\frac{s}{D}}h\right)}$$

The solution in the s-domain is

$$y(s,x) = C_1 \left\{ \frac{\left[ \sinh\left(\sqrt{\frac{s}{D}}h\right)\cosh\left(\sqrt{\frac{s}{D}}h\right) - \cosh\left(\sqrt{\frac{s}{D}}h\right)\sinh\left(\sqrt{\frac{s}{D}}x\right) \right]}{s\sinh\left(\sqrt{\frac{s}{D}}h\right)} \right\}$$

and the inverse transform is derived using the residue theorem:

$$C(t,x) = C_1 L^{-1} \left\{ \frac{\sinh(h-x)\sqrt{s/D}}{s\sinh\left(\sqrt{\frac{s}{D}}h\right)} \right\} = C_1 \sum_{n=0}^{\infty} \frac{P(s_1,x)}{Q'(s_n)} \exp(s_n t)$$

For $s_0 = 0$, the residue $= \lim_{s \to 0} \dfrac{s\sinh(h-x)\sqrt{s/D}}{s\sinh\left(\sqrt{\frac{s}{D}}h\right)} = \dfrac{h-x}{h}$

using l'Hopital's rule. For $s_n \neq 0$

$$Q'(s_n) = \sinh\left(\sqrt{\frac{s}{D}}h\right) + s\left[ \frac{h}{2\sqrt{sD}}\cosh\left(\sqrt{\frac{s}{D}}h\right) \right]$$

Also, substituting

$$\sqrt{\frac{s}{D}} = i\lambda$$

simplifies the results and we get

$$\lambda = n\frac{\pi}{h}, \ n = 1,2,\ldots$$

from $\sinh\sqrt{\dfrac{s}{D}}h = 0$. Finally,

$$\frac{P(s_n,x)}{Q'(s_n)} = -\frac{2}{n\pi}\sin\left(\frac{n\pi x}{h}\right)$$

and the concentration profile is

$$C(t,x) = C_1 \left\{ \frac{h-x}{h} - \frac{2}{\pi} \sum_{n=1}^{\infty} \frac{1}{n} \exp\left( \frac{n^2 \pi^2 tD}{h^2} \right) \sin\left( \frac{n\pi x}{h} \right) \right\}$$

The rate of penetrant across the surface $x = h$ is given by

$$J(t) = -DA \left( \frac{\partial C}{\partial x} \right)_{x=h}$$

$$= \frac{DAC_1}{h} \left\{ 1 + 2 \sum_{n=1}^{\infty} (-1)^n \exp\left( -\frac{n^2 \pi^2 Dt}{h^2} \right) \right\}$$

for a flat membrane with a surface area A (cm$^2$). Also, notice that the steady-state rate $J_{ss}$, is given by

$$J_{ss} = D \frac{AC_1}{h}$$

Cylinder

The differential equation is Laplace transformed to

$$\frac{d^2 u}{dr^2} + \frac{1}{r} \frac{du}{dr} - \frac{s}{D} u = 0$$

subject to

$$u(s,a) = 0$$

$$u(s,b) = \frac{C_1}{s}$$

The general solution of the transformed differential equation is

$$u(s,r) = k_1 J_0\left( i\sqrt{s/D}\, r \right) + k_2 Y_0\left( i\sqrt{s/D}\, r \right)$$

where $J_0(.)$ are $Y_0(.)$ the zero-order Bessel functions of the first and second kind, respectively. Following the use of the transformed boundary conditions

$$k_1 = \frac{C_1 Y_0(\alpha b)}{s \left[ J_0(\alpha a) Y_0(\alpha b) - J_0(\alpha b) Y_0(\alpha a) \right]}$$

and

$$k_2 = -\frac{C_1 J_0(\alpha b)}{s\left[J_0(\alpha a)Y_0(\alpha b)-J_0(\alpha b)Y_0(\alpha a)\right]}$$

where the substitution

$$\alpha = i\sqrt{s/D}$$

has been used. Therefore,

$$u(s,r)=C_1 \frac{\left[Y_0(\alpha b)J_0(\alpha r)-J_0(\alpha b)Y_0(\alpha r)\right]}{s\left[J_0(\alpha a)Y_0(\alpha b)-J_0(\alpha b)Y_0(\alpha a)\right]}$$

The inverse transform gives

$$C(t,r)=\rho_0(t,r)+\sum_{n=1}^{\infty} \frac{P(s_n,r)}{Q'(s_n)}\exp(s_n t)$$

where

$$\rho_0(t,r)=\lim_{s\to 0} s\frac{P(s,r)}{Q(s)} \equiv \lim_{s\to 0} s\, u(s,r)=C_1\frac{\ln(b/r)}{\ln(b/a)}.$$

The results for small arguments of the Bessel functions were employed where appropriate (*where?*).

Finally, the concentration profile for the cylindrical geometry is

$$C(t,r)=C_1\left\{ \begin{array}{l} \dfrac{\ln(b/r)}{\ln(b/a)}+\pi\displaystyle\sum_{n=1}^{\infty} J_0(\alpha_n b)J_0(\alpha_n a) \\[2ex] \times\dfrac{\left[Y_0(\alpha_n b)J_0(\alpha_n r)-J_0(\alpha_n b)Y_0(\alpha_n r)\right]}{J_0^{\,2}(\alpha_n a)-J_0^{\,2}(\alpha_n b)} \end{array}\right\} \exp\left(-\alpha_n^2 Dt\right)$$

where

$$J_0(\alpha_n a)Y_0(\alpha_n b)-J_0(\alpha_n b)Y_0(\alpha_n a)=0$$

defines the eigenvalues, $\alpha_n \neq 0$, $n = 1, 2, \ldots$

The rate of penetrant across the surface at $r = a$, for a cylinder of length L is

$$J(t) = 2\pi D L \left(\frac{\partial C}{\partial r}\right)_{r=a}$$

$$= \frac{2\pi D L C_1}{\ln(b/a)}\left\{1 + 2\ln(b/a)\sum_{n=1}^{\infty}\frac{J_0(\alpha_n a)J_0(\alpha_n b)}{J_0^2(\alpha_n a) - J_0^2(\alpha_n b)}\exp(-\alpha_n^2 D t)\right\}$$

Again, the steady-state permeation rate is

$$J_{ss} = \frac{2\pi D L C_1}{\ln(b/a)}$$

for the cylinder.

## Example 7.2.3

This model describes the release of relatively small molecules such as benzoic acid (Ramraj, Farrell, and Loney[6]). Initially, the solute to be released is present in solution in a reservoir. A microporous membrane ($a \le x \le b$) without non-porous coating bounds the reservoir. The pores of the membrane are filled with a liquid that is immiscible with the reservoir phase ($0 \le x < a$).

In addition, we assumed the following:

1. Diffusion of the agent (benzoic acid) is Fickian.
2. Interfacial boundary layers are not influential here.
3. Diffusivities of the agent are independent of concentration.

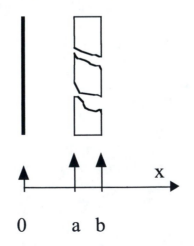

**FIGURE 7.4**
Cross-section of membrane device.

4. The aqueous-organic partition coefficient for the agent is independent of concentration.

5. Reservoir and bath solutions are ideal.

6. Uniform temperature exists throughout the system.

7. The aqueous and organic phases are immiscible.

8. Initially, the agent in the reservoir is at some concentration below its saturation value in the solvent.

Following previous works in this area (Jost[7] and Bird[8]), the governing equation for the agent concentration in the reservoir is

$$\frac{\partial C_1}{\partial t} = D_1 \frac{\partial^2 C_1}{\partial x^2} \tag{7.2.1}$$

and

$$\frac{\partial C_2}{\partial t} = D_2 \frac{\partial^2 C_2}{\partial x^2} \tag{7.2.2}$$

is the governing equation for the agent in the membrane. The subscripts refer to the respective regions [0,a] and [a,b] as shown in Figure 7.4. $D_2$ is an effective diffusivity and is defined as

$$D_2 = \frac{D\varepsilon}{\tau} \tag{7.2.3}$$

where D is the agent diffusivity in the pore liquid, $\varepsilon$ is the membrane porosity, and $\tau$ is the membrane tortuosity. Both the agent diffusivity in the pore liquid (D) and in the reservoir liquid ($D_1$) are calculated (Wilke-Chang correlation[9]). The quantity $\tau$ was defined and measured for various systems (Prasad and Sirkar[10]); here we adopt the value given for a hydrophobic membrane. The porosity value (0.38) is a manufacturer supplied quantity.

Equations 7.2.1 and 7.2.2 are subject to the following boundary conditions:

$$C_1(a,t) = m_{1,2} C_2(a,t) \tag{7.2.4}$$

$$\frac{\partial C_1(0,t)}{\partial x} = 0 \tag{7.2.5}$$

$$D_1 \frac{\partial C_1(a,t)}{\partial x} = D_2 \frac{\partial C_2(a,t)}{\partial x} \tag{7.2.6}$$

$$V_w \frac{\partial C_2(b,t)}{\partial t} = -D_2 \alpha_2 m_{2,w} \frac{\partial C_2(b,t)}{\partial x} \qquad (7.2.7)$$

Further, since the entire agent is initially present in the reservoir phase, then

$$C_1(x,0) = C_1^0 \qquad (7.2.8)$$

$$C_2(x,0) = 0 \qquad (7.2.9)$$

Equation 7.2.4 is a statement of the equilibrium partitioning at the reservoir/pore liquid interface with $m_{1,2}$ the partition coefficient. Equation 7.2.5 indicates that the solute concentration is expected to be finite at the bottom of the reservoir. Equation 7.2.6 displays the continuity of the agent flux across the reservoir/pore interface, while Equation 7.2.7 accounts for the material leaving the membrane and entering the surrounding water bath. The quantity $\alpha_2$ is the membrane area at the outer wall (cm²).

In deriving the solution to the model, we recast the model in a dimensionless form by introducing the following quantities:

$$u_1(\xi,\theta) = \frac{C_1(x,t)}{C_1^0} \qquad (7.2.10)$$

$$u_2(\xi,\theta) = \frac{C_2(x,t)}{C_1^0} \qquad (7.2.11)$$

where

$$\xi = \frac{x}{b}, \quad \theta = \frac{D_1 t}{b^2} \qquad (7.2.12)$$

The dimensionless model now consists of the following eight equations:

$$\frac{\partial u_1}{\partial \theta} = \frac{\partial^2 u_1}{\partial \xi^2} \qquad (7.2.13)$$

$$\frac{D_1}{D_2} \frac{\partial u_2}{\partial \theta} = \frac{\partial^2 u_2}{\partial \xi^2} \qquad (7.2.14)$$

$$u_1(a/b,\theta) = m_{1,2} u_2(a/b,\theta) \qquad (7.2.15)$$

$$\frac{\partial u_1(0,\theta)}{\partial \xi} = 0 \tag{7.2.16}$$

$$\frac{D_1}{D_2}\frac{\partial u_1(a/b,\theta)}{\partial \xi} = \frac{\partial u_2(a/b,\theta)}{\partial \xi} \tag{7.2.17}$$

$$\beta\frac{\partial u_2(1,\theta)}{\partial \theta} = -\frac{\partial u_2(1,\theta)}{\partial \xi} \tag{7.2.18}$$

$$u_1(\xi,0) = 1 \tag{7.2.19}$$

$$u_2(\xi,0) = 0 \tag{7.2.20}$$

where

$$\beta = \frac{V_w D_1}{bm_{2,w}D_2\alpha_2} \tag{7.2.21}$$

This type of coupled system of equations is very amenable to the technique of Laplace transform. As such, we let

$$\bar{u}_1(\xi,s) = \int_0^\infty u_1(\xi,\theta)e^{-s\theta}d\theta \quad \text{and} \quad \bar{u}_2(\xi,s) = \int_0^\infty u_2(\xi,\theta)e^{-s\theta}d\theta \tag{7.2.22}$$

such that Equations 7.2.13, 7.2.14, 7.2.19, and 7.2.20 transforms to the second order linear differential equations:

$$s\bar{u}_1 - 1 = \frac{d^2\bar{u}_1}{d\xi^2} \tag{7.2.23}$$

and

$$\frac{D_1}{D_2}s\bar{u}_2 = \frac{d^2\bar{u}_2}{d\xi^2} \tag{7.2.24}$$

subject to the transformed boundary conditions:

$$\bar{u}_1(a/b,s) = m_{1,2}\bar{u}_2(a/b,s) \tag{7.2.25}$$

$$\frac{d\bar{u}_1(0,s)}{d\xi} = 0 \tag{7.2.26}$$

$$\frac{D_1}{D_2}\frac{d\bar{u}_1(a/b,s)}{d\xi} = \frac{d\bar{u}_2(a/b,s)}{d\xi} \tag{7.2.27}$$

$$\beta s\bar{u}_2(1,s) = -\frac{d\bar{u}_2(1,s)}{d\xi} \tag{7.2.28}$$

The solution to the dimensionless model, Equations 7.2.23 to 7.2.28, are

$$\bar{u}_1 = \frac{1}{s} + \frac{\sqrt{\dfrac{D_2}{D_1}}\begin{bmatrix}\lambda\beta D_2/D_1\cos\lambda(1-a/b)\\ +\sin\lambda(1-a/b)\end{bmatrix}\cos\lambda\sqrt{D_2/D_1}\,\xi}{\mathrm{sq}(\lambda)} \tag{7.2.29}$$

and

$$\bar{u}_2 = \frac{\left[\beta\lambda D_2/D_1\sin\lambda(1-\xi)-\cos\lambda(1-\xi)\right]}{\mathrm{sq}(\lambda)}\sin\left(\lambda a/b\sqrt{D_2/D_1}\right) \tag{7.2.30}$$

where

$$q(\lambda) = m_{1,2}\left[\lambda\beta D_2/D_1\sin\lambda(1-a/b)-\cos\lambda(1-a/b)\right]\sin\left(\lambda\frac{a}{b}\sqrt{D_2/D_1}\right)$$

$$-\sqrt{D_2/D_1}\left[\lambda\beta D_2 \Big/ \begin{array}{c}D_1\cos\lambda(1-a/b)\\ +\sin\lambda(1-a/b)\end{array}\right]\cos\left(\lambda\frac{a}{b}\sqrt{D_2/D_1}\right) \tag{7.2.31}$$

The substitution

$$i\lambda = \sqrt{\frac{D_1 s}{D_2}} \tag{7.2.32}$$

where *i* is the imaginary unit, was used in Equations 7.2.29 to 7.2.31.
    Equations 7.2.29 and 7.2.30 are inverted by the residue theorem (Loney[11]) to give

$$u_1(\xi,\theta) = \cfrac{m_{1,2}\, a/b}{1+\beta\cfrac{D_2}{D_1}+\cfrac{a}{b}\left(m_{1,2}-1\right)}$$

$$+\sum_{n=1}^{\infty}\cfrac{\sqrt{D_2/D_1}\begin{bmatrix}\lambda_n\beta D_2/D_1\cos\lambda_n\left(1-a/b\right)\\[4pt]+\sin\lambda_n\left(1-a/b\right)\end{bmatrix}\cos\left(\lambda_n\,\dfrac{D_2}{D_1}\xi\right)e^{-\left(\lambda_n^2\frac{D_2}{D_1}\theta\right)}}{\left(\dfrac{\lambda}{2}\dfrac{dq}{d\lambda}\right)_{\lambda=\lambda_n}}$$

and

$$u_2(\xi,\theta) = \cfrac{a/b}{1+\beta\cfrac{D_2}{D_1}+\cfrac{a}{b}\left(m_{1,2}-1\right)} +$$

$$\sum_{n=1}^{\infty}\cfrac{\begin{bmatrix}\lambda_n\beta D_2/D_1\sin\lambda_n\left(1-\xi\right)\\[4pt]-\cos\lambda_n\left(1-\xi\right)\end{bmatrix}\sin\left(\lambda_n\,a/b\sqrt{D_2/D_1}\right)e^{-\left(\lambda_n^2\frac{D_2}{D_1}\theta\right)}}{\left(\dfrac{\lambda}{2}\dfrac{dq}{d\lambda}\right)_{\lambda=\lambda_n}} \quad (7.2.34)$$

where the eigenvalues, $\lambda_n$, are defined by

$$\tan\lambda_n\left(1-a/b\right) = \frac{m_{1,2}\tan\left(\lambda_n\,a/b\sqrt{D_2/D_1}\right)+D_2/D_1\sqrt{D_2/D_1}\beta\lambda_n}{\lambda_n m_{1,2}D_2/D_1\beta\tan\left(\lambda_n\,a/b\sqrt{D_2/D_1}\right)-\sqrt{D_2/D_1}} \quad (7.2.35)$$

Then using Equation 7.2.11, the concentration profile of the agent at $x = b$ can be determined.

## 7.3 Comparison Between Heat and Mass Transfer Results

Quite frequently heat transfer problems have their analogs in mass transfer. This can be fortuitous if recognized, since time and labor can be saved when solving a given problem. One approach that can be very helpful in exposing similarities between two problems requires the use of dimensionless quantities. That is, recast the given differential equation and its conditions into a dimensionless form. Then derive a dimensionless solution. At this juncture the solution is not tied to either heat transfer or mass transfer and can be interpreted as needed.

In two previously worked-out applications, Example 7.1.1 and Example 7.2.1, both situations were reduced to the identical dimensionless differential equations:

$$\frac{\partial Y}{\partial \tau} = \frac{\partial^2 Y}{\partial \xi^2}$$

subject to

$$\frac{\partial Y(0,\tau)}{\partial \xi} = 0$$

$$Y(1,\tau) = 0 \quad \text{and} \quad Y(\xi,0) = 1$$

This dimensionless model has the solution

$$Y(\xi,\tau) = \sum_{n=1}^{\infty} \frac{4}{(2n-1)\pi} \exp\left[-\left(\frac{2n-1}{2}\pi\right)^2 \tau\right] \cos\left(\frac{2n-1}{2}\pi\xi\right)$$

This result may now be interpreted in terms of a heat transfer application (Example 7.1.10) or a mass transfer application (Example 7.2.1).

As a second illustration, consider a solid body occupying the space from y = 0 to y = ∞ that is initially at a temperature $T_0$. At time t = 0, the surface at y = 0 is suddenly raised to a temperature $T_1$ and is maintained at that temperature for some t > 0. Determine the temperature profile T(y, t), if the mathematical statement of the problem is given by

$$\frac{\partial T}{\partial t} = \alpha \frac{\partial^2 T}{\partial y^2} ; y > 0$$

$$T(y,0) = T_0$$

$$T(0,t) = T_1$$

$$T(\infty,t) = T_0$$

Conveniently, the dimensionless temperature profile can be taken as

$$\theta(y,t) = \frac{T - T_0}{T_1 - T_0}$$

which recasts the differential equation into

$$\frac{\partial \theta}{\partial t} = \alpha \frac{\partial^2 \theta}{\partial y^2}$$

subject to

$$\theta = 0 \quad \text{at} \quad t \le 0 \quad \forall y$$

$$\theta = 1 \quad \text{at} \quad y = 0 \quad \forall t > 0$$

$$\theta = 0 \quad \text{at} \quad y = \infty \quad \forall t > 0$$

Then, if we let

$$\theta = f(\eta), \eta = \frac{y}{\sqrt{4\alpha t}}$$

the dimensionless differential equation can be reduced to an ordinary differential equation

$$f''(\eta) + 2\eta f'(\eta) = 0$$

This latter differential equation has the general solution

$$f(\eta) = k_1 \int_0^\eta e^{-\eta^2} \, d\eta + k_2$$

with arbitrary constants $k_1$ and $k_2$. By employing the conditions

$$f(0) = 1 \quad \text{and} \quad f(\infty) = 0$$

there results

$$\theta = \frac{T - T_0}{T_1 - T_0} = 1 - \frac{2}{\sqrt{\pi}} \int_0^\eta e^{-\xi^2} d\xi = 1 - \text{erf}\left(y/\sqrt{4\alpha t}\right)$$

By comparing this dimensionless temperature profile with the dimensionless concentration profile (Example 7.4.2) results in identical form and substance.

$$\frac{c_A - c_{A\infty}}{c_{A0} - c_{A\infty}} = 1 - \text{erf}\left(\frac{x}{\sqrt{4D_{AB} z/v_0}}\right)$$

Notice the argument of the error functions. In the heat transfer interpretation, $\alpha$ represents thermal diffusivity while $D_{AB}$ represents mass diffusivity in the mass transfer case. Also, the quantity $z/v_0$ represents time while x and y are coordinates in their respective systems. If one is able to anticipate analogies between conduction and diffusion in certain specific situations, then a given issue may be resolved in a fraction of the time it would otherwise take.

A procedure on how to convert dimensional differential equations to their dimensionless forms is given in Chapter 8.

## 7.4   Simultaneous Diffusion and Convection

By extending our definition of diffusion to include the process of heat transfer by conduction, the examples that follow demonstrate how some problems involving diffusion and flow can be treated effectively with the mathematical tools previously discussed. For example, consider the problem involving heat transfer to a flowing fluid. This problem was solved (Huang, Matlosz, Pan, and Snyder[12]) with the use of the Confluent Hypergeometric function (Slater[13]).

### Example 7.4.1

An example of heat transfer to a laminar flow fluid in a circular tube, this, the so-called Graetz problem, involves the determination of the temperature profile in a fully developed laminar flow fluid inside a circular tube.

The governing equation for the Graetz problem may be obtained from an energy balance in cylindrical coordinates. Alternatively, one can start with the equation of energy in terms of transport properties for Newtonian fluids of constant density $\rho$ and thermal conductivity k (Bird, Stewart, and Lightfoot[14]). In cylindrical coordinates, the steady-state equation of energy excluding the r and $\theta$ component of velocities and neglecting viscous dissipation is given by

$$\rho C_p v_z(r) \frac{\partial T}{\partial z} = k\left[\frac{1}{r}\frac{\partial}{\partial r}\left(r\frac{\partial T}{\partial r}\right) + \frac{1}{r^2}\frac{\partial^2 T}{\partial \theta^2} + \frac{\partial^2 T}{\partial z^2}\right] \tag{7.4.1}$$

Equation 7.4.1 can be further simplified if both the terms

$$\frac{\partial^2 T}{\partial \theta^2} \quad \text{and} \quad \frac{\partial^2 T}{\partial z^2}$$

are neglected on the basis that the amount of heat conducted in these directions are negligible in comparison to that conducted in the r-direction as well as that conducted by convection. The dimensionless form of the reduced Equation 7.4.1 is

$$\left(1-\xi^2\right)\frac{\partial u}{\partial \zeta} = \frac{1}{\xi}\frac{\partial}{\partial \xi}\left(\xi \frac{\partial u}{\partial \xi}\right) \tag{7.4.2}$$

subject to

$$u(0,\zeta) \text{ is finite} \tag{7.4.3}$$

$$u(1,\zeta) = 0 \tag{7.4.4}$$

$$u(\xi,0) = 1 \tag{7.4.5}$$

where the dimensionless quantities

$$u(\xi,\zeta) = \frac{T_w - T}{T_w - T_0}, \quad \xi = \frac{r}{r_1}, \quad \zeta = \frac{kz}{\rho C_p v_{zmax} r_1^2}$$

have been used. Here, the quantities $T_0$, $T_w$, $r_1$, and $v_{zmax}$ are entering fluid temperature, wall temperature, radius of tube, and maximum axial fluid velocity, respectively. Then, applying the method of separation of variables to Equation 7.4.2, by assuming that

$$u(\xi,\zeta) = f(\xi)g(\zeta) \tag{7.4.6}$$

such that Equation 7.4.2 reduces to

$$g' + \lambda^2 g = 0, \quad \Rightarrow g(\zeta) = c_1 e^{-\lambda^2 \zeta} \tag{7.4.7}$$

and

$$\xi \frac{d^2 f}{d\xi^2} + \frac{df}{d\xi} + \lambda^2 \xi \left(1-\xi^2\right) f = 0 \tag{7.4.8}$$

where $\lambda$ is real. In order to solve Equation 7.4.8, the Frobenius series method is appropriate. However, the following substitutions will be made in order to recast the differential equation into a more recognizable form:

(i) Let $w = \lambda \xi^2$ and (ii) $f(w) = e^{-w/2} y(w)$
Then Equation 7.4.8 becomes

$$w\frac{d^2y}{dw^2}+(1-w)\frac{dy}{dw}-\left(\frac{1}{2}-\frac{\lambda}{4}\right)y=0 \qquad (7.4.9)$$

following application of the chain rule. Equation 7.4.9 is a confluent hypergeometric equation for which there are tabulated solutions (Slater[13]). That is, Equation 7.4.9 has two linearly independent solutions:

$$y_1(w)={}_1F_1[a;b;w]=1+\frac{a}{b}w+\frac{a(a+1)w^2}{b(b+1)2!}+\frac{a(a+1)(a+2)w^3}{b(b+1)(b+2)3!}+\cdots$$

where, in this case

$$b=1,\quad a=\frac{1}{2}-\frac{\lambda}{4}$$

and

$$y_2(w)=y_1\log w+\frac{aw}{1!1!}\left(\frac{1}{a}-\frac{2}{1}\right)+\cdots$$

Therefore, the general solution to Equation 7.4.9 is

$$y(w)=c_2y_1(w)+c_3y_2(w)$$

The boundary condition given by Equation 7.4.3 requires that $f(0)$ be finite, which means that $y(w)$ must also be finite. However, $y_2(w)$ is unbounded at $w=0$, thus $c_3$ must be chosen as zero. For a nontrivial solution, $c_2\neq0$ such that

$$f(\xi)=c_2e^{\frac{-\lambda\xi^2}{2}}\,{}_1F_1\left[\frac{1}{2}-\frac{\lambda}{4};1;\lambda\xi^2\right] \qquad (7.4.10)$$

gives

$$f(1)=c_2e^{\frac{-\lambda}{2}}\,{}_1F_1\left[\frac{1}{2}-\frac{\lambda}{4};1;\lambda\right]=0$$

following application of the boundary condition given by Equation 7.4.4 resulting in

$${}_1F_1\left[\frac{1}{2}-\frac{\lambda_n}{4};1;\lambda_n\right]=0\quad\text{for}\quad n\geq1 \qquad (7.4.11)$$

Equation 7.4.11 defines the n eigenvalues while Equation 7.4.10 defines the corresponding eigenfunctions. Then, by the principle of superposition

$$u(\xi,\zeta) = \sum_{n=1}^{\infty} C_n e^{-\lambda_n^2 \zeta} \, e^{\frac{-\lambda_n \xi^2}{2}} \, {}_1F_1\left[\frac{1}{2} - \frac{\lambda_n}{4};1;\lambda_n\xi^2\right] \tag{7.4.12}$$

is the solution to Equations 7.4.2 to 7.4.4. The condition given by Equation 7.4.5 reduces Equation 7.4.12 to

$$u(\xi,0) = 1 = \sum_{n=1}^{\infty} C_n e^{\frac{-\lambda_n \xi^2}{2}} \, {}_1F_1\left[\frac{1}{2} - \frac{\lambda_n}{4};1;\lambda_n\xi^2\right]$$

which is a generalized Fourier series. Then, by making use of the orthogonal properties of a Sturm-Liouville problem, $C_n$ is given by

$$C_n = \frac{\left(\frac{1}{2} - \frac{1}{\lambda_n}\right)e^{-\lambda_n/2} \, {}_1F_1\left[\frac{3}{2} - \frac{\lambda_n}{4};2;\lambda_n\right]}{\int_0^1 \xi\left(1-\xi^2\right)e^{-\lambda_n\xi^2}\left({}_1F_1\left[\frac{1}{2} - \frac{\lambda_n}{4};1;\lambda_n\xi^2\right]\right)^2 d\xi} \tag{7.4.13}$$

From the temperature profile given by Equation 7.4.12, the heat flux at the wall, the total rate of heat transfer, and the bulk temperature of the fluid at the exit can be evaluated. Also, using the arithmetic mean of terminal temperature differences together with the definition of heat transfer coefficient, the arithmetic mean Nusselt number can be expressed as a function of the Graetz number. Similarly, the logarithmic Nusselt number can also be obtained as a function of the Graetz number (Huang, Matlosz, Pan, and Snyder[12]).

## Example 7.4.2

Diffusion into a Falling Film (Bird[8]). Consider the diffusion of a solute A into a moving liquid film B. The liquid is in laminar flow. Assuming that (1) the film moves with a flat velocity profile $v_0$, (2) the film may be taken to be infinitely thick with respect to the penetration of the absorbed material, and (3) the concentration at the interface $x = 0$ is $c_{A0}$, then the mathematical statement of this problem is

$$v_0 \frac{\partial c_A}{\partial z} = D_{AB} \frac{\partial^2 c_A}{\partial x^2}$$

subject to

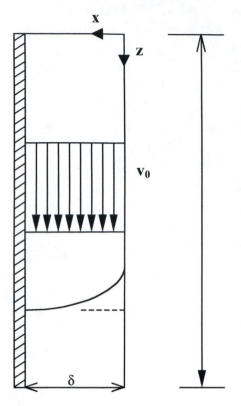

**FIGURE 7.5**
Diffusion into a falling film.

$$c_A = c_{A0} \quad \text{at} \quad x = 0$$

$$c_A = c_{A\infty} \quad \text{at} \quad x = \infty$$

$$c_A = c_{A\infty} \quad \text{at} \quad z = 0$$

(a) Derive the concentration profile for the solute A.

(b) Derive a liquid-film mass transfer coefficient based on the driving force $c_{A0} - c_{A\infty}$.

*Solution*
Let

$$\frac{c_A - c_{A\infty}}{c_{A_0} - c_{A\infty}} = f(\eta), \quad \eta = \frac{x}{\sqrt{4D_{AB}\, z/v_0}}$$

Then

$$\frac{\partial c_A}{\partial z} = \left(c_{A_0} - c_{A_\infty}\right) f'(\eta) \frac{d\eta}{dz}$$

$$\frac{\partial c_A}{\partial x} = \left(c_{A_0} - c_{A_\infty}\right) f'(\eta) \frac{d\eta}{dx}$$

$$\frac{\partial^2 c_A}{\partial x^2} = \left(c_{A_0} - c_{A_\infty}\right) f''(\eta) \left(\frac{d\eta}{dx}\right)^2 + f'(\eta) \frac{d\eta}{dx}$$

but

$$\frac{d\eta}{dz} = -\frac{2D_{AB}}{v_0} \frac{x}{\sqrt{4D_{AB}\, z/v_0}} \frac{1}{4D_{AB}\, z/v_0}\; ; \frac{d\eta}{dx} = \frac{1}{\sqrt{4D_{AB}\, z/v_0}}$$

Therefore, the differential equation reduces to

$$\frac{f''}{f'} = -2\eta$$

and the boundary conditions become

$$f(\eta) = 1 \quad \text{at} \quad \eta = 0; \quad f(\eta) = 0 \quad \text{at} \quad \eta = \infty$$

Integration of this new ordinary differential equation results in

$$f(\eta) = k_1 \int_0^\eta e^{-\xi^2}\, d\xi + k_2$$

where $k_1$ and $k_2$ are arbitrary constants. Imposing the boundary conditions results in

$$f(\eta) = -\frac{2}{\sqrt{\pi}} \int_0^\eta e^{-\xi^2}\, d\xi + 1 = 1 - \text{erf}\left(\frac{x}{\sqrt{4D_{AB}\, z/v_0}}\right).$$

That is

$$\frac{c_A - c_{A\infty}}{c_{A0} - c_{A\infty}} = 1 - \text{erf}\left(\frac{x}{\sqrt{4D_{AB}\, z/v_0}}\right)$$

(c) For short contact times for absorption in wetted-wall towers, the following expression for the liquid-phase mass transfer coefficient was developed in the literature (Higbie[15]). The total moles of A transferred per unit time per unit cross-sectional transfer area is

$$N_A = \frac{v_0}{L} \int_0^\infty \left( c_A - c_{A\infty} \right)_{z=L} dx = \sqrt{\frac{4D_{AB}v_0}{\pi L}} \left( c_{A0} - c_{A\infty} \right) = k_L \left( c_{A0} - c_{A\infty} \right)$$

such that

$$k_L = \sqrt{\frac{4D_{AB}v_0}{\pi L}}$$

Note that the quantity $L/v_0$ is the time required for the liquid film to traverse the length L.

## 7.5  Simultaneous Diffusion and Chemical Reaction

There are numerous practical situations in which both diffusion and chemical reaction may be occurring, for instance, during low pressure chemical vapor deposition, solid catalytic processes, or decoking a Pyrolysis furnace, to name a few.

### Example 7.5.1

This example involves the derivation of the differential equation and boundary conditions for a process step that is integral to Microelectronics processing.

Consider the process of low pressure chemical vapor deposition (LPCVD) in a horizontal cylindrical reactor heated from the outside (Loney and Huang[16]). That is, the reactor is surrounded by a furnace that supplies heat to the reacting contents. This heating strategy is the so-called hot wall process, and it is usually assumed that heats of reaction are small in comparison to the supplied heat. In this assumed isothermal process, thin circular disks called wafers are supported in a special wafer holder consisting of prearranged slots of equal distance apart so that the disks are located axisymmetric with the cylindrical tube. Material flows from one end of the reactor in the annulus created by the wafers and the reactor wall, but diffusion is anticipated to be the dominant mass transfer mechanism in the region bounded by any two wafers. Derive the mathematical model using the following assumptions:

1. Diffusion is the mass transfer mechanism in the region between any two wafers.
2. The gap between two wafers (inter-wafer region) is $2\delta$ long.
3. Surface reaction dominates over homogeneous reaction.
4. Azimuthal flow effects can be neglected.
5. A pseudo-binary mixture prevails.
6. Homogeneous gas phase reactions are negligible.

Also solve the derived system of equations for the concentration profile as a function of r (radial) and z (along the axis).

*Solution*
Conservation of mass applied to the reactant species A in the inter-wafer region:

$$\frac{1}{r}\frac{\partial}{\partial r}(rN_{Ar}) + \frac{\partial}{\partial z}N_{Az} = 0 \tag{7.5.1}$$

where $N_A$ is the molar flux relative to stationary coordinates (Bird, Stewart, and Lightfoot[14]) and is defined by

$$\overline{N}_A = -CD_{AB}\overline{\nabla}X_A + X_A\left(\overline{N}_A + \overline{N}_B\right) \tag{7.5.2}$$

Then the molar flux in the r-direction for a dilute system is

$$N_{Ar} = -D_{AB}\frac{\partial C_A}{\partial r} + X_A\left(N_{Ar} + N_{Br}\right) \tag{7.5.3}$$

which consists of a diffusion part and a bulk flow part. For a diffusion dominated mechanism, Equation 7.5.3 becomes

$$N_{Ar} = -D_{AB}\frac{\partial C_A}{\partial r} \tag{7.5.4}$$

and the z-directed flux becomes

$$N_{Az} = -D_{AB}\frac{\partial C_A}{\partial z} \tag{7.5.5}$$

Therefore, substitution of Equations 7.5.4 and 7.5.5 into Equation 7.5.1 gives

$$\frac{1}{r}\frac{\partial}{\partial r}\left(r\frac{\partial C_A}{\partial r}\right) + \frac{\partial^2 C_A}{\partial z^2} = 0 \tag{7.5.6}$$

subject to

$$\frac{\partial C_A(0,z)}{\partial r} = 0 \qquad (7.5.7)$$

that is, no concentration gradient exists along the axis of the cylinder. By symmetry,

$$\frac{\partial C_A(r,0)}{\partial x} = 0 \qquad (7.5.8)$$

Also

$$C_A(R_W,z) = C_{Ab} \qquad 0 < z < \delta \qquad (7.5.9)$$

where $R_W$ is the radius of the wafer and $C_{Ab}$ is the bulk concentration of species A which varies along the length of the reactor. In addition, only half the inter-wafer region need be considered.

The surface reaction can be taken as a first order reaction; that is

$$-D_{AB}\frac{\partial C_A}{\partial z} = kC_A \qquad \text{at } z = \delta \qquad (7.5.10)$$

Therefore, the mathematical model for the outlined process is

$$\frac{1}{r}\frac{\partial}{\partial r}\left(r\frac{\partial C_A}{\partial r}\right) + \frac{\partial^2 C_A}{\partial z^2} = 0$$

$$\frac{\partial C_A(0,z)}{\partial r} = 0$$

$$\frac{\partial C_A(r,0)}{\partial z} = 0$$

$$C_A(R_W,z) = C_{Ab} \qquad 0 < z < \delta$$

$$-D_{AB}\frac{\partial C_A}{\partial z} = kC_A \qquad \text{at } z = \delta$$

To solve this system of equations, the method of separation of variables will be employed. *Note that this is a homogeneous problem that requires special handling.* At this point, we need to broaden the definition of homogeneous to include *a linear condition or equation that when satisfied by a particular function F is also satisfied by cF, where c is an arbitrary constant.* This means that Equation 7.5.10 is homogeneous, but cannot be treated in the usual way. In this problem, Equation 7.5.9 plays the role of the initial condition.

Let $C_A(r, z) = R(r) Z(z)$, then substitute into the differential equation to get

$$R'' + \frac{1}{r}R' - \lambda R = 0 \tag{I}$$

and

$$Z'' + \lambda R = 0 \tag{II}$$

where $\lambda$ is the separation constant (three cases to be considered). The homogeneous boundary conditions become

$$R'(0) = 0$$

$$Z'(0) = 0$$

The other two nonhomogeneous boundary conditions will be dealt with later.

Case 1: $\lambda = 0$
   Equation (I) solves to

$$R(r) = \frac{c_2}{r} + c_3$$

and $R'(0) \rightarrow \infty$ unless $c_2$ is choosen as zero.
   Further, Equation (II) becomes

$$Z'' = (0) \Rightarrow Z(z) = m_1 z + m_2$$

and

$$Z'(0) = 0 = m_1$$

Therefore, the implication is

$$C_A(r, z) = \text{a constant.}$$

However, Equation 7.5.10 would be violated; therefore, on physical grounds, the case $\lambda$ is zero must be rejected.
   Suppose $\lambda > 0$, say $\lambda = \alpha^2$, then we get from Equation (I)

$$R'' + \frac{1}{r}R' - \alpha^2 R = 0$$

a modified Bessel's equation (Spiegel[3] and Watson[17]). This equation can be solved by comparing with Equation 3.6.4 to give

$$R(r) = c_4 I_0(\alpha r) + c_5 K_0(\alpha r)$$

where $I_0(\alpha r)$ is the zero order modified Bessel function of the first kind and $K_0(\alpha r)$ is the zero order modified Bessel function of the second kind, which contains a logarithmic term.

The constant $c_5$ must be chosen as zero, since $K_0(.)$ and its derivative become unbounded as r approaches zero. Therefore,

$$R(r) = c_4 I_0(\alpha r)$$

Also, for this $\lambda$, Equation (II) gives $Z(z) = m_3 \cos \alpha z + m_4 \sin \alpha z$ and

$$Z'(0) = 0 = \alpha m_4 \Rightarrow m_4 = 0$$

such that

$$Z(z) = m_3 \cos \alpha z$$

Then for each n, we expect that

$$C_{A,n}(r,z) = R_n Z_n = A_n I_0(\alpha_n r) \cos \alpha_n z$$

and by the principle of superposition

$$C_A(r,z) = \sum_{n=1}^{\infty} A_n I_0(\alpha_n r) \cos \alpha_n z$$

which must now satisfy

$$-D_{AB} \frac{\partial C_A}{\partial z} = kC_A \quad \text{at } z = \delta$$

from which we get

$$D_{AB} \sum_{n=1}^{\infty} A_n \alpha_n I_0 (r,z) \sin \alpha_n \delta = k \sum_{n=1}^{\infty} A_n I_0 (\alpha_n r) \cos \alpha_n \delta$$

Then, for $n \geq 1$, the $\alpha_n$ are defined by

$$\alpha_n \tan \alpha_n \delta = \frac{k}{D_{AB}}$$

Now, Equation 7.5.9 will be useful in defining the quantity $A_n$, that is

$$C_A (R_w, z) = C_{Ab} = \sum_{n=1}^{\infty} A_n I_0 (\alpha_n R_w) \cos \alpha_n z$$

which is recognizable as a generalized Fourier series. Therefore, for $n \geq 1$,

$$\int_0^{\delta} C_{Ab} \cos \alpha_n z \, dz = A_n I_0 (\alpha_n R_w) \int_0^{\delta} \cos^2 \alpha_n z \, dz$$

$$= A_n I_0 (\alpha_n R_w) \int_0^{\delta} \left[ \frac{1}{2} + \frac{1}{2} \cos 2\alpha_n z \right] dz$$

$$= A_n I_0 (\alpha_n R_w) \left[ \frac{\delta}{2} + \frac{1}{2} \left( \frac{1}{2\alpha_n} \right) \sin 2\alpha_n \delta \right]$$

or

$$A_n = \left( \frac{1}{I_0 (\alpha_n R_w) \left[ \dfrac{\delta}{2} + \dfrac{\sin 2\alpha_n \delta}{4\alpha_n} \right]} \right) \left( \int_0^{\delta} C_{Ab} \cos^2 \alpha_n z \, dz \right)$$

For the case $\lambda < 0$, say $\lambda = -\beta^2$, $\beta > 0$, Equation (I) becomes

$$R'' + \frac{1}{2} R' + \beta^2 R = 0$$

a Bessel differential equation (Spiegel[3] and Watson[17]), which can be solved by the same process of comparison used above. That is, compare this new R-equation with

$$y'' - \left(\frac{2a-1}{x}\right)y' + \left(b^2c^2x^{2c-2} + \frac{a^2 - \gamma^2c^2}{x^2}\right)y = 0$$

whose general solution is

$$y = x^a J_\gamma\left(bx^c\right) + x^a J_{-\gamma}\left(bx^c\right)$$

Then, by comparison to the R-equation above

$$2a - 1 = -1 \;\Rightarrow a = 0$$

$$2c - 2 = 0 \;\Rightarrow c = 1$$

$$a^2 - \gamma^2 c^2 = 0 \;\Rightarrow \gamma = 0$$

$$b^2 c^2 = \beta^2 \;\Rightarrow b = \beta$$

such that the R-equation has general solution

$$R(r) = c_6 J_0\left(\beta r\right) + C_7 Y_0\left(\beta r\right)$$

where $J_0(\beta r)$ is the zero order Bessel function of the first kind and $Y_0(\beta r)$ is the zero order Bessel function of the second kind, which contains a logarithmic term. As a result of the logarithmic term, $Y_0(\beta r)$ and its derivative become unbounded as their argument approaches zero. Therefore, the constant $c_7$ must be chosen as zero. Such that

$$R(r) = c_6 J_0\left(\beta r\right)$$

Also, Equation (II) gives the general solution

$$Z(z) = m_5 e^{\beta z} + m_6 e^{-\beta z}$$

and

$$Z'(0) = 0 = \beta m_5 - \beta m_6 \quad \Rightarrow m_5 = m_6$$

such that

$$Z(z) = m_5\left(e^{\beta z} + e^{-\beta z}\right) = 2m_5 \cosh \beta z$$

Then, for each n, all we can conclude at this point is

$$C_{A,n} = H_n J_0(\beta_n r) \cosh \beta_n z$$

where the $\beta_n$ are to be defined. Then, by the principle of superposition,

$$C_A(r,z) = \sum_{n=1}^{\infty} H_n J_0(\beta_n r) \cosh \beta_n z$$

and

$$\frac{\partial C_A(r,\delta)}{\partial z} = \sum_{n=1}^{\infty} H_n \beta_n J_0(\beta_n r) \cosh \beta_n \delta$$

Therefore, Equation 7.5.10 becomes

$$-D_{AB} \sum_{n=1}^{\infty} H_n \beta_n J_0(\beta_n r) \cosh \beta_n \delta = k \sum_{n=1}^{\infty} H_n J_0(\beta_n r) \cosh \beta_n \delta$$

or

$$\beta_n \tanh \beta_n \delta = -\frac{k}{D_{AB}}, \quad n \geq 1$$

which defines $\beta_n$. However, the argument of the hyperbolic tangent is positive which means that the left-hand side is always positive, while the right-hand side is negative. This is impossible and therefore $\lambda < 0$ has to be rejected.

Finally, the solution to this model is given as

$$C_A(r,z) = \sum_{n=1}^{\infty} A_n I_0(\alpha_n r) \cos \alpha_n z$$

where the Fourier coefficient is given by

$$A_n = \left( \frac{1}{I_0(\alpha_n R_w)\left[\dfrac{\delta}{2} + \dfrac{\sin 2\alpha_n \delta}{4\alpha_n}\right]} \right) \left( \int_0^\delta C_{Ab} \cos \alpha_n z \, dz \right)$$

and the eigenvalues are defined by

$$\alpha_n \tan \alpha_n \delta = \frac{k}{D_{AB}}$$

for $n \geq 1$.

## Example 7.5.2

The following system of equations have been used in the modeling of biodegradation processes:

$$\frac{1}{D_{se}} \frac{\partial C_A}{\partial t} = \frac{\partial^2 C_A}{\partial r^2} + \frac{2}{r} \frac{\partial C_A}{\partial r} - \left(\frac{1-\varepsilon_s}{\varepsilon_s}\right) \frac{K}{D_{se}} C_A \tag{7.5.11}$$

$$C_A(0,t) \text{ is finite} \tag{7.5.12}$$

$$\frac{\partial C_A}{\partial r} = \frac{k}{D_{se}}(C_A - C_b), \quad \text{at} \quad r = R \tag{7.5.13}$$

$$C_A(r,0) = C_{A0} \tag{7.5.14}$$

Determine the concentration profile for species A.

*Solution*
In order to simplify the problem let

$$C_A(r,t) = u(r,t) + F(r) \tag{7.5.15}$$

where $u(r, t)$ represents the transient solution and $F(r)$ represents the steady-state solution. Further, we require that

$$F'' + \frac{2}{r} F' - \beta F = 0 \tag{7.5.16}$$

$$F \text{ is to be finite at } r = 0 \tag{7.5.17}$$

and

$$\frac{dF(R)}{dr} = \frac{k_m}{D_{se}}(F(R) - C_b) \tag{7.5.18}$$

Then Equations 7.5.11, 7.5.12, and 7.5.13 reduce to

$$\frac{1}{D_{se}}\frac{\partial u}{\partial t} = \frac{\partial^2 u}{\partial r^2} + \frac{2}{r}\frac{\partial u}{\partial r} - \beta u \tag{7.5.19}$$

subject to

$$u(0,t) \text{ is finite} \tag{7.5.20}$$

$$\frac{\partial u}{\partial r} = \frac{k_m}{D_{se}} u \text{ at } r = R \tag{7.5.21}$$

where

$$\beta = \frac{1-\varepsilon_s}{\varepsilon_s}\frac{K}{D_{se}}$$

As a further simplification to Equations 7.5.19 to 7.5.21, let

$$u(r,t) = e^{-D_{se}\beta t}w(r,t) \tag{7.5.22}$$

such that Equations 7.5.19 to 7.5.22 become

$$\frac{1}{D_{se}}\frac{\partial w}{\partial t} = \frac{\partial^2 w}{\partial r^2} + \frac{2}{r}\frac{\partial w}{\partial r} \tag{7.5.23}$$

subject to

$$W(0,t) \text{ is finite} \tag{7.5.24}$$

$$\frac{\partial w}{\partial r} = \frac{k_m}{D_{se}} w \text{ at } r = R \tag{7.5.25}$$

Then, by separation of variables, that is for w = f(r)T(t), we get

$$\frac{1}{D_{se}}\frac{T'}{T} = \frac{f''}{f} + \frac{2}{r}\frac{f'}{f} = \lambda$$

For the case $\lambda = 0$, we get f(r) = a constant, which implies that zero is an eigenvalue for this problem. This solution must be rejected, however, since

Equation 7.5.25 could only be satisfied for the particular case of w being identically zero.

The case $\lambda > 0$, say $\lambda = \alpha^2$ produces

$$T(t) = c_1 e^{\lambda D_{se} t}$$

and

$$f'' + \frac{2}{r} f' - \alpha^2 f = 0$$

which has

$$f(r) c_2 r^{-\frac{1}{2}} J_{\frac{1}{2}}(i\alpha r) + c_2 r^{-\frac{1}{2}} J_{-\frac{1}{2}}(i\alpha r)$$

as its general solution. However, w = f(r)T(t) is required to remain finite, even as t gets larger and larger. This condition cannot be met for $\lambda > 0$.

For $\lambda < 0$, say $\lambda = -\eta^2$ then we have

$$f'' + \frac{2}{r} f' + \eta^2 f = 0$$

whose general solution is

$$f(r) = c_4 \frac{J_{\frac{1}{2}}(\eta r)}{\sqrt{r}} + c_5 \frac{J_{-\frac{1}{2}}(\eta r)}{\sqrt{r}}$$

Again, for a bounded solution, $c_5$ must be chosen as zero, since

$$\frac{J_{-1/2}(\eta r)}{\sqrt{r}}$$

becomes unbounded as r approaches zero. However, for a nontrivial solution to exist, $c_4 \neq 0$.

Recall now that

$$J_{\frac{1}{2}}(\eta r) = \sqrt{\frac{2}{n\pi r}} \sin \eta r$$

such that

$$\frac{J_{\frac{1}{2}}(\eta r)}{\sqrt{r}} = \sqrt{\frac{2}{n\pi}} \frac{\sin \eta r}{r}$$

Therefore, the solution up to now is

$$w(r,t) = \sum_{i=1}^{\infty} A_i e^{-\eta^2 D_{se} t} \sqrt{\frac{2}{\eta\pi}} \frac{\sin \eta r}{r} \qquad (7.5.26)$$

with the use of the superposition principle. That is, Equation 7.5.26 solves Equations 7.5.23 and 7.5.24. Equation 7.5.25 will define the eigenvalues for this system in the following way:

$$\frac{\partial w}{\partial r} = \sum_{n=1}^{\infty} A_i e^{-\eta^2 D_{se} t} \sqrt{\frac{2}{\eta\pi}} \left( \frac{\eta \cos \eta r}{r} - \frac{\sin \eta r}{r^2} \right)$$

then, at r = R

$$\frac{\eta \cos \eta R}{R} - \frac{\sin \eta R}{R^2} = \frac{k_m}{D_{se}} \frac{\sin \eta R}{R}, \quad i \geq 1$$

or

$$\eta_i R = \left( \frac{k_m R}{D_{se}} + 1 \right) \tan \eta_i R \quad i \geq 1 \qquad (7.5.27)$$

The initial condition is

$$u(r,0) = w(r,0) = C_{A0} - F(r) \qquad (7.5.28)$$

where F(r) is the solution to Equations 7.5.16 to 7.5.18. That is,

let $F(r) = \dfrac{q(r)}{r}$, then $F'(r) = \dfrac{q'(r)}{r} \dfrac{-q(r)}{r^2}$, and $F''(r) = \dfrac{q''(r)}{r} - \dfrac{2}{r} q' + \dfrac{2q(r)}{r^3}$

Then these substitutions reduce Equation 7.5.16 to

$$q''(r) - \beta q(r) = 0$$

whose general solution is

$$q(r) = k_1 \cosh \sqrt{\beta} r + k_2 \sinh \sqrt{\beta} r$$

Therefore,

$$F(r) = \frac{k_1 \cosh \sqrt{\beta} r}{r} + \frac{k_2 \sinh \sqrt{\beta} r}{r}$$

However, $k_1$ must be chosen as zero, since $F(0)$ is required to be finite. Also, Equation 7.5.18 gives

$$k_2 \frac{R^2 C_b k_m}{D_{se}\left\{\left(1 + \dfrac{R k_m}{D_{se}}\right) \sinh \sqrt{\beta} R - R\sqrt{\beta} \cosh \sqrt{\beta} R\right\}}$$

such that

$$F(r) = \frac{R^2 C_b k_m \sinh \sqrt{\beta} r}{D_{se}\left\{\left(1 + \dfrac{R k_m}{D_{se}}\right) \sinh \sqrt{\beta} R - R\sqrt{\beta} \cosh \sqrt{\beta} R\right\} r} \tag{7.5.29}$$

Finally,

$$u(r,t) = e^{-D_{se}\beta t} w(r,t) = \sum_{i=1}^{\infty} A_i e^{-\left(D_{se}\beta + \eta_i^2 D_{se}\right)t} \sqrt{\frac{2}{\eta_i \pi}} \frac{\sin \eta_i \pi}{r}$$

or

$$C_A(r,t) = F(r) + \sum_{i=1}^{\infty} A_i e^{-\left(D_{se}\beta + \eta_i^2 D_{se}\right)t} \sqrt{\frac{2}{\eta_i \pi}} \frac{\sin \eta_i \pi}{r}$$

then at $t = 0$,

$$C_A(r,0) = F(r) + \sum_{i=1}^{\infty} A_i \sqrt{\frac{2}{\eta_i \pi}} \frac{\sin \eta_i \pi}{r} = C_{A0}$$

or

$$C_{A0} - F(r) = \sum_{i=1}^{\infty} A_i \sqrt{\frac{2}{\eta_i \pi}} \frac{\sin \eta_i \pi}{r}$$

defines $A_i$. That is, for $i \geq 1$

$$\int_0^R \left(C_{A0} + F(r)\right) r^2 \frac{\sin \eta_i r}{r} \, dr = A_i \sqrt{\frac{2}{\eta_i \pi}} \int_0^R r^2 \left[\frac{\sin \eta_i r}{r}\right]^2 dr$$

where $r^2$ is the weighting function.

---

## 7.6  Simultaneous Diffusion, Convection, and Chemical Reaction

This type of phenomena is typified by flowing systems, in which both surface and homogeneous chemical reactions are occurring. However, sometimes the mathematics associated with these phenomena may be too nontractable and a simpler model may be sought. Simplifying assumptions that are used to reduce a model would still require some measure of justification, at least in theory. In the very next example an approach is taken that can be used to determine the effect of diffusion and surface reaction on a chemical vapor deposition process.

### Example 7.6.1

In multiwafer low pressure chemical vapor deposition (LPCVD) systems, the convection contribution to the transport of reactants to surface sites in the Interwafer region is usually neglected (Loney[18]). Consider laminar flow with a velocity profile in the annulus, between the wafers and the reactor walls. This velocity profile is assumed constant and contains the ratio of wafer radius to that of the reactor denoted by $\kappa$. Then in this case, an average velocity, $v_{avg}$ can be described by (Bird, Stewart, and Lightfoot[14])

$$v_{avg} = \frac{(P_0 - P_L) R^2}{8 \mu L} \left[\frac{1 - \kappa^4}{1 - \kappa^2} - \frac{1 - \kappa^2}{\ln\left(\frac{1}{\kappa}\right)}\right] \tag{7.6.1}$$

R is the reactor radius. A mathematical description of the process occurring in this annulus region is given as

$$v_{avg} \frac{\partial C_A}{\partial z} = D_{AB} \frac{1}{r} \frac{\partial}{\partial r}\left(r \frac{\partial C_A}{\partial r}\right) - k_b C_A \tag{7.6.2}$$

subject to

$$C_A = C_{A0} \quad \text{at} \quad z = 0 \tag{7.6.3}$$

$$C_A \text{ is finite at } r = 0 \tag{7.6.4}$$

and

$$-D_{AB} \frac{\partial C_A}{\partial r} = k_s C_A \quad \text{at} \quad r = R \tag{7.6.5}$$

The differential equation is the equation of continuity of species A for constant density $\rho$ and binary diffusivity $D_{AB}$ (Bird, Stewart, and Lightfoot[3]). Also, the r and $\theta$ components of velocity are neglected. Develop an equation that can be used to justify the assumption of transport by diffusion in the interwafer region of a multiwafer LPCVD reactor.

*Solution*
Equations 7.6.2 to 7.6.5 can be recasted in dimensionless form as

$$\frac{\partial F}{\partial \zeta} = \frac{\partial^2 F}{\partial \xi^2} + \frac{1}{\xi} \frac{\partial F}{\partial \xi} - \alpha F \tag{7.6.2A}$$

subject to

$$F = 1 \quad \text{at} \quad \zeta = 0 \tag{7.6.3A}$$

$$F \text{ is finite at } \xi = 0 \tag{7.6.4A}$$

$$-\frac{\partial F}{\partial \xi} = \beta F \tag{7.6.5A}$$

where the dimensionless quantities

$$F(\xi, \zeta) = \frac{C_A}{C_{A0}}; \quad \xi = \frac{r}{R}; \quad \zeta = \frac{D_{AB} z}{v_{avg} R^2}; \quad \alpha = \frac{R^2 k_b}{D_{AB}}; \quad \beta = \frac{k_s R}{D_{AB}}$$

were used. Equation 7.6.2A is nonhomogeneous and can be made homogeneous by the substitution

$$w(\xi, \zeta) = F e^{\alpha \zeta} \tag{7.6.6}$$

Therefore, the system

$$\frac{\partial w}{\partial \zeta} = \frac{\partial^2 w}{\partial \xi^2} + \frac{1}{\xi}\frac{\partial w}{\partial \xi}$$

$$w = 1 \quad \text{at} \quad \zeta = 0$$

$$w \text{ is finite at } \xi = 0$$

$$-\frac{\partial w}{\partial \xi} = \beta w \quad \text{at} \quad \xi = 1$$

can now be solved using the method of separation of variables to give

$$F = \sum_{n=1}^{\infty} B_n J_0(\lambda_n \xi) e^{-(\lambda_n^2 + \alpha)\zeta} \tag{7.6.7}$$

where the eigenvalues $\lambda_n$ are defined by

$$\beta J_0(\lambda_n) = \lambda_n J_1(\lambda_n), \quad n \geq 1 \tag{7.6.8}$$

and

$$B_n = \frac{2 J_1(\lambda_n)}{\lambda_n \left[ \left( J_1(\lambda_n) \right)^2 + \left( J_0(\lambda_n) \right)^2 \right]} \tag{7.6.9}$$

The axial concentration profile can be predicted from

$$\frac{\partial F}{\partial \zeta} = -\left[ \sum_{n=1}^{\infty} \left\{ B_n \lambda_n^2 J_0(\lambda_n \xi) + \alpha B_n J_0(\lambda_n \xi) \right\} e^{-(\lambda_n^2 + \alpha)\zeta} \right] \tag{7.6.10}$$

$$= -\lambda_n^2 F - \alpha F$$

The left hand side of Equation 7.6.10) can now be written in dimensional form as $dC_A/dt$. That is

$$\frac{dC_A}{dt} = -\left( \lambda_n^2 \frac{D_{AB}}{R^2} + k_b \right) C_A \tag{7.6.11}$$

or by the ideal gas law

$$\frac{dp_A}{dt} = -\left( \lambda_n^2 \frac{D_{AB}}{R^2} + k_b \right) p_A \tag{7.6.12}$$

where $t = z/v_{avg}$ has been used.

Equation 7.6.11 or 7.6.12 can be used to argue the effect of diffusion and surface reaction on the overall deposition process. Notice that the eigenvalues, $\lambda_n$, are defined in relation to the overall surface rate coefficient ks, and the term $(\lambda_n^2 D_{AB}/R^2)$ grows as n increases while $k_b$ tends to remain relatively constant.

## Example 7.6.2

This is an example of rapid chemical reaction in the laminar boundary-layer on a flat plate. Consider the steady-state dissolution of a slightly soluble solid in a flowing dilute solution, and suppose the solid is acidic while the flowing solution is basic (Friedlander and Litt[19]). The geometry under consideration is a flat plate consisting of acid, located at zero incidence to the flow. The process may be assumed *isothermal*, and the fluid properties are to be treated as independent of position.

Species A dissolves from the plate and diffuses into the reaction zone while species B diffuses into the reaction zone from the main body of the solution. In the negligibly thin reaction region, the two substances undergo a fast irreversible reaction:

$$aA + bB \rightarrow \text{products}. \tag{7.6.13}$$

In the region between the reaction zone and the surface, the diffusional process can be described by

$$u \frac{\partial C_A}{\partial x} + v \frac{\partial C_A}{\partial y} = D_A \frac{\partial^2 C_A}{\partial y^2} \tag{7.6.14}$$

while

$$u \frac{\partial C_B}{\partial x} + v \frac{\partial C_B}{\partial y} = D_B \frac{\partial^2 C_B}{\partial y^2} \tag{7.6.15}$$

describes the diffusional process in the region outside the reaction zone. $D_A$ and $D_B$ represent the diffusivities of species A and B in the fluid, while $C_A$ and $C_B$ are molar concentrations. The quantities u and v are respectively the x and y components of the boundary-layer velocity and satisfy the continuity equation (Bird, Stewart, and Lightfoot[14] and Schlichting[20])

$$\frac{\partial u}{\partial x} + \frac{\partial v}{\partial y} = 0 \tag{7.6.16}$$

and the x-component equation of motion (Bird, Stewart, and Lightfoot[14] and Schlichting[20])

$$u\frac{\partial u}{\partial x} + v\frac{\partial u}{\partial y} = v\frac{\partial^2 u}{\partial y^2} \tag{7.6.17}$$

Then, following Schlichting,[12] the velocity components are given by

$$u = \frac{\partial \Psi}{\partial y} = \frac{\partial \Psi}{\partial \eta}\frac{\partial \eta}{\partial y} = U_\infty f'(\eta) \tag{7.6.18}$$

and

$$v = -\frac{\partial \Psi}{\partial x} = \frac{1}{2}\sqrt{\frac{vU_\infty}{x}}\left(\eta f' - f\right) \tag{7.6.19}$$

with $U_\infty$ being the free stream velocity (far away from the surface) and $\eta$ is the combined variable defined as

$$\eta = y\sqrt{\frac{U_\infty}{vx}} \tag{7.6.20}$$

where $v$ is the kinematic viscosity. The dependent variable $\psi$ is the stream function (Bird, Stewart, and Lightfoot[14]).

Upon substitution of Equations 7.6.18 to 7.6.20 into Equation 7.6.17, there results

$$ff'' + 2f''' = 0 \tag{7.6.21}$$

the so-called Blasius equation (Schlichting[20]).

Now, one would like to know the rate at which dissolution takes place as a function of distance from the leading edge of the plate and the concentration of reactant in the mainstream (far from the surface).

*Solution*
Let

$$c_A = \frac{C_A}{C_{AS}}, \quad c_B = \frac{C_B}{C_{BM}} \tag{7.6.22}$$

where $C_{AS}$ is the surface concentration (moles/L) of A, and $C_{BM}$ is the mainstream concentration (moles/L) of B. Then, for

$$c_A(x,y) = f(\eta) \qquad (7.6.23)$$

$$\frac{\partial c_A}{\partial x}dx + \frac{\partial c_A}{\partial y}dy = \frac{df}{d\eta}d\eta \qquad (7.6.24)$$

but

$$\eta = \eta(x,y) \qquad (7.6.25)$$

Therefore,

$$d\eta = \frac{\partial \eta}{\partial x}dx + \frac{\partial \eta}{\partial y}dy \qquad (7.6.26)$$

Substituting Equation 7.6.26 into Equation 7.6.24 gives

$$\frac{\partial c_A}{\partial x}dx + \frac{\partial c_A}{\partial y}dy = \frac{df}{d\eta}\left[\frac{\partial \eta}{\partial x}dx + \frac{\partial \eta}{\partial y}dy\right]$$

Equating coefficients of dx and dy:

$$dx: \frac{\partial c_A}{\partial x} = \frac{df}{d\eta}\frac{\partial \eta}{\partial x} = f'(\eta)\frac{\partial \eta}{\partial x}$$

and

$$dy: \frac{\partial c_A}{\partial y} = \frac{df}{d\eta}\frac{\partial \eta}{\partial y} = f'(\eta)\frac{\partial \eta}{\partial y}$$

By letting

$$H(x,y) = \frac{\partial c_A}{\partial y} \quad \text{and} \quad \phi(\eta y) = f'(\eta)\frac{\partial \eta}{\partial y}$$

suggest that

$$dH(x,y) = d\phi(\eta,y)$$

such that

$$\frac{\partial H}{\partial x}dx + \frac{\partial H}{\partial y}dy = \frac{\partial \phi}{\partial \eta}d\eta + \frac{\partial \phi}{\partial y}dy$$

$$= \frac{\partial \phi}{\partial \eta}\left[\frac{\partial \eta}{\partial x}dx + \frac{\partial \eta}{\partial y}dy\right] + \frac{\partial \phi}{\partial y}dy$$

and equating coefficients give

$$dx : \frac{\partial H}{\partial x} = \frac{\partial \phi}{\partial \eta}\frac{\partial \eta}{\partial x}$$

$$dy : \frac{\partial H}{\partial y} = \frac{\partial \phi}{\partial \eta}\frac{\partial \eta}{\partial y} + \frac{\partial \phi}{\partial y}$$

that is

$$\frac{\partial H}{\partial y} = \frac{\partial^2 c_A}{\partial y^2} = 2f''(\eta)\left(\frac{\partial \eta}{\partial y}\right)^2 + f'(\eta)\frac{\partial}{\partial \eta}\left(\frac{\partial \eta}{\partial y}\right) + f'(\eta)\frac{\partial^2 \eta}{\partial y^2}$$

But

$$\frac{\partial \eta}{\partial y} = \sqrt{\frac{U_\infty}{vx}}, \quad \frac{\partial}{\partial \eta}\left(\sqrt{\frac{U_\infty}{vx}}\right) = 0 = \frac{\partial^2 \eta}{\partial y^2}$$

and

$$\frac{\partial \eta}{\partial x} = -\frac{\eta}{2x}$$

Therefore, Equation 7.6.14 becomes

$$-\frac{u\eta f'(\eta)}{2x} + v\sqrt{\frac{U_\infty}{vx}}f'(\eta) = 2D_A \frac{U_\infty}{vx}f''(\eta)$$

Then with Equations 7.6.18 and 7.6.19

$$-\frac{u\left[f'(\eta)\right]^2}{2x} + \frac{1}{2}\left(\frac{U_\infty v}{x}\right)^{\frac{1}{2}}\left(\frac{U_\infty}{vx}\right)^{\frac{1}{2}}f'(\eta)\left(\eta f' - f\right) = 2D_A \frac{U_\infty}{vx}f''(\eta)$$

which reduces to

$$-\frac{ff'}{2} = 2\frac{D_A}{v}f''$$

That is

$$\frac{2}{S_A}\frac{d^2c_A}{d\eta^2} + \frac{f}{2}\frac{dc_A}{d\eta} = 0 \qquad (7.6.27)$$

Equation 7.6.15 can be similarly transformed to

$$\frac{2}{S_B}\frac{d^2c_B}{d\eta^2} + \frac{f}{2}\frac{dc_B}{d\eta} = 0 \qquad (7.6.28)$$

The quantities $S_A$ and $S_B$ are *Schmidt numbers* $\left(\dfrac{v}{D}\right)$ for the respective specie. Equations 7.6.27 and 7.6.28 are subject to the conditions

$$c_A = 1 \quad \text{at} \quad \eta = 0$$

$$c_A = c_B = 0 \quad \text{at} \quad \eta = \eta_R$$

$$c_B = 1 \quad \text{at} \quad \eta = \infty$$

and

$$c_B = 1 \text{ at } \eta = \infty$$

Substituting Equation 7.6.21 into Equation 7.6.27 and integrating results in

$$c_A = 1 - \frac{\displaystyle\int_0^\eta (f'')^{S_A/2}\,d\eta}{\displaystyle\int_0^{\eta_R} (f'')^{S_A/2}\,d\eta}, \qquad 0 < \eta < \eta_R \qquad (7.6.29)$$

Similarly, Equation 7.6.28 gives

$$c_B = 1 - \frac{\displaystyle\int_{\eta_R}^\eta (f'')^{S_B/2}\,d\eta}{\displaystyle\int_{\eta_R}^{\infty} (f'')^{S_B/2}\,d\eta}, \qquad \eta_R < \eta < \infty \qquad (7.6.30)$$

## Example 7.6.3

This is an example of rapid chemical reaction in the laminar boundary layer on a flat plate. Reconsider the previous example with non-isothermal condition and constant fluid properties. In this case, assume that the surface temperature $T_S$ and the main stream temperature $T_M$ are constant. This implies that the temperature of the reaction zone $T_R$ is also constant. Under these conditions, determine the temperature profile for species A.

*Solution*
The equation of energy (Bird, Stewart, and Lightfoot[14]) can be modified to

$$u \frac{\partial T_A}{\partial x} + v \frac{\partial T_A}{\partial y} = \alpha \frac{\partial^2 T_A}{\partial y^2} \tag{7.6.31}$$

for the problem under consideration. The temperature of species A can be made dimensionless by

$$t_A = \frac{T_R - T_A}{T_R - T_S} \tag{7.6.32}$$

Equation 7.6.31 becomes

$$u \frac{\partial t_A}{\partial x} + v \frac{\partial t_A}{\partial y} = \alpha \frac{\partial^2 t_A}{\partial y^2} \tag{7.6.33}$$

Then, following the procedure of Example 7.6.2, Equation 7.6.33 can be reduced to

$$\frac{2}{P_r} \frac{d^2 t_A}{d\eta^2} + \frac{f}{2} \frac{dt_A}{d\eta} = 0 \tag{7.6.34}$$

subject to the conditions

$$t_A = 1 \quad \text{at} \quad \eta = 0$$
$$t_A = 0 \quad \text{at} \quad \eta = \eta_R \tag{7.6.35}$$

where the quantity $P_r$ is the Prandtl number $\left( \frac{c_p \rho v}{k} \right)$

Equations 7.6.34 and 7.6.35 solve to

$$t_A = \frac{T_R - T_A}{T_R - T_S} = 1 - \frac{\displaystyle\int_0^{\eta} \left(f''\right)^{P_r/2} d\eta}{\displaystyle\int_0^{\eta_R} \left(f''\right)^{P_r/2} d\eta}, \quad 0 < \eta < \eta_R$$

## Example 7.6.4

In this example of carrier-facilitated transport in membrane separation, consider a bundle of parallel hollow fibers through which a fluid flows (Kim and Stroeve[21]). One can examine the concentration profile in a single fiber and then predict the performance of the separation device. Further, consider a fluid (Newtonian) from which the solute is to be extracted while entering the reactive section of the hollow fiber in fully developed, one-dimensional laminar flow. As shown in the schematic, Figure 7.6, at z equals zero, the fluid contacts the reactive membrane. At such a location the solute concentration $C_A$ is uniform and has the value $C_{A0}$. As the fluid flows further into the reactive section, the solute diffuses through the membrane by carrier-facilitated transport and emerges into the second fluid (shell side) which surrounds the hollow fiber. $C_A$ is negligible axially on the shell side, since the incoming shell side fluid is devoid of the solute, and is at a much higher flow rate. An alternative to this condition is a constant solute concentration on the shell side.

Following the development in the literature (Way and Noble[22]), a reversible equilibrium reaction of the form

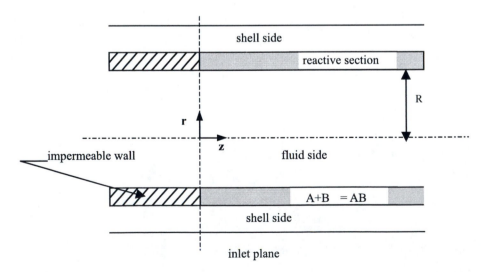

**FIGURE 7.6**
Hollow fiber with reactive walls containing the supported carrier.

$$A + B \underset{k_2}{\overset{k_1}{=}} AB \qquad (7.6.36)$$

occurs inside the membrane, where A is the solute, B is the carrier, and AB is the solute-carrier complex. The quantities $k_1$ and $k_2$ are the forward and reverse rate coefficients.

The equation of continuity for the solute (species A) in this system is:

$$2v_{avg}\left(1 - \frac{r^2}{R^2}\right)\frac{\partial C_A}{\partial z} = D_A \frac{1}{r}\frac{\partial}{\partial r}\left(r\frac{\partial C_A}{\partial r}\right) \qquad (7.6.37)$$

subject to

$$C_A = C_{A0} \text{ at } z = 0 \qquad (7.6.38)$$

$$\frac{\partial C_A}{\partial r} = 0 \text{ at } r = 0 \qquad (7.6.39)$$

$$-D_A \frac{\partial C_A}{\partial r} = k_w Sf(C_A) \text{ at } r = R \qquad (7.6.40)$$

where $D_A$ is the solute diffusivity in the fluid, $k_w$ is the membrane mass transfer coefficient for the solute, and $S$ is the shape factor based on the inside radius, R, of the hollow fiber. This shape factor (Kim and Stroeve [21] and Noble[23]) behaves as a correction factor for geometry. It permits the translation of experimental data derived in flat membranes to nonflat membrane configurations for the same experimental conditions.

Equation 7.6.40 is the boundary condition that incorporates the relevant information given in Equation 7.6.36 as to the rate of disappearance of species A. Usually, $f(C_A)$ is a quotient of two polynomials, and is typically of the form (Kim and Stroeve[21]):

$$f(C_A) = \frac{D_B' C_T K_{eq}}{D_A'\left(1 + K_{eq} C_A H\right)} C_A \qquad (7.6.6)$$

where the prime indicates diffusivity of that substance in the membrane, $K_{eq}$ is $k_1/k_2$. The equilibrium (thermodynamic) distribution coefficient, H, is defined as a proportionality factor relating the concentration of species A in the two phases (membrane and fluid). The quantity $C_T$ accounts for the concentration of species B and AB. Determine the concentration profile for species A.

*Solution*

Introducing the dimensionless variables:

$$C = \frac{C_A}{C_{A0}}; \; \xi = \frac{r}{R}; \; \zeta = \frac{zD_A}{v_{avg}R^2}$$

into Equations 7.6.37 to 7.6.39, we get:

$$2(1-\xi^2)\frac{\partial C}{\partial \zeta} = \frac{\partial^2 C}{\partial \xi^2} + \frac{1}{\xi}\frac{\partial C}{\partial \xi} \qquad (7.6.37A)$$

$$C = 1 \text{ at } \zeta = 0 \qquad (7.6.38A)$$

$$\frac{\partial C}{\partial \xi} = 0 \text{ at } \xi = 0 \qquad (7.6.39A)$$

Then, recognizing that $f(C_A)$ can be recasted as an infinite series in dimensionless form to be

$$(1+\alpha)C + \alpha \sum_{n=1}^{\infty} (-1)^n \varepsilon^n C^{n+1}$$

Equation 7.6.40 becomes:

$$-\frac{\partial C}{\partial \xi} = w[(1+\alpha)C - \varepsilon\alpha C^2 + \varepsilon^2\alpha C^3 - \varepsilon^3\alpha C^4 + ...] \text{ at } \xi = 1 \qquad (7.6.41)$$

where $w$ is defined as

$$w = \frac{Rk_w S}{D_A} \qquad (7.6.41A)$$

a dimensionless group called the wall Sherwood number. The dimensionless quantities, $\alpha$ and $\beta$ are defined in the literature (Kim and Stroeve[21]). In this work $\varepsilon$ is less than one, and can be associated with the quantity, $\beta$, in two cases. In one case, let $\varepsilon$ be $\beta$, while in another let $\varepsilon$ be the reciprocal of $\beta$ (for $\beta > 1$).

The system consisting of Equations 7.6.37A to 7.6.39A and 7.6.41 can be recasted by using the following form of the dimensionless concentration:

$$C = F_0 + \varepsilon F_1 + \varepsilon^2 F_2 + ... \qquad (7.6.42)$$

such that:

$$\frac{\partial C}{\partial \zeta} = \frac{\partial F_0}{\partial \zeta} + \varepsilon \frac{\partial F_1}{\partial \zeta} + \varepsilon^2 \frac{\partial F_2}{\partial \zeta} + \ldots$$

$$\frac{\partial C}{\partial \xi} = \frac{\partial F_0}{\partial \xi} + \varepsilon \frac{\partial F_1}{\partial \xi} + \varepsilon^2 \frac{\partial F_2}{\partial \xi} + \ldots$$

$$\frac{\partial^2 C}{\partial \xi^2} = \frac{\partial^2 F_0}{\partial \xi^2} + \varepsilon \frac{\partial^2 F_1}{\partial \xi^2} + \varepsilon^2 \frac{\partial^2 F_2}{\partial \xi^2} + \ldots$$

which then results in:

$$2(1-\xi^2)\left[\frac{\partial F_0}{\partial \zeta} + \varepsilon \frac{\partial F_1}{\partial \zeta} + \varepsilon^2 \frac{\partial F_2}{\partial \zeta} + \ldots\right]$$

$$= \frac{\partial^2 F_0}{\partial \xi^2} + \varepsilon \frac{\partial^2 F_1}{\partial \xi^2} \ldots + \frac{1}{\xi}\left(\frac{\partial F_0}{\partial \xi} + \varepsilon \frac{\partial F_1}{\partial \xi} + \ldots\right) \tag{7.6.43}$$

$$F_0 + \varepsilon F_1 + \varepsilon^2 F_2 + \ldots = 1 \text{ at } \zeta = 0 \tag{7.6.44}$$

$$\frac{\partial F_0}{\partial \xi} + \varepsilon \frac{\partial F_1}{\partial \xi} + \varepsilon^2 \frac{\partial F_2}{\partial \xi} + \ldots = 0 \text{ at } \xi = 0 \tag{7.6.45}$$

$$-\left(\frac{\partial F_0}{\partial \xi} + \varepsilon \frac{\partial F_1}{\partial \xi} + \ldots\right) = w\left\{(1+\alpha)F_0 + \varepsilon\left((1+\alpha)F_1 - \alpha F_0^2\right) + O(\varepsilon^2)\right\} \tag{7.6.46}$$

Equating coefficients of like powers of $\varepsilon$ gives:

$$\varepsilon^0: 2(1-\xi^2)\frac{\partial F_0}{\partial \zeta} = \frac{\partial^2 F_0}{\partial \xi^2} + \frac{1}{\xi}\frac{\partial F_0}{\partial \xi} \tag{7.6.47}$$

$$F_0 = 1 \text{ at } \zeta = 0 \tag{7.6.48}$$

$$\frac{\partial F_0}{\partial \xi} = 0 \text{ at } \xi = 0 \tag{7.6.49}$$

$$-\frac{\partial F_0}{\partial \xi} = w(1+\alpha)F_0 \text{ at } \xi = 1 \tag{7.6.50}$$

$$\varepsilon: 2(1-\xi^2)\frac{\partial F_1}{\partial \zeta} = \frac{\partial^2 F_1}{\partial \xi^2} + \frac{1}{\xi}\frac{\partial F_1}{\partial \xi} \tag{7.6.51}$$

$$F_1 = 0 \text{ at } \zeta = 0 \tag{7.6.52}$$

$$\frac{\partial F_1}{\partial \xi} = 0 \text{ at } \xi = 0 \tag{7.6.53}$$

$$-\frac{\partial F_1}{\partial \xi} = w\left[(1+\alpha)F_1 - \alpha F_0^2\right] \text{ at } \xi = 1 \tag{7.6.54}$$

$$\varepsilon^2: 2(1-\xi^2)\frac{\partial F_2}{\partial \zeta} = \frac{\partial^2 F_2}{\partial \xi^2} + \frac{1}{\xi}\frac{\partial F_2}{\partial \xi} \tag{7.6.55}$$

$$F_2 = 0 \text{ at } \zeta = 0 \tag{7.6.56}$$

$$\frac{\partial F_2}{\partial \xi} = 0 \text{ at } \xi = 0 \tag{7.6.57}$$

$$-\frac{\partial F_2}{\partial \xi} = w\left[(1+\alpha)F_2 - 2\alpha F_0 F_1 + \alpha F_0^3\right] \text{ at } \xi = 1 \tag{7.6.58}$$

The system of Equations 7.6.47 to 7.6.50 is linear, and can be solved using separation of variables to give:

$$F_0 = \sum_{n=0}^{\infty} B_n \exp\left\{-\frac{\lambda_n}{2}\left(\lambda_n\zeta + \xi^2\right)\right\}{}_1F_1\left[\frac{1}{2} - \frac{\lambda_n}{4}; 1; \lambda_n\xi^2\right] \tag{7.6.59}$$

where the $\lambda_n$ are defined by:

$$\left\{\lambda_n - w(1+\alpha)\right\}{}_1F_1\left[\frac{1}{2} - \frac{\lambda_n}{4}; 1; \lambda_n\right]$$

$$-2\lambda_n\left(\frac{1}{2} - \frac{\lambda_n}{4}\right){}_1F_1\left[\frac{3}{2} - \frac{\lambda_n}{4}; 2; \lambda_n\right] = 0 \tag{7.6.60}$$

and $B_n$ by:

$$B_n = \dfrac{\displaystyle\int_0^1 \xi(1-\xi)\exp\left\{-\frac{\lambda_n}{2}\xi^2\right\}{}_1F_1\left[\frac{1}{2}-\frac{\lambda_n}{4};\,1;\,\lambda_n\xi^2\right]d\xi}{\displaystyle\int_0^1 \xi(1-\xi^2)\exp\left\{-\lambda_n\xi^2\right\}{}_1F_1^2\left[\frac{1}{2}-\frac{\lambda_n}{4};\,1;\,\lambda_n\xi^2\right]d\xi}\,,\quad n \geq 0 \quad (7.6.61)$$

where

$${}_1F_1\left[\frac{1}{2}-\frac{\lambda_n}{4};\,1;\,\lambda_n\xi^2\right]$$

is one solution of the confluent hypergeometric equation or Kummer's equation (Slater[13]). Values of both $\lambda_n$ and $B_n$ can be derived as described in the literature (Brown[24]), or by employing standard software packages such as *Mathematica* (Wolfram[25]).

   *The system described by Equations 7.6.51 to 7.6.54 is also linear, but separation of variables is not the appropriate solution technique. Here Laplace transform is more suitable.* That is, by transforming the $\zeta$ variable while treating $\xi$ as a parameter, one get Kummer's equation as shown below:

$$\text{Let } L\{F_1(\zeta,\xi)\} \equiv \int_0^\infty e^{-s\zeta}F_1(\zeta,\xi)d\zeta = \overline{F}_1(s,\xi) \quad (7.6.62)$$

then Equations 7.6.51 and 7.6.52 become

$$\frac{d}{d\xi}\left(\xi\frac{d\overline{F}_1}{d\xi}\right) - 2s\xi(1-\xi^2)\overline{F}_1 = 0 \quad (7.6.63)$$

where $s$ is a complex number. Upon solving Equation 7.6.63 and applying the transformed Equation 7.6.53, results in:

$$\overline{F}_1(s,\xi) = \eta(s)\exp\left\{-i\frac{\sqrt{2s}}{2}\xi^2\right\}{}_1F_1\left[\frac{1}{2}=\frac{i\sqrt{s}}{2\sqrt{2}};\,1;\,i\sqrt{2s}\xi^2\right] \quad (7.6.64)$$

and one can see that replacing $i\sqrt{2s}$ with $\lambda$ results in

$$F_1\left[\frac{1}{2}-\frac{\lambda_n}{4};\,1;\,\lambda_n\xi^2\right]$$

the identical solution of Kummer's equation previously obtained. Where $\eta(s)$ is defined by:

$$\eta(s) = \frac{-\alpha \bar{F}_0^2 \exp\left\{i\frac{\sqrt{s}}{2}\right\}}{\left[i\sqrt{2s} - w(1+\alpha)\right] {}_1F_1\left[\frac{1}{2} - \frac{i\sqrt{2s}}{2\sqrt{2}}; 1; i\sqrt{2s}\right]}$$

$$-2i\sqrt{2s}\left(\frac{1}{2} - \frac{i\sqrt{s}}{2\sqrt{2}}\right) {}_1F_1\left[\frac{3}{2} - \frac{i\sqrt{2s}}{2\sqrt{2}}; 2; i\sqrt{2s}\right]$$

following the Laplace transformation of Equation 7.6.54. Here, the quantity $\bar{F}_0^2$ represents the Laplace transform of $F_0^2$, where $F_0$ is the solution of Equations 7.6.47 to 7.6.50. Equation 7.6.64 can now be expressed in terms of $\lambda$ as:

$$\bar{F}_1(s,\xi) = \frac{-\alpha w \bar{F}_0^2 \exp\left\{\lambda/2(1-\xi^2)\right\} {}_1F_1\left[\frac{1}{2} - \frac{\lambda}{4}; 1; \lambda\xi^2\right]}{\left[\lambda - w(1+\alpha)\right] {}_1F_1\left[\frac{1}{2} - \frac{\lambda}{4}; 1; \lambda\right] - 2\lambda\left(\frac{1}{2} - \frac{\lambda}{4}\right) {}_1F_1\left[\frac{3}{2} - \frac{\lambda}{4}; 2; \lambda\right]} \quad (7.6.65)$$

where $i\sqrt{2s}$ is replaced by $\lambda$. Equation 7.6.65 can be inverted with the use of the residue theorem (Mickley et al.[26]). That is,

$$L^{-1}\{\bar{F}_1(s,\xi)\} = \sum_{n=0}^{\infty} \frac{p(s_n)}{q'(s_n)} \exp\{s_n\zeta\} \quad (7.6.66)$$

where $p(s_n)$ and $q(s_n)$ are the numerator and denominator, respectively, of Equation 7.6.65.

This latter example demonstrates that several mathematical techniques may be required to achieve an analytical solution to a complicated transport model. In addition, the regular perturbation technique is used to reduce the problem to a set of linear problems that we know how to solve.

## 7.7  Viscous Flow

### Example 7.7.1
A fluid of constant density ($\rho$) and viscosity ($\mu$) is contained in a very long horizontal pipe of length L and radius R (Bird, Stewart, and Lightfoot[14]). The fluid is at rest initially. At t = 0, a pressure gradient

$$\left(\frac{P_0 - P_L}{L}\right)$$

is impressed on the system. Determine how the velocity profiles change with time.

*Solution*
Cylindrical coordinates are convenient.

Assumptions:

1. Both the r and θ-component of velocity are zero.

2. $v_z = v_z(r, t)$

*Problem Setup:*
The equations of motion and continuity (Bird, Stewart, and Lightfoot[14]) are combined to give

$$\rho\frac{\partial v_z}{\partial t} = \frac{P_0 - P_L}{L} + \mu\frac{1}{r}\frac{\partial}{\partial r}\left(r\frac{\partial v_z}{\partial r}\right)$$

subject to

$$v_z(r,0) = 0 \quad 0 \le r \le R$$

$$v_z(0,t)\text{ is finite,} \quad t > 0$$

$$v_z(R,t) = 0, \quad t > 0$$

Then it is convenient to introduce the following dimensionless variables:

$$\phi(\xi,\tau) = \frac{v_z}{(P_0 - P_L)R^2/4\mu L}; \quad \xi = \frac{r}{R}; \quad \tau = \frac{\mu t}{\rho R^2}$$

Substitution of the dimensionless variables into the differential equation gives

$$\frac{\partial\phi}{\partial\tau} = 4 + \frac{1}{\xi}\frac{\partial}{\partial\xi}\left(\xi\frac{\partial\phi}{\partial\xi}\right)$$

and the conditions become

$$\phi(\xi,0)=0, \quad \phi(1,\tau)=0$$

$$\text{and} \quad \phi(0,\tau) \text{ is finite}$$

This is a case in which the differential equation is nonhomogeneous. However, the fact that the system is expected to reach a steady state as $\tau \to \infty$ can be used to reduce the differential equation to a homogeneous one. That is, assume a solution of the dimensionless system to be

$$\phi(\xi,\tau)=\phi_\infty - w(\xi,\tau)$$

where $\phi_\infty$ is the steady-state solution satisfying

$$0 = 4 + \frac{1}{\xi}\frac{d}{d\xi}\left(\xi \frac{d\phi_\infty}{d\xi}\right)$$

$$\phi_\infty(1)=0 \quad \text{and} \quad \phi_\infty(0) \text{ is finite}$$

and $w(\xi, \tau)$ is the transient solution satisfying

$$\frac{\partial w}{\partial \tau} = \frac{1}{\xi}\frac{\partial}{\partial \xi}\left(\xi \frac{\partial w}{\partial \xi}\right)$$

$$w(\xi,0)=\phi_\infty$$

$$w(1,\tau)=0$$

$$w(0,\tau) \quad \text{is finite}$$

Further, assuming a solution of the form

$$w(\xi,\tau)=Z(\xi)T(\tau)$$

gives

$$\frac{T'}{T}=\frac{Z''}{Z}+\frac{1}{\xi}\frac{Z'}{Z}=-\alpha^2$$

or

$$\frac{T'}{T} = -\alpha^2 \Rightarrow T(\tau) = c_1 e^{-\alpha^2 \tau}$$

and

$$Z'' + \frac{1}{\xi} Z' + \alpha^2 Z = 0$$

subject to

$$Z(0) \text{ is finite and } Z(1) = 0$$

which is a singular Sturm-Liouville problem involving Bessel's differential equation. As in previous examples, the general solution to this differential equation is

$$Z(\xi) = c_2 J_0(\alpha\xi) + c_3 Y_0(\alpha\xi)$$

where $J_0(.)$ and $Y_0(.)$ are zero-order Bessel functions. Applying the boundary conditions, we see that the constant $c_3$ must be chosen as zero, since $Y_0(\alpha\xi)$ becomes unbounded as $\xi \to 0$ and $Z(\xi)$ is to be finite. The second boundary condition gives

$$J_0(\alpha) = 0$$

for nontrivial solutions to exist. Further, since $J_0(\alpha)$ crosses the $\alpha$-axis infinitely many times, then for each n

$$Z_n = c_{2,n} J_0(\alpha_n \xi)$$

is the eigenfunction corresponding to the eigenvalue $\alpha_n$ satisfying

$$J_0(\alpha_n) = 0$$

Also for each n,

$$w_n(\xi, \tau) = B_n e^{-\alpha_n^2 \tau} J_0(\alpha_n \xi)$$

where the constants $c_1$ and $c_{2,n}$ are combined as $B_n$. Then, by the principle of superposition

$$w_n(\xi, \tau) = \sum_{n=1}^{\infty} B_n e^{-\alpha_n^2 \tau} J_0(a_n \xi)$$

satisfies the differential equation and the boundary conditions. Application of the initial condition gives

$$(1-\xi^2) = \sum_{n=1}^{\infty} B_n J_0(a_n \xi)$$

which is a generalized Fourier series. Then the Fourier coefficient can be determined by (Sturm-Liouville theory of Chapter 4 )

$$\int_0^1 \xi(1-\xi^2) J_0(\alpha_n \xi) d\xi = B_n \int_0^1 \xi [J_0(\alpha_n \xi)]^2 d\xi, \quad n \geq 1$$

Here, the weight function is $\xi$, for the case of a singular Sturm-Liouville problem as discussed in Chapter 4. Then with the aid of integral tables for Bessel functions (Spiegel[3] and Watson[17]), we get

$$B_n = \frac{8}{\alpha_n^3 J_1(\alpha_n)} \quad \text{for } n \geq 1$$

Therefore,

$$\phi(\xi, \tau) = (1-\xi^2) - 8 \sum_{n=1}^{\infty} \frac{J_0(\alpha_n \xi)}{\alpha_n^3 J_1(\alpha_n)} e^{-\alpha_n^2 \tau}$$

is the solution.

An alternative to the steady-state hypothesis is to solve the dimensionless problem using Laplace transform. That is, reconsider

$$\frac{\partial \phi}{\partial \tau} = 4 + \frac{1}{\xi} \frac{\partial}{\partial \xi} \left( \xi \frac{\partial \phi}{\partial \xi} \right)$$

and the conditions

$$\phi(\xi, 0) = 0, \quad \phi(1, \tau) = 0$$

$$\phi(0, \tau) \text{ is finite}$$

Then let

$$L\{\phi(\xi, \tau)\} = u(\xi, s) = \int_0^{\infty} \phi(\xi, \tau) e^{-s\tau} d\tau$$

The differential equation transforms to

$$\frac{d^2u}{ds^2} + \frac{1}{\xi}\frac{du}{d\xi} - su = \frac{-4}{s}$$

subject to

$$\begin{cases} L\{\phi(0,\tau)\} = u(0,s) \\ L\{\phi(0,\tau)\} = u(1,s) = 0 \end{cases}$$

The general solution to the nonhomogeneous differential equation is

$$u(\xi,s) = c_1 J_0\left(i\sqrt{s}\xi\right) + c_2 Y_0\left(i\sqrt{s}\xi\right) + \frac{4}{s^2}$$

Since $\phi(0,\tau)$ is finite, then its Laplace transform is also expected to be finite. This means that $c_2$ must be chosen as zero, since $Y_0(i\sqrt{s}\,\xi) \to \infty$ as $\xi \to 0$. The second boundary condition gives

$$c_1 = \frac{-4}{s^2 J_0(i\sqrt{s})}$$

Therefore,

$$u(\xi,s) = \frac{-4J_0(i\sqrt{s}\xi)}{s^2 J_0(i\sqrt{s})} + \frac{4}{s^2}$$

The inverse transform for $u(\xi, s)$ can be located in a table of Laplace transforms, for example (Spiegel[3]):

$$L^{-1}\left\{\frac{J_0(i\sqrt{s}\xi)}{s^2 J_0(i\sqrt{s})}\right\} = \frac{1}{4}(\xi^2 - 1) + \tau + 2\sum_{n=1}^{\infty}\frac{e^{-\lambda_n^2\tau}J_0(\lambda_n\xi)}{\lambda_n^3 J_1(\lambda_n)}$$

where $\lambda_1, \lambda_2, \ldots$ are the positive roots of $J_0(\lambda) = 0$. Therefore,

$$\phi(\xi,\tau) = (1 - \xi^2) - 8\sum_{n=1}^{\infty}\frac{e^{-\lambda_n^2\tau}J_0(\lambda_n\xi)}{\lambda_n^3 J_1(\lambda_n)}$$

## Example 7.7.2

Consider now the problem of tangential Newtonian flow in annuli. Suppose one is interested in studying the velocity profiles of an isothermal, incompressible viscous fluid in the annular space between two cylinders, with either one or both cylinders rotating (Figure 7. 7). Then, following the literature (Bird and Curtiss[27]),

$$\rho \frac{\partial v_\theta}{\partial t} = \mu \frac{\partial}{\partial r}\left[\frac{1}{r}\frac{\partial}{\partial r}(rv_\theta)\right]$$

$$v_\theta = \kappa R\Omega_i \text{ at } r = \kappa R$$

$$v_\theta = R\Omega_o \text{ at } r = R$$

$$v_\theta = 0 \text{ at } t < 0$$

models the phenomena of interest. If the following dimensionless variables are substituted

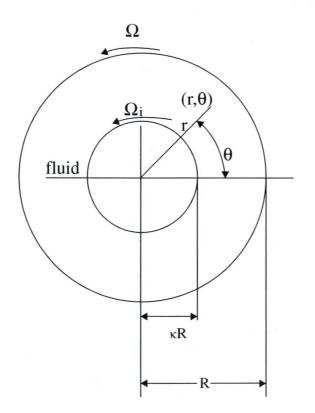

**FIGURE 7.7**
Fluid in annular space between rotating cylinders.

$$\xi = r/R: \text{ radial coordinate}$$

$$\tau = \mu t/\rho R^2: \text{ dimensionless time}$$

$$\phi = \frac{V_\theta}{R(\Omega_o - \Omega_i)}: \text{ tangential velocity}$$

$$-\alpha = \frac{\Omega_i}{(\Omega_o - \Omega_i)}: \text{ angular velocity}$$

the model can be reduced to

$$\frac{\partial \phi}{\partial \tau} = \frac{\partial}{\partial \xi}\left[\frac{1}{\xi}\frac{\partial}{\partial \xi}(\xi\phi)\right]$$

subject to

$$\phi = -\alpha\kappa \text{ at } \xi = \kappa$$

$$\phi = 1 - \alpha \text{ at } \xi = 1$$

$$\phi = 0 \text{ at } \tau = 0 \tag{7.7.1}$$

The condition given by Equation 7.7.1 motivates the use of the Laplace transform method to determine $\phi(\tau, \xi)$. If

$$L\{\phi(\tau,\xi)\} = \int_0^\infty e^{-st}\phi(\tau,\xi)d\tau \equiv y(s,\xi) \tag{7.7.2}$$

then the dimensionless differential equation and Equation 7.7.1 become

$$sy(s,\xi) - \phi(0,\xi) = \frac{d^2y(s,\xi)}{d\xi^2} + \frac{1}{\xi}\frac{dy(s,\xi)}{d\xi} - \frac{y}{\xi^2}$$

subject to the Laplace transformed boundary conditions:

$$y(s, k) = -\kappa\alpha/s$$

$$y(s, 1) = (1 - \alpha)/s$$

Rewriting, the second order ordinary differential equation is recognizable as a Bessel equation where s is a parameter and $\xi$ is the independent variable:

$$\frac{d^2 y(s,\xi)}{d\xi^2} + \frac{1}{\xi}\frac{dy(s,\xi)}{d\xi} - \left(s + \frac{1}{\xi^2}\right)y = 0$$

Then, by comparison to

$$y'' - \left(\frac{2a-1}{x}\right)y' + \left(b^2 c^2 x^{2c-2} + \frac{a^2 - v^2 c^2}{x^2}\right)y = 0$$

whose solution is

$$y = c_1 x^a J_v(bx^c) + c_2 x^a J_{-v}(bx^c)$$

$$a = 0,\ c = 1,\ b = i\sqrt{s},\ \text{and}\ v = 1$$

we get

$$y(s,\xi) = c_1 J_1(i\sqrt{s}\,\xi) + c_2 Y_1(i\sqrt{s}\,\xi)$$

where the second solution, $J_{-1}(.)$, is denoted by $Y_1(.)$. Then at $\xi = \kappa$ and at

$$c_1 J_1(i\sqrt{s}\,\kappa) + c_2 Y_1(i\sqrt{s}\,\kappa) = -\kappa\alpha/s$$

and at $\xi = 1$,

$$c_1 J_1(i\sqrt{s}) + c_2 Y_1(i\sqrt{s}) = (1-\alpha)/s$$

Therefore,

$$c_1 = \frac{\begin{vmatrix} -\dfrac{\alpha\kappa}{s} & Y_1(i\sqrt{s}\kappa) \\[2mm] \dfrac{1-\alpha}{s} & Y_1(i\sqrt{s}) \end{vmatrix}}{\begin{vmatrix} J_1(i\sqrt{s}\kappa) & Y_1(i\sqrt{s}\kappa) \\[1mm] J_1(i\sqrt{s}) & Y_1(i\sqrt{s}) \end{vmatrix}} = \frac{-\dfrac{\alpha\kappa}{s}Y_1(i\sqrt{s}) - \dfrac{1-\alpha}{s}Y_1(i\sqrt{s}\kappa)}{J_1(i\sqrt{s}\kappa)Y_1(i\sqrt{s}) - J_1(i\sqrt{s})Y_1(i\sqrt{s}\kappa)}$$

and

$$c_2 = \frac{\dfrac{1-\alpha}{s}J_1(i\sqrt{s}\kappa) + \dfrac{\alpha\kappa}{s}J_1(i\sqrt{s})}{J_1(i\sqrt{s}\kappa)Y_1(i\sqrt{s}) - J_1(i\sqrt{s})Y_1(i\sqrt{s}\kappa)}$$

Therefore,

$$y(s,\xi) = \frac{Y_1(i\sqrt{s})\alpha\kappa + (1-\alpha)Y_1(i\sqrt{s}\kappa)}{s\left(J_1(i\sqrt{s})Y_1(i\sqrt{s}\kappa) - J_1(i\sqrt{s}\kappa)Y_1(i\sqrt{s})\right)} J_i(i\sqrt{s}\xi)$$

$$- \frac{(1-\alpha)J_1(i\sqrt{s}\kappa) + \alpha\kappa J_1(i\sqrt{s})}{s\left(J_1(i\sqrt{s})Y_1(i\sqrt{s}\kappa) - J_1(i\sqrt{s}\kappa)Y_1(i\sqrt{s})\right)} Y_i(i\sqrt{s}\xi)$$

Let

$$\lambda_j = i\sqrt{s_j} \Rightarrow s_j = -\lambda_j^2 \text{ and } \frac{d\lambda_j}{ds} = \frac{-1}{2\lambda_j}$$

then in terms of $\lambda$, $y(s, \xi)$ becomes

$$y(\lambda,\xi) = \frac{\left[Y_1(\lambda)\alpha\kappa + (1-\alpha)Y_1(\lambda\kappa)\right]J_1(\lambda\xi)}{-\lambda^2\left(J_1(\lambda)Y_1(\lambda\kappa) - J_1(\lambda\kappa)Y_1(\lambda)\right)} - \frac{\left[(1-\alpha)J_1(\lambda\kappa) + \alpha\kappa J_1(\lambda)\right]Y_1(i\sqrt{s}\xi)}{-\lambda^2\left(J_1(\lambda)Y_1(\lambda\kappa) - J_1(\lambda\kappa)Y_1(\lambda)\right)}$$

Now, by applying the formula:

$$L^{-1}\{y(s,\xi\} = \phi(\tau,\xi) = \sum_{n=0}^{\infty} \rho_n(\tau,\xi)$$

where

$$\rho_n(\tau,\xi) = \frac{P(s_n,\xi)}{Q'(s_n)} e^{s_n\tau}$$

the inverse Laplace transform of $y(s, \xi)$ can be derived.
First, recall that

$$\frac{P(s_n)}{Q'(s_n)} = \lim_{s\to s_n} \frac{P(s)}{\left[\dfrac{Q(s)-Q(s_n)}{s-s_n}\right]} = \lim_{s\to s_n}(s-s_n)\frac{P(s)}{Q(s)}$$

such that for $s_0 = 0$,

$$\rho_0(\tau) = \frac{P(0)}{Q'(0)} = \lim_{s\to 0} s\frac{P(s)}{Q(s)}$$

Then for

$$y(\lambda, \xi) = \frac{P(\lambda, \xi)}{Q(\lambda)}$$

where

$$P(\lambda, \xi) = [\alpha\kappa Y_1(\lambda) + (1 - \alpha)Y_1(\lambda\kappa)]J_1(\lambda\xi) - [(1 - \alpha)J_1(\lambda\kappa) + \alpha\kappa J_1(\lambda)]Y_1(\lambda\xi)$$

and

$$Q(\lambda) = -\lambda^2[J_1(\lambda)Y_1(\lambda\kappa) - J_1(\lambda\kappa)Y_1(\lambda)]$$

$$\rho_0(\tau) = \lim_{s \to 0} s\frac{P(s)}{Q(s)} = \lim_{\lambda \to 0}(-\lambda^2)\frac{P(\lambda, \xi)}{Q(\lambda)}$$

But

$$\frac{-\lambda^2 P(\lambda, \xi)}{Q(\lambda)} = \frac{[\alpha\kappa Y_1(\lambda) + (1 - \alpha)Y_1(\lambda\kappa)]J_1(\lambda\xi) - [(1 - \alpha)J_1(\lambda\kappa) + \alpha\kappa J_1(\lambda)]Y_1(\lambda\xi)}{J_1(\lambda)Y_1(\lambda\kappa) - J_1(\lambda\kappa)Y_1(\lambda)}$$

and for small values of $\lambda$:

$$J_1(\lambda\kappa) = \lambda\kappa/2$$

and

$$Y_1(\lambda\kappa) = -2/\pi\lambda\kappa$$

such that

$$\lim_{\lambda \to 0} \frac{[\alpha\kappa Y_1(\lambda) + (1 - \alpha)Y_1(\lambda\kappa)]J_1(\lambda\xi) - [(1 - \alpha)J_1(\lambda\kappa) + \alpha\kappa J_1(\lambda)]Y_1(\lambda\xi)}{J_1(\lambda)Y_1(\lambda\kappa) - J_1(\lambda\kappa)Y_1(\lambda)}$$

$$= \frac{1 - \alpha(1 - \kappa^2)}{1 - \kappa^2}\xi - \frac{k^2}{1 - \kappa^2}\xi^{-1}$$

That is

$$\rho_0(\tau) = \frac{1 - \alpha(1 - \kappa^2)}{1 - \kappa^2}\xi - \frac{k^2}{1 - \kappa^2}\xi^{-1}$$

is the steady-state solution.

$$\text{Then for } s_n \neq 0 \Leftrightarrow \lambda_n \neq 0, n = 1, 2, \ldots$$

$$J_1(\lambda_n)Y_1(\lambda_n\kappa) - J_1(\lambda_n\kappa)Y_1(\lambda_n) = 0$$

(7.7.3)

With the use of Equation 7.7.3, $P(\lambda, \xi)$ becomes

$$P(\lambda,\xi) = \frac{1}{J_1(\lambda\kappa)}\left[(1-\alpha)J_1(\lambda\kappa) + \alpha\kappa J_1(\lambda)\right]\left[Y_1(\lambda\kappa)J_1(\lambda\xi) - J_1(\lambda\kappa)Y_1(\lambda\xi)\right]$$  (7.7.4)

Differentiating $Q(\lambda)$ and using Equation 7.7.3 results in

$$\frac{dQ}{d\lambda} = -\lambda^2\left[J_1'(\lambda)Y_1(\lambda\kappa) + J_1(1)Y_1'(\lambda\kappa) - J_1'(\lambda\kappa)Y_1(\lambda) - J_1(\lambda\kappa)Y_1'(\lambda)\right]$$

But

$$J_1'(z) = J_0(z) - \frac{1}{z}J_1(z)$$

and

$$Y_1'(z) = Y_0(z) - \frac{1}{z}Y_1(z)$$

therefore,

$$\frac{dQ}{d\lambda} = -\lambda^2\left\{J_0(\lambda)Y_1(\lambda\kappa) - J_1(\lambda\kappa)Y_0(\lambda) + \kappa\left[J_1(\lambda)Y_0(\lambda\kappa) - \frac{J_0(\lambda\kappa)J_1(\lambda)Y_1(\lambda\kappa)}{J_1(\lambda\kappa)}\right]\right\}$$

which reduces to

$$\frac{dQ}{d\lambda} = \frac{-2\lambda}{\pi J_1(\lambda\kappa)J_1(\lambda)}\left[J_1^2(\lambda) - J_1^2(\lambda\kappa)\right]$$

following substitution of

$$J_1(\lambda)Y_0(\lambda) - J_0(\lambda)Y_1(\lambda) = 2/\pi\lambda$$

and

$$J_1(\lambda\kappa)Y_0(\lambda\kappa) - J_0(\lambda\kappa)Y_1(\lambda\kappa) = 2/\pi\lambda\kappa$$

Therefore,

$$Q'(s_n) = \frac{-1}{2\lambda}\frac{dQ}{d\lambda}\bigg]_{\lambda=\lambda_n} = \frac{1}{\pi J_1(\lambda_n\kappa)J_1(\lambda)}\left[J_1^2(\lambda_n) - J_1^2(\lambda_n\kappa)\right]$$

Finally,

$$\rho_n(\tau,\xi) = \frac{P(\lambda_n,\xi)}{Q'(s_n)}e^{-\lambda_n^2\tau}$$

$$= \frac{\pi J_1(\lambda_n)\left[(1-\alpha)J_1(\lambda_n\kappa) + \alpha\kappa J_1(\lambda_n)\right]\left[Y_1(\lambda_n\kappa)J_1(\lambda_n\xi) - J_1(\lambda_n\kappa)Y_1(\lambda_n\xi)\right]}{J_1^2(\lambda_n) - J_1^2(\lambda_n\kappa)}e^{-\lambda_n^2\tau}$$

and

$$\phi(\tau,\xi) = \frac{1-\alpha(1-\kappa^2)}{1-\kappa^2}\xi - \frac{\kappa}{1-\kappa^2}\xi^{-1}$$

$$+ \sum_{n=1}^{\infty}\frac{\pi J_1(\lambda_n)\left[(1-\alpha)J_1(\lambda_n\kappa) + \alpha\kappa J_1(\lambda_n)\right]\left[Y_1(\lambda_n\kappa)J_1(\lambda_n\xi) - J_1(\lambda_n\kappa)Y_1(\lambda_n\xi)\right]}{J_1^2(\lambda_n) - J_1^2(\lambda_n\kappa)}e^{-\lambda_n^2\tau}$$

Using these dimensionless results, now back substitute to recover the dimensioned variables in which the problem was stated.

## 7.8   Problems

1. (a)  Use Equations 7.6.16, 7.6.17, and 7.6.20 to derive Equation 7.6.21.
   (b)  Solve Equation 7.6.21 subject to:

$$y = 0;\ u = v = 0$$

$$y = \infty;\ u = U_\infty$$

2. Show all the intermediate steps to derive Equation 7.6.29 and Equation 7.6.30 using Equation 7.6.21 and the given conditions in Example 2 of Section 7.6.

3. Solve:

$$\frac{2}{S_A}f'' + \frac{f\,f'}{2} = 0; \quad \text{subject to:} \begin{cases} f(0) = 1 \\ f(\eta_R) = 0 \end{cases}$$

4. For the region $\eta_R < \eta < \infty$, reduce the equation of energy to:

$$\frac{2}{\rho}\frac{d^2 T_B}{d\eta^2} + \frac{f}{2}\frac{dT_B}{d\eta} = 0$$

Solve the reduced equation subject to:
$$T_B = 0 \text{ at } \eta = \eta_R$$
$$T_B = 1 \text{ at } \eta = \infty$$

5. In a porous medium, the continuity equation can be written as

$$\varepsilon \frac{\partial \rho}{\partial t} = -\overline{\nabla} \bullet \rho \overline{u}$$

where $\varepsilon$ is the porosity, and the equation of motion (Darcy's law) can be written as

$$\overline{u} = -\frac{\kappa}{\mu}\overline{\nabla}p$$

where gravity is neglected, $\kappa$ is the permeability and $\mu$ the viscosity (Bird, Stewart, and Lightfoot[14] and Jenson and Jeffreys[4]).

(a) Show that for an incompressible fluid, $\nabla^2 p = 0$.

(b) Show that for isothermal flow of a compressible gas

$$\frac{2\varepsilon\mu\rho_0}{\kappa}\frac{\partial \rho}{\partial t} = \nabla^2 \rho^2$$

6. Two concentric spherical metallic shells of radii a and b cm (a < b) are separated by a solid of thermal diffusivity $\alpha$ (cm²/sec). The outer surface of the inner shell is maintained at $T_0$°C and the inner surface of the outer shell at $T_1$°C. Derive the differential equation governing the unsteady state temperature distribution in the solid as a function of time and radial coordinate.

Show that the solution takes the form: where $\beta = n\pi/(b - a)$. Demonstrate how $B_n$ can be determined from any initial temperature distribution.

7. If a liquid is flowing through a fixed bed at a rate v cm/sec when a pulse of a tracer is added, show that the later distribution of the tracer is given by the solution of

$$E\frac{\partial^2 C}{\partial x^2} - v\frac{\partial C}{\partial x} = \frac{\partial C}{\partial t}$$

where E is the mixing coefficient which operates in the axial direction (x) only. Use the substitution $z = x - vt$, and assume infinite bed length to find C.

8. The steady laminar flow of a liquid through a heated cylindrical pipe has a parabolic velocity profile if natural convection effects, and variation of physical properties with temperature are neglected (Jenson and Jeffreys[4]). If the fluid entering the heated section is at a uniform temperature $(T_1)$ and the wall is maintained at a constant temperature $(T_W)$, develop Graetz's solution by neglecting the thermal conductivity in the axial direction.

9. The sudden closure of a valve generates a pressure wave within the liquid flowing in the pipe leading to the valve. The passage of this wave causes compression of the liquid and expansion of the pipe. Show that the velocity of the liquid and the pressure are related by

$$-\frac{\partial p}{\partial x} = \frac{\rho}{g}\frac{\partial v}{\partial t}$$

and

$$-\frac{\partial^2 p}{\partial t^2} = c^2\frac{\partial^2 p}{\partial x^2}$$

where c is the velocity of propagation of the pressure wave.

If a uniform pipe of length L connects a reservoir at $x = 0$ to the valve at $x = L$, show that

$$p(x,t) = p_0 + \frac{4c\rho v_0}{\pi g}\sum_{n-0}^{\infty}\frac{(-1)^n}{(2n+1)}\sin(2n+1)\frac{\pi x}{2L}\sin(2n+1)\frac{\pi ct}{2L}$$

10. Consider the case of plug flow with homogeneous chemical reaction and reaction at the wall of a tubular reactor (Smith, Krieger and Herzog[28]). For an average velocity, a first order homogeneous

reaction and a first order wall reaction, show that a reasonable model is

$$V_{avg}\frac{\partial C_A}{\partial z} = D_{AB}\left[\frac{1}{r}\frac{\partial}{\partial r}\left(r\frac{\partial C_A}{\partial r}\right)\right] - kC_A$$

subject to

$$\frac{\partial C_A}{\partial r} = 0 \text{ at } r = 0$$

$$C_A = C_{A0} \text{ at } z = 0$$

$$\text{and } -D_{AB}\frac{\partial C_A}{\partial r} = k_w C_A \text{ at } r = R$$

Find the concentration profile and use it to show that

$$\frac{dC_A}{dr} = -(k_d + k)C_A, k_d = D_{AB}\frac{(\lambda_1^2 - 1)}{\alpha R^2}, \alpha = \frac{D_{AB}}{kR^2}$$

11. Consider a tubular flow reactor modeled by

$$v_0\left[1 - \left(\frac{r}{R}\right)^2\right]\frac{\partial C_A}{\partial z} = D_{AB}\left[\frac{1}{r}\frac{\partial}{\partial r}\left(r\frac{\partial C_A}{\partial r}\right) + \frac{\partial^2 C_A}{\partial z^2}\right] + kC_A^n$$

where $D_{AB}$ is the binary diffusivity, $v_0$ is the axial velocity at the center of the tube of radius R. The differential equation is subject to the following boundary conditions:

$$C_A = C_{A0} \text{ at } z = 0$$

$$\frac{\partial C_A}{\partial r} = 0 \text{ at } r = 0$$

$$-D_{AB}\frac{\partial C_A}{\partial r} = k_w C_A \text{ at } r = R$$

Further, suppose that n = 1 and that axial diffusion is negligible in comparison to bulk flow, determine the "cup mixing" concentration:

$$\overline{C}_A = 4 \int_0^1 [1 - u^2] \theta \, u \, du$$

$$\text{where} \quad u = \frac{r}{R}, \theta = \frac{C_A}{C_{A0}}$$

12. Consider evaluating the diffusion coefficient, D, for the resin phase in an ion exchange system modeled by:

$$\frac{\partial q_i}{\partial t} = D\left[\frac{\partial^2 q_i}{\partial r^2} + \frac{2}{r}\frac{\partial q_i}{\partial r}\right]$$

where $q_i$ represents point concentration (meq/g dry resin), and t is time (sec).

The differential equation is subject to:

$$q_i = 0 \text{ for } t < 0, 0 < r < b$$

and

$$q_i = q_s \text{ at } r = b, t > 0$$

where $q_s = A + Bt + Ct^2$, A, B, and C are experimentally determined constants. Find the total amount of acid, w, adsorbed per particle up to time t where w is defined as:

$$w = \int_0^t D\rho\left(\frac{\partial q_i}{\partial r}\right)_{r=b} (4\pi b^2) dt$$

in which $\rho$ is the bulk density of the resin (g dry resin/cm$^3$).

13. Bounded Equimolal Counterdiffusion (Bird[8]). Consider a system containing A and B with a partition at z = 0. At the boundaries, z = ±L, $\partial C_A/\partial z = 0$ for all t and at t = 0, $c_A = c_A^-$ for −L < z < 0 and $c_A = c_A^+$ for 0 < z < L. The differential equation describing this system is given by

$$\frac{\partial c_A}{\partial t} = \frac{\partial}{\partial z}\left(D_{AB}\frac{\partial c_A}{\partial z}\right)$$

Derive the concentration profile for the case when the diffusion coefficient is independent of concentration.

Answer:

$$\frac{c_A - c_A^-}{c_A^+ - c_A^-} = \frac{1}{2}\left[1 + \frac{2}{\pi}\sum_{n=0}^{\infty}\frac{\sin(n+1/2)\pi z/L}{(n+1/2)}\exp-\left[(n+1/2)\pi/L\right]^2 D_{AB}t\right]$$

14. Diffusion in a Two-Phase System (Bird[8]).Consider the system consisting of two immiscible phases, I ($-\infty < z < 0$) and II ($0 < z < +\infty$), separated by a partition. Initially, the solute concentration in phase I is $c_I^0$ and $c_{II}^0$ in phase II. At time $t = 0$ the partition at $z = 0$ is removed and diffusion is allowed to take place. Assuming that the diffusion can be described by Fick's second law in both phases, and there is equilibrium at the interface ($c_{II} = m\ c_I$), derive the concentration profile for each phase in terms of the respective initial concentration. The mathematical statement of the problem is

Diffusion in phase I: $\quad \dfrac{\partial c_I}{\partial t} = D_I \dfrac{\partial^2 c_I}{\partial z^2}$

Diffusion in phase II: $\quad \dfrac{\partial c_{II}}{\partial t} = D_{II}\dfrac{\partial^2 c_{II}}{\partial z^2}$

Boundary conditions:

At $z = 0 \quad c_{II} = m\ c_I$

At $z = 0 \quad D_I(\partial c_I/\partial z) = D_{II}(\partial c_{II}/\partial z)$

At $z = -\infty \quad \partial c_I/\partial z = 0$

At $z = +\infty \quad \partial c_{II}/\partial z = 0$

Initial conditions: At $t = 0 \quad\begin{aligned} c_I &= c_I^0 \\ c_{II} &= c_{II}^0 \end{aligned}$

Answer:

$$\frac{c_I - c_I^0}{c_{II}^0 - m\,c_I^0} = \frac{1 + \text{erf}\,z/\sqrt{4D_I t}}{m + \sqrt{D_I/D_{II}}}$$

$$\frac{c_{II} - c_{II}^0}{c_I^0 - (1/m)c_{II}^0} = \frac{1 + \text{erf}\,z/\sqrt{4D_{II}t}}{(1/m) + \sqrt{D_{II}/D_I}}$$

where m is the distribution coefficient, $D_I$ and $D_{II}$ are the diffusion coefficients in the respective phase.

15. If the flat velocity profile assumed in Example 7.4.2 is incorrect, use the Navier-Stokes equations of motion to show that

$$0 = \mu \frac{d^2 v_z}{dx^2} + \rho g$$

For the boundary conditions $v_z = 0$ at $x = \delta$, and $dv_z/dx = 0$ at $x = 0$ show that

$$v_z = \frac{\rho g \delta^2}{2\mu}\left[1 - \left(\frac{x}{\delta}\right)^2\right] = v_{max}\left[1 - \left(\frac{x}{\delta}\right)^2\right]$$

where the maximum velocity occurs at the film surface. Also show that the film thickness is

$$\delta = \sqrt[3]{\frac{3\mu\Gamma}{\rho^2 g}} = \sqrt[3]{\frac{3\mu^2}{\rho^2 g}\,Re}$$

where $R_e = 4\Gamma/m$ and $\Gamma = \rho\, v_{avg}\delta$ is the mass rate of flow in the z-direction per unit with of wetted wall in the y direction.

Use the derived velocity profile together with the boundary conditions:

$$c_A = c_{A0} \text{ at } x = 0$$

$$\partial c_A/\partial x = 0 \text{ at } x = \delta$$

$$c_A = c_{A1} \text{ at } z = 0$$

to derive the concentration profile for $c_A$.

16. Reconsider the model discussed in Example 7.2.3, but with cylindrical geometry (Ramraj, Farrell, and Loney[29]).

Following the development in the literature (Jost[7] and Bird[8]), the governing equation for the agent concentration in the reservoir is

$$\frac{\partial C_1}{\partial t} = D_1\left(\frac{\partial^2 C_1}{\partial r^2} + \frac{1}{r}\frac{\partial C_1}{\partial r}\right) \tag{1}$$

and

$$\frac{\partial C_2}{\partial t} = D_2 \left( \frac{\partial^2 C_2}{\partial r^2} + \frac{1}{r}\frac{\partial C_2}{\partial r} \right) \tag{2}$$

is the governing equation for the agent in the membrane. The subscripts refer to the respective region as shown in Figure 1. $D_2$ is an effective diffusivity and is defined as

$$D_2 = \frac{D\varepsilon}{\tau} \tag{3}$$

where $D$ is the agent diffusivity in the pore liquid, $\varepsilon$ is the membrane porosity, and $\tau$ is the membrane tortuosity. Both the agent diffusivity in the pore liquid ($D$) and in the reservoir liquid ($D_1$) are calculated with the Wilke-Chang correlations (Reid, Prausnitz, and Sherwood[9]). The quantity $\tau$ was defined and measured for various systems (Prasad and Sirkar[10]), we use the values given for hydrophobic membrane. The porosity value is a manufacturer supplied quantity.

Equations (1) and (2) are subject to the following boundary conditions:

$$C_1(a,\ t) = m_{1,2}C_2(a,\ t) \tag{4}$$

$$\frac{\partial C_1(0,t)}{\partial r} = 0 \tag{5}$$

$$D_1 \frac{\partial C_1(a,t)}{\partial r} = D_2 \frac{\partial C_2(a,t)}{\partial r} \tag{6}$$

$$V_w \frac{\partial C_2(b,t)}{\partial t} = -D_2\alpha_2 m_{2,w} \frac{\partial C_2(b,t)}{\partial r} \tag{7}$$

Further, since the entire agent is initially present in the reservoir phase, then

$$C_1(r,\ 0) = C_1^{\ 0} \tag{8}$$

$$C_2(r,\ 0) = 0 \tag{9}$$

Equation (4) is a statement of the equilibrium partitioning at the reservoir/pore liquid interface with $m_{1,2}$ being the partition coefficient. Equation (5) indicates that the solute concentration is expected to be finite at the bottom of the reservoir. Equation (6) displays the continuity of the agent flux across the reservoir/pore interface, while equation (7) accounts for the material leaving the membrane and entering the surrounding water bath. The quantity $\alpha_2$ is the membrane area at the outer wall (cm²).

Determine the concentration profile exiting the membrane (region 2).

*Hint:* In deriving the solution to the model, first recast the model in dimensionless form by introducing the following quantities:

$$u_1(\xi, \theta) = \frac{C_1(r, t)}{C_1^0} \tag{10}$$

$$u_2(\xi, \theta) = \frac{C_2(r, t)}{C_1^0} \tag{11}$$

where

$$\xi = \frac{r}{b}, \quad \theta = \frac{D_1 t}{b^2} \tag{12}$$

---

# References

1. Bennett, C.O. and Myers, J.E., *Momentum, Heat and Mass Transfer*, McGraw-Hill, New York, 1962.
2. Myers, G.E., *Analytical Methods in Conduction Heat Transfer*, McGraw-Hill, New York, 1971.
3. Spiegel, M.R., *Mathematical Handbook*, McGraw-Hill, New York, 1968.
4. Jenson, V.G. and Jeffreys, G.V., *Mathematical Methods in Chemical Engineering*, Academic Press, London, 1963.
5. Felder, R.M., Spence, R.D., and Ferrell, J.K., A method for the dynamic measurement of diffusivities of gases in polymers, *J. Appl. Polym. Scie.*, 19, 3193, 1975.
6. Ramraj, R., Farrell, S., and Loney, N.W., Analytical solution to controlled release using microporous membrane, *Separation Sci. Technol.*, 34, 225, 1999.
7. Jost, W., *Diffusion in Solids, Liquids and Gases*, Academic Press, New York, 1960.
8. Bird, R.B., Theory of diffusion, in *Advances in Chemical Engineering*, Vol. I, Drew, T.B. and Hoopes, J.W., eds., Academic Press Inc., 1956, 156.
9. Reid, R., Prausnitz, J., and Sherwood, T., *Properties of Gases and Liquids*, McGraw-Hill, New York, 1977.
10. Prasad, R. and Sirkar, K., Dispersion-free solvent extraction with microporous hollow-fiber modules, *AIChE J.*, 34, 177, 1988.
11. Loney, N.W., Analytical solution to mass transfer in laminar flow in hollow fiber with heterogeneous chemical reaction, *Chem. Eng. Sci.*, 51, 3995, 1996.

12. Huang, C.R., Matlosz, M., Pan, W.D. and Snyder, W., Heat transfer to a laminar flow fluid in a circular tube, *AIChE J.*, 30, 833, 1984.

13. Slater, L.C.J., *Confluent Hypergeometric Functions*, Cambridge University Press, London, 1960.

14. Bird, R.B., Stewart, W.E., and Lightfoot, E.N., *Transport Phenomena*, John Wiley, New York, 1960.

15. Higbie, R., The rate of absorption of a pure gas into a still liquid during short periods of exposure, *Trans. Amer. Inst. Chem. Eng.*, 31, 365, 1935.

16. Loney, N.W. and Huang, C.R., Analytical solution of an LPCVD reactor inter-wafer region, *Thin Solid Films*, 226, 15, 1993.

17. Watson, G.N., *A Treatise on the Theory of Bessel Functions*, 2nd ed., Cambridge University Press, London, 1966.

18. Loney, N.W., Justification of the assumption of transport by diffusion in the interwafer region of a multiwafer CVD reactor, *J. Mater. Sci. Lett.*, 15, 1219, 1996.

19. Friedlander, S.K. and Litt, M., Diffusion controlled reaction in a laminar boundary layer, *Chem. Eng. Sci.*, 7, 229, 1958.

20. Schlichting, H., *Boundary-Layer Theory*, 7th ed., McGraw-Hill, New York, 1979.

21. Kim, J. I. and Stroeve, P., Mass transfer in separation devices with reactive hollow fibers, *Chem. Eng. Sci.*, 43, 247, 1988.

22. Noble, D.R. and Way, D.J., Facilitated transport, in *Membrane Handbook*, Ho, W.S.W. and Sirkar, K., eds., Van Nostrand Reinhold, New York, 1992, chap. 44.

23. Noble R.D., Shape Factors in Facilitated Transport through Membranes, *Indust. Chem. Fundam.*, 22, 139, 1983.

24. Brown, G.M., Heat or mass transfer in a fluid in laminar flow in a circular or flat conduit, *AIChE J.*, 6, 179, 1960.

25. Wolfram, S., *Mathematica: a System for Doing Mathematics by Computer*, Addison-Wesley, Redwood City, 1991.

26. Mickley, H.S., Sherwood, T.K., and Reed, C.E., *Applied Mathematics in Chemical Engineering*, McGraw-Hill, New York 1957.

27. Bird, R.B. and Curtiss, C.F., Tangential Newtonian flow in annuli-I, *Chem. Eng. Sci.*, 11, 108, 1959.

28. Smith, G.K., Krieger, B.B., and Herzog, P.M., Experimental and analytical study of wall reaction and transport effects in fast reaction systems, *AIChE J.*, 26, 567, 1980.

29. Ramraj, R., Farrell, S., and Loney, N.W., Mathematical modeling of controlled release from a hollow fiber, *J. Membrane Sci.*, 162, 73, 1999.

# 8

---

## Dimensional Analysis and Scaling
## of Boundary Value Problems

---

### 8.1  Introduction

In the practice of chemical engineering, dimensional analysis and scaling are techniques that can be employed to emphasize similarities between a proto-type and its model. Dimensionless quantities are developed which can serve as new variables. Usually, the number of new variables is much smaller than the original number of physical variables of the system under consideration. For example, suppose one was interested in conducting an investigation to determine the power required to drive an ordinary house fan (Taylor[1]). More specifically, suppose one wants to relate the size and shape of the fan to the rotational speed and torque, where Newton's second law relates the torque (t) to the forces generated when the fan accelerates the air which passes through it. Suppose the torque is chosen as the dependent variable and the physical variables are the following:

- Fan diameter (d)
- Fan design or shape of fan (R)
- Medium characteristics (air viscosity, density, sound velocity, and ratio of specific heats)
- Rotative speed (n)

Since this house fan is to be uncomplicated, one can neglect the effects of the air viscosity, sound velocity, and the ratio of specific heats. Then, using the mass (M), length (L), and time (T) system, Table 8.1 can be generated:
If the torque is divided by the density, Table 8.2 results:
If $t/\rho$ is divided by the product $D^5n^2$, there results the quantity

$$\left[\frac{t}{\rho D^5 n^2}\right] = 1$$

**TABLE 8.1**

|              |        | Dimensions |    |     |
|--------------|--------|------------|----|-----|
| Quantity     | Symbol | M          | L  | T   |
| Torque       | t      | 1          | 2  | -2  |
| Fan diameter | D      | 0          | 1  | 0   |
| Fan design   | R      | 0          | 0  | 0   |
| Air density  | ρ      | 1          | -3 | 0   |
| Rotative speed | n    | 0          | 0  | -1  |

*Source:* Adapted from Taylor, E.S., *Dimensional Analysis for Engineers*, Oxford University Press, London, UK, 1974. With permission.

**TABLE 8.2**

| Symbol | M | L | T  |
|--------|---|---|----|
| t/ρ    | 0 | 5 | -2 |
| D      | 0 | 1 | 0  |
| n      | 0 | 0 | -1 |

Therefore, the final analysis yields

$$f\left(\frac{t}{\rho D^5 n^2}, R\right) = 0$$

which means that the torque for a given design R is proportional to the dimensionless product

$$\left[\frac{t}{\rho D^5 n^2}\right]$$

Therefore, instead of five variables, only one need be considered.

Another example, more familiar to chemical engineers, involves fluid flow. Consider a very long cylindrical pipe having a uniform cross-section area (A) in which an incompressible fluid is flowing. Assume that the motion is steady and that viscosity ($\mu$) and density ($\rho$) are not negligible. One may want to characterize the motion of the fluid. In this case, the pressure drop along the pipe, the average velocity (v) of the fluid, or the fluid discharge per unit time through A is needed. Hence, the motion in the pipe can be determined by $\rho$, $\mu$, v and D, where D is the pipe inside diameter. That is, if one forms the dimensionless quantity (drag):

$$\phi = \frac{P_1 - P_2}{l\left(\frac{\rho v^2}{D}\right)}$$

Then, the resistance of a section of pipe of length $\ell$ is

$$\left(p_1 - p_2\right)A = \Delta pA = \phi \frac{\ell}{D} A\rho v^2$$

Among the four physical variables, only one dimensionless quantity can be formed, namely

$$N_{Re} = \frac{\rho v D}{\mu}$$

where $N_{Re}$ is the Reynolds number. Therefore,

$$\phi = \phi\left(N_{Re}\right)$$

Again, instead of 4 variables, the Reynolds number can be used to characterize the phenomenon.

## 8.2   A Classical Approach to Dimensional Analysis

The *pi theorem* is a generalized method of dimensional analysis and detailed discussions can be found in (Taylor,[1] Massey,[2] Massey,[3] Pankhurst,[4] Hansen,[5] Murphy,[6] and Isaacson and Isaacson[7]). Below is a brief review of the pi theorem.

Consider the magnitude $f_1$ of some physical quantity depending on other, independent magnitudes $g_1, g_2, ..., g_n$, then

$$f_1 = \phi\left(g_1, g_2, \cdots, g_n\right)$$

or alternatively

$$\psi\left(f_1, g_1, g_2, \cdots, g_n\right) = 0 \qquad (8.2.1)$$

Equation 8.2.1 is required to be dimensionally homogeneous. The pi theorem says that if the number of distinct reference quantities required to express the dimensional formula of all n magnitudes is r, then the n magnitudes may be grouped in n − r independent dimensionless Π terms, resulting in the relation

$$\phi\left(\Pi_1, \Pi_2, \cdots, \Pi_{n-r}\right) = 0 \qquad (8.2.2)$$

An immediate advantage is demonstrated if Equation 8.2.1 is compared with Equation 8.2.2. That is, the number of independent variables to be studied is reduced to n − r.

Usually, it is not difficult to determine the value of n; just add up the variables. However, determining r requires more effort. Above, r is stated as *distinct reference quantities*. The term *distinct* is significant. That is, in the linear algebra sense, r is the rank of the dimensional matrix. For example, suppose a problem involving force (F), velocity (v), density (ρ), viscosity (μ), and length (L) is to be restated in terms of dimensionless groups. Then, the rank of the dimensional matrix must be determined. In solving such problems, a usual first step is the construction of a table, listing the variables and their associated dimensions. Such a table is shown as follows in Table 8.3.

**TABLE 8.3**

| Variable | Symbol | Dimensions |
|----------|--------|------------|
| Force | F | $M\,L/t^2$ |
| Velocity | v | $L/t$ |
| Density | ρ | $ML^3$ |
| Viscosity | μ | $M/L\,t$ |
| Length | L | $L$ |

The next step is to define the dimensional matrix whose elements are the exponents of the fundamental dimensions M, L, and t appearing in Table 8.3.

Dimensional Matrix

|   | F | v | ρ | μ | L |
|---|---|---|---|---|---|
| M | 1 | 0 | 1 | 1 | 0 |
| L | 1 | 1 | −3 | −1 | 1 |
| t | −2 | −1 | 0 | −1 | 0 |

In order to determine r, recall that the definition of the rank, r, of a matrix is the number of rows (columns) in the largest nonzero determinant which can be formed from the matrix. In our situation, we have a matrix that contains 3 rows and 5 columns. However, determinants are only defined for square matrices (see Appendix). Therefore, the largest possible determinant would contain 3 rows and 3 columns (determinant of order 3). It turns out that, for this problem, the largest nonzero determinant is of order 3; therefore, the rank is 3. However, in other situations, one should verify that the determinant is, in fact, nonzero.

As a second example of the determination of the rank, r, consider the dimensional matrix:

| | P | Q | R | S |
|---|---|---|---|---|
| M | 2 | 1 | 3 | 4 |
| L | −1 | 6 | −3 | 0 |
| T | 1 | 20 | −3 | 8 |

Here, the largest determinant will be that which contains 3 rows and 3 columns. However, all such determinants are zero. That is,

$$\begin{vmatrix} 2 & 1 & 3 \\ -1 & 6 & -3 \\ 1 & 20 & -3 \end{vmatrix} = \begin{vmatrix} 2 & 1 & 4 \\ -1 & 6 & 0 \\ 1 & 20 & 8 \end{vmatrix} = \begin{vmatrix} 2 & 3 & 4 \\ -1 & -3 & 0 \\ 1 & -3 & 8 \end{vmatrix} = \begin{vmatrix} 1 & 3 & 4 \\ 6 & -3 & 0 \\ 20 & -3 & 3 \end{vmatrix} = 0$$

It turns out here that only second order determinants have nonzero values. For example, the determinant

$$\begin{vmatrix} 2 & 1 \\ -1 & 6 \end{vmatrix} = 14 \neq 0$$

Therefore, the rank of this dimensional matrix is 2.

## 8.3 Finding the Πs

In this section, two ways of deriving the Πs are discussed. The first is based on the Pi theorem, while the second is a more practical approach. Recall the example of fluid flow through a long, smooth pipe of circular cross-section. There are five variables representing the n magnitudes, namely $\Delta p/\ell$, $\rho$, $\mu$, $v$, and D. The number of distinct reference quantities (r) is three (rank of the dimensional matrix). Therefore there should be $5 - 3 = 2$ independent dimensionless products ($\Pi_1$ and $\Pi_2$). Further, the theorem requires that, from among the n original variables, r of them are to be selected to form a recurring set. This recurring set further requires that the r variables, together, must involve the r distinct reference quantities. Here $v$, $\mu$, and $\rho$ form a suitable recurring set. Each Π will be formed from the recurring set and one of the remaining variables. For example, $\Pi_1$ is constructed from $v$, $\rho$, $\mu$, and $\Delta p/\ell$, while $\Pi_2$ from $\rho$, $\mu$, $v$, and D. For the formulation of $\Pi_1$, let the dimensionless number be $v^a \mu^b \rho^c \Delta p/\ell$, where a, b, and c are constants to be determined. Then, the dimensional formula of this product is

$$\left[LT^{-1}\right]^a \left[ML^{-1}T^{-1}\right]^b \left[ML^{-3}\right]^c \left[ML^{-2}T^{-2}\right] = \left[M^0 L^0 T^0\right]$$

Equating exponents of like magnitudes gives

$$
\begin{aligned}
\text{L:} \quad & a - b - 3c - 2 = 0 \\
\text{M:} \quad & b + c + 1 = 0 \\
\text{T:} \quad & -a - b - 2 = 0
\end{aligned}
$$

resulting in $a = -3$, $b = 1$, and $c = -2$. Therefore, the dimensionless number $\Pi_1$ is

$$
\frac{\mu \Delta p / \ell}{v^3 \rho^2}
$$

Similarly, for the formation of $\Pi_2$, let $v^a \mu^b \rho^c D$ be the dimensionless number. Then, the dimensional formula of this product is

$$
\left[LT^{-1}\right]^a \left[ML^{-1}T^{-1}\right]^b \left[ML^{-3}\right]^c [L] = \left[M^0 L^0 T^0\right]
$$

giving $a = 1$, $b = -1$, and $c = 1$. Therefore, Equation 8.2.2 for this example is

$$
\phi\left(\Pi_1, \Pi_2\right) = 0
$$

That is,

$$
\phi\left( \frac{\mu \dfrac{\Delta p}{\ell}}{v^3 \rho^2} \quad \frac{vD\rho}{\mu} \right) = 0
$$

or, in terms of $\Delta p / \ell$:

$$
\Delta p / \ell = \frac{v^3 \rho^2}{\mu} \phi_1\left(N_{Re}\right)
$$

It is important to note that the recurring set must provide the opportunity for canceling of any dimensional components that are involved in the other magnitudes. Therefore, *its constituent magnitudes must involve each of the reference magnitudes at least once*. However, *it must not be possible to form a dimensionless group from the recurring set alone*.

As a second illustration on finding the $\Pi$s, consider a screw propeller operating in a fluid of constant density, such as seawater. Here, one is interested in the thrust produced by the propeller, and the goal is to determine how the magnitude of this force depends on other associated quantities. Following Massey,[2] the fluid is assumed to be homogeneous (no air bubbles) and the

**TABLE 8. 4**

| Quantity | Symbol | Dimensional Formula |
|----------|--------|---------------------|
| Thrust (force) | F | $[MLT^{-2}]$ |
| Fluid density | $\rho$ | $[ML^{-3}]$ |
| Propeller diameter | D | $[L]$ |
| Speed of advance | v | $[LT^{-1}]$ |
| Angular speed | $\varpi$ | $[T^{-1}]$ |
| Fluid viscosity | $\mu$ | $[ML^{-1}T^{-1}]$ |
| Weight per unit mass | g | $[LT^{-2}]$ |

*Source:* Adapted from Massey, B.S., *Measures in Science and Engineering*, Ellis Horwood Limited, Chichester West, UK, 1986. With permission.

propeller is remote from all other surfaces (this latter assumption excludes ships, etc.). Listed in Table 8.4 are the relevant quantities:

The quantity, g, is included to account for work done against gravity in moving the fluid surface vertically. However, surface tension is being neglected, since the propeller is assumed to be large. From Table 8.4, it can be seen that the n magnitudes are 7 and the number of reference magnitudes (mass, length, and time interval) is 3. Therefore, 7 – 3 = 4 dimensionless quantities (products) are expected. That is,

$$\phi(\Pi_1, \Pi_2, \Pi_3, \Pi_4) = 0 \qquad (8.3.1)$$

is the appropriate form to expect.

Since the number of reference magnitudes is 3, then 3 magnitudes (variables) must be selected from among the 7 for use as a recurring set. The variables F, $\rho$, and D are suitable, while the set D, v, $\varpi$ is not suitable because the reference magnitude M is not included. Also the set consisting of F, $\rho$, and $\mu$ is an unsuitable choice, even though all the reference magnitudes are included. This latter set is unsuitable because a dimensionless group, $F\rho/\mu^2$, can be formed from these variables alone.

Using F, $\rho$, and D as the recurring set, the dimensionless numbers ($\Pi$s) take the forms:

$$F^{a1}\rho^{b1}D^{c1}v; \quad F^{a2}\rho^{b2}D^{c2}\varpi; \quad F^{a3}\rho^{b3}D^{c3}\mu; \quad F^{a4}\rho^{b4}D^{c4}g$$

For the first $\Pi$, the dimensional formula is

$$\left[MLT^{-2}\right]^{a1}\left[ML^{-3}\right]^{b1}[L]^{c}\left[LT^{-1}\right] = \left[M^0L^0T^0\right]$$

such that al = –1/2, b1 = 1/2, c1 = 1. Therefore, the first $\Pi$ may be written as $F^{-1/2}\rho^{1/2}D$ v. However, the fractional exponents may prove to be inconvenient and can be removed by squaring to get

$$\Pi_1 = \frac{\rho D^2 v^2}{F}$$

Similarly, the other $\Pi$s are derived:

$$\Pi_2 = \frac{\rho D^4 \varpi^2}{F}, \quad \Pi_3 = \frac{\mu^2}{F\rho}, \quad \Pi_4 = \frac{\rho D^3 g}{F}$$

Then, Equation 8.3.1 may be expressed as

$$\phi\left(\frac{\rho D^2 v^2}{F}, \frac{\rho D^4 \varpi^2}{F}, \frac{\mu^2}{F\rho}, \frac{\rho D^3 g}{F}\right) = 0 \tag{8.3.2}$$

It is important to note that the recurring set F, $\rho$, and D is not unique. However, if analysis is performed with any other suitable recurring set, the same information, appearing in different forms, would result. For example, the recurring set consisting of $\rho$, D, v results in

$$\phi\left(\frac{\rho D^2 v^2}{F}, \frac{D\varpi}{v}, \frac{\rho Dv}{\mu}, \frac{Dg}{v}\right) = 0 \tag{8.3.3}$$

It appears that only the first of the arguments in Equation 8.3.2 corresponds directly with any in Equation 8.3.3. However, the first two arguments of Equation 8.3.2 may be divided to yield

$$\frac{\Pi_1}{\Pi_2} = \left(\frac{D\varpi}{v}\right)^{-2}$$

which is a power of the second $\Pi$ in Equation 8.3.3. The other two products in Equation 8.3.3 may also be obtained from combinations of those in Equation 8.3.2. This type of transformation of $\Pi$s may be useful, but none of the original magnitudes must be completely removed from the set of $\Pi$s. Also, the number of independent $\Pi$s must be that specified by the Pi theorem, n − r. That is, given

$$\phi\left(\Pi_1, \Pi_2, \Pi_3, \Pi_4\right) = 0$$

then

$$\phi\left(\Pi_1, \Pi_2, \frac{\Pi_1}{\Pi_2}, \Pi_4\right) = 0$$

is not appropriate, while the following is

$$\phi\left(\Pi_1 \middle/ \Pi_2, \Pi_3, \Pi_4\right) = 0$$

The following example illustrates the key steps in the application of the Pi theorem (Welty, Wicks, and Wilson[8]).

## Example 8.3.1

Determine the dimensionless groups formed from the variables involved in the flow of a fluid external to a solid body. The force exerted on the body is a function of $v$, $\rho$, $\mu$, and $L$ (Welty, Wicks, and Wilson[8]).

*Solution*
Typically, a table of the variables and their dimensions is constructed.

| Variable | Symbol | Dimensions |
|----------|--------|------------|
| Force | F | $M L/t^2$ |
| Velocity | v | $L/t$ |
| Density | $\rho$ | $M/L^3$ |
| Viscosity | $\mu$ | $M/L\,t$ |
| Length | L | $L$ |

Then, a dimensional matrix is formed whose elements are the exponents of the fundamental dimensions M, L, and t appearing in each of the variables. In this case, the dimensional matrix is

|   | F | v | $\rho$ | $\mu$ | L |
|---|---|---|--------|-------|---|
| M | 1 | 0 | 1 | 1 | 0 |
| L | 1 | 1 | -3 | -1 | 1 |
| t | -2 | -1 | 0 | -1 | 0 |

From the above matrix, the rank is determined. This was done in the previous section, and there, as well as in this case, the rank r is 3. In order to determine the number of independent dimensionless groups, the number of variables is determined. In this case, n = 5. Therefore, the number of $\Pi$s is n − r = 2. Next, a recurring set of r variables is selected. This recurring set will be made up of those variables, which will appear in each pi group, and among them, contain all of the fundamental dimensions. One approach in choosing this set is to exclude from it those variables whose effect is being investigated. In this problem, one would like to isolate the drag force effect; therefore, it will not be part of the recurring set. For the remaining exclusion, the viscosity is chosen. The remaining variables are v, $\rho$, and L, which include the fundamental dimensions M, L, and t among them. That is, the recurring set consists of v, $\rho$, and L.

Each of the two $\Pi$s will include the recurring set and one of the previously excluded variables. That is,

$$\Pi_1 = v^a \rho^b L^c F \quad \text{and} \quad \Pi_2 = v^d \rho^e L^f \mu$$

Notice that each $\Pi$ is required to be dimensionless. As such, the variables in the recurring set are raised to certain exponents, which will satisfy this dimensionless condition. Each pi group is evaluated independently. Therefore,

$$\Pi_1 = v^a \rho^b L^c F$$

Dimensionally,

$$\left[M^0 L^0 t^0\right] = \left[\frac{L}{t}\right]^a \left[\frac{M}{L^3}\right]^b [L]^c \frac{ML}{t^2}$$

Equating exponents of M, L, and t, on both sides, results in

$$M: 0 = b + 1$$

$$L: 0 = a - 3b + c + 1$$

$$t: 0 = -a - 2$$

giving $a = -2$, $b = -1$ and $c = -2$. Therefore,

$$\Pi_1 = \frac{F/L^2}{\rho v^2}$$

which is the Euler number. Similarly, evaluating the exponents for the second pi group results in

$$\Pi_2 = \mu/\rho v L = 1/N_{Re}$$

where $N_{Re}$ is the Reynolds number. Therefore, the application of dimensional analysis has reduced a 5-variable problem to a 2-parameter problem

$$Eu = \varphi_1 (N_{Re})$$

The second method for finding the $\Pi$s in this section is demonstrated using the following example. Reconsider the illustration involving the flow of a fluid through a very long smooth pipe. This method involves the construction of a series of tables containing the variables and their dimensions as follows:

**TABLE 8.5A**

|          | [M] | [L] | [T] |
|----------|-----|-----|-----|
| $\Delta p/\ell$ | 1 | -2 | -2 |
| D        | 0 | 1 | 0 |
| v        | 0 | 1 | -1 |
| $\mu$    | 1 | -1 | -1 |
| $\rho$   | 1 | -3 | 0 |

*Following the construction of Table 8.5A, the dependence of each variable on the reference magnitudes is systematically eliminated:* Eliminate the dependence on M by dividing each variable having dimensions in respect to M by $\rho$. This results in Table 8.5B:

**TABLE 8.5B**

|          | [M] | [L] | [T] |
|----------|-----|-----|-----|
| $\Delta p/\ell\rho$ | 0 | 1 | -2 |
| D        | 0 | 1 | 0 |
| v        | 0 | 1 | -1 |
| $\mu/\rho$ | 0 | 2 | -1 |
| $\rho/\rho = 1$ | 0 | 0 | 0 |

Eliminate the dependence on T. Each variable with dimensions involving T is multiplied or divided by an appropriate power of v, so as to make the T exponent zero, as shown in Table 8.5C below.

**TABLE 8.5C**

|          | [M] | [L] | [T] |
|----------|-----|-----|-----|
| $\Delta p/\ell\rho v^2$ | 0 | -1 | 0 |
| D        | 0 | 1 | 0 |
| $v/v = 1$ | 0 | 0 | 0 |
| $\mu/\rho v$ | 0 | 1 | 0 |

Finally, eliminate the dependence on L by using appropriate powers of D, as shown in Table 8.5D:

**TABLE 8.5D**

|          | [M] | [L] | [T] |
|----------|-----|-----|-----|
| $D\Delta p/\ell\rho v^2$ | 0 | 0 | 0 |
| $D/D = 1$ | 0 | 0 | 0 |
| $\mu/\rho v D$ | 0 | 0 | 0 |

*Source:* Tables 8.5A–D adapted from Massey, B.S., *Measures in Science and Engineering*, Ellis Horwood Limited, Chichester West, UK, 1986. With permission.

The resulting two dimensionless groups ($\Pi$s) are $D\Delta p/\ell\rho v^2$ and $\mu/\rho\,v\,D$. If a particular variable is to be used as the dependent variable, then that variable (or any power of it) must be excluded as a multiplier or divisor in the elimination steps ($\Delta p/\ell$). Also, the order in which reference magnitudes are eliminated is only important in so far as expediency is concerned. Even though this latter technique may involve more labor, it is more straightforward to apply.

So far, the discussion has been centered on developing dimensionless groups from a set of given process variables. These dimensionless groups are available as design parameters instead of the individual process variables. This is especially helpful when relationships are to be established between a model and an as yet designed larger scale process.

Below, scaling of the independent and dependent variables in given differential equations will establish dimensionless groups.

## 8.4  Scaling Boundary Value Problems

In studying transport phenomena one is guaranteed to encounter situations involving fluid-solid or gas-liquid interfaces. These problems usually become more complex when fluid motion is involved. Traditionally, such complex problems are analyzed with the aid of the concept of a *boundary layer*. The boundary layer equations are well established for fluid flow, heat transfer, and mass transfer (Bird, Stewart, and Lightfoot[9], Incropera and DeWitt,[10] Kays and Crawford,[11] White,[12] Schlichting,[13] and Welty, Wicks, and Wilson[8]). This set of equations is a convenient starting point for scaling of the independent and dependent variables. However, before we begin it is helpful to briefly review descriptions of the three types of boundary layers.

In fluid flow, one may consider a flat plate over which the fluid is flowing (Figure 8.1). Then fluid particles making contact with the surface are

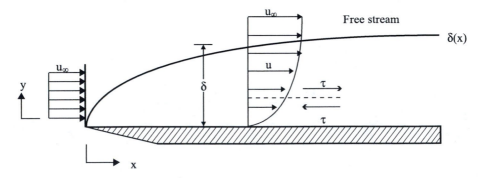

**FIGURE 8.1**
Velocity boundary layer.

presumed to have zero velocity. These particles further act to retard the motion of particles in the adjacent fluid layers until, at some distance, y = δ from the solid surface, the fluid velocity is no longer influenced by the surface. The quantity δ is the boundary layer thickness. This retardation of fluid motion is associated with shear stresses, τ, acting in planes parallel to the fluid velocity (see Figure 8.1). Also, as one moves away from the solid surface in the y-direction, the x-component of the fluid velocity, u, increases until it approaches the free stream value $u_\infty$. This is the velocity boundary layer and is expected to develop whenever there is fluid flow over a surface.

The velocity boundary layer is important in establishing surface frictional effects, such as local friction coefficient $C_f$:

$$C_f = \frac{\tau_s}{\rho u_\infty^2 / 2} \tag{8.4.1}$$

where the subscript, s, refers to the surface. For Newtonian fluids, the surface shear stress is

$$\tau_s = \mu \left( \frac{\partial u}{\partial y} \right)_{y=0} \tag{8.4.2}$$

where $\mu$ is the fluid viscosity.

Similar to a velocity boundary layer developing whenever there is fluid flow over a surface, a thermal boundary layer develops whenever there is a temperature gradient between the surface and the free stream fluid. That is, flow over an isothermal flat plate (see Figure 8.2) is expected to have a uniform temperature profile $T(y) = T_\infty$ at its leading edge. However, those fluid particles that come into contact with the plate achieve thermal equilibrium at the surface temperature. These equilibrated particles exchange energy with those in the adjacent layers, thus developing a temperature gradient in a region of the fluid. The region in which such temperature gradients develop is the thermal boundary layer whose thickness is $\delta_t$.

A relation between conditions in the thermal boundary layer and the convection heat transfer coefficient, h, is

$$h = -\frac{k(\partial T / \partial y)_{y=0}}{T_s - T_\infty} \tag{8.4.3}$$

since at the surface (y = 0), there is no fluid motion and energy transfer occurs by conduction.

Analogous to velocity and thermal boundary layers is the concentration boundary layer. That is, it determines convection mass transfer, similar to the

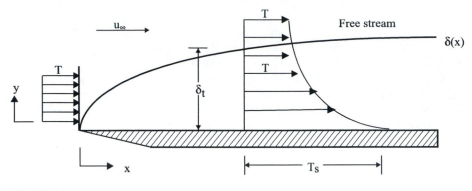

**FIGURE 8.2**
Thermal boundary layer.

velocity boundary layer determining wall friction or the thermal boundary layer determining convection heat transfer.

When a binary mixture of species A and B flows over a surface (see Figure 8.3) and the concentration of species A at the surface, $C_{A,S}$, differs from $C_{A,\infty}$, the free stream concentration, a concentration boundary layer is expected to develop. That is, the region of the fluid in which concentration gradients exist is the concentration boundary layer. The thickness of the concentration boundary layer is $\delta_c$.

Similar to Equation 8.4.3, a relationship between conditions in the concentration boundary layer and the mass transfer coefficient, $h_m$, is given by

$$h_m = -D_{AB} \frac{(\partial C_A / \partial y)_{y=0}}{C_{A,S} - C_{A,\infty}} \qquad (8.4.4)$$

For steady, two-dimensional velocity, thermal and concentration boundary layers, the equations of change apply. These equations are developed elsewhere (Welty, Wicks, and Wilson,[8] Bird, Stewart, and Lightfoot,[9] Incropera

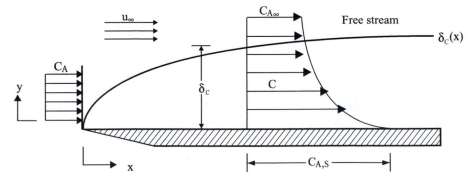

**FIGURE 8.3**
Species concentration boundary layer.

and DeWitt,[10] and Schlichting[13]). However, it is unusual for all the terms in the equations of change to be needed to resolve a given situation. Therefore, it is customary to simplify these equations by assuming fluids with constant physical properties ($\rho$, k, and $\mu$), negligible body forces, no chemical reaction, and no heat generation. In addition, the usual boundary layer approximations are incorporated to further reduce the equations of change (Incropera and DeWitt[10]) to:

$$\text{Continuity} \quad \frac{\partial u}{\partial x} + \frac{\partial v}{\partial y} = 0 \tag{8.4.5}$$

$$\text{X-momentum} \quad u\frac{\partial u}{\partial x} + v\frac{\partial u}{\partial y} = -\frac{1}{\rho}\frac{\partial p}{\partial x} + v\frac{\partial^2 u}{\partial y^2} \tag{8.4.6}$$

$$\text{Energy} \quad u\frac{\partial T}{\partial x} + v\frac{\partial T}{\partial y} = \alpha\frac{\partial^2 T}{\partial y^2} + \frac{v}{C_p}\left(\frac{\partial u}{\partial y}\right)^2 \tag{8.4.7}$$

$$\text{Species continuity} \quad u\frac{\partial C_A}{\partial x} + v\frac{\partial C_A}{\partial y} = D_{AB}\frac{\partial^2 C_A}{\partial y^2} \tag{8.4.8}$$

Equations 8.4.5, 8.4.6, 8.4.7, and 8.4.8 are the *convection transfer equations*. When appropriate boundary conditions are included, these equations can be solved to determine spatial variations of u, v, T, and $C_A$ in the boundary layers. In this section, these reduced equations will be scaled to establish important analogies between momentum, heat, and mass transfer, while identifying key design parameters.

In order to achieve the stated purpose, we will define dimensionless independent and dependent variables in the following way:

$$\text{Let} \quad x^* = \frac{x}{L} \quad \text{and} \quad y^* = \frac{y}{L} \tag{8.4.9}$$

where L is a *characteristic length* for the surface of interest such as the length of a flat plate. Also, define the dimensionless velocities $\mu^*$ and $y^*$ as

$$u^* = \frac{u}{u_\infty} \quad \text{and} \quad v^* = \frac{v}{u_\infty} \tag{8.4.10}$$

and the dimensionless temperature, $T^*$, and species concentration, as $C_A^*$,

$$T^* = \frac{T - T_s}{T_\infty - T_s} \quad \text{and} \quad C^*_A = \frac{C_A - C_{A,S}}{C_{A,\infty} - C_{A,S}} \tag{8.4.11}$$

Substitution of these dimensionless quantities into the convection transfer equations result in

$$\text{Continuity} \quad \frac{\partial u^*}{\partial x^*} + \frac{\partial v^*}{\partial y^*} = 0 \tag{8.4.12}$$

$$\text{Velocity} \quad u^* \frac{\partial u^*}{\partial x^*} + v^* \frac{\partial u^*}{\partial y^*} = -\frac{dp^*}{dx^*} + \frac{1}{N_{ReL}} \frac{\partial^2 u^*}{\partial y^{*2}} \tag{8.4.13}$$

where $p^* = p/\rho\, u_\infty^2$ and $N_{reL} = u_\infty L/v$ is the Reynolds number, based on length as opposed to diameter.

$$\text{Thermal} \quad u^* \frac{\partial T^*}{\partial x^*} + v^* \frac{\partial T^*}{\partial y^*} = \frac{1}{N_{ReL}\, Pr} \frac{\partial^2 T^*}{\partial y^{*2}} \tag{8.4.14}$$

where $Pr = v/\alpha$ is the Prandtl number and the viscous dissipation term $(\partial u/\partial y)^2$ neglected.

$$\text{Concentration} \quad u^* \frac{\partial C_A^*}{\partial x^*} + v^* \frac{\partial C_A^*}{\partial y^*} = \frac{1}{N_{ReL}\, Sc} \frac{\partial^2 C_A^*}{\partial y^{*2}} \tag{8.4.15}$$

where $Sc = v/D_{AB}$ is the Schmidt number.

Notice the similarity between Equations 8.3.14 and 8.3.15. Either one of these equations can now be used to model heat or mass transfer, and only the interpretation would change for heat or mass transfer.

The dimensionless differential equations suggest how important results may be generalized. For example, in functional form, the dimensionless velocity is

$$u^* = f_1\left( x^*, y^*, N_{ReL}, \frac{dp^*}{dx^*} \right) \tag{8.4.16}$$

such that the friction coefficient may be expressed as

$$C_f = \frac{2}{N_{ReL}} f_2\left( x^*, N_{ReL} \right) \tag{8.4.17}$$

for a given geometry. Similarly, the dimensionless temperature is

$$T^* = f_3\left( x^*, y^*, N_{ReL}, Pr, \frac{dp^*}{dx^*} \right) \tag{8.4.18}$$

and can be used to establish the Nusselt number

$$\text{Nu} = +\frac{\partial T^*}{\partial y^*} = f_4\left(x^*, N_{ReL}, Pr\right) \tag{8.4.19}$$

An average Nusselt number may be derived based on an average heat transfer coefficient, which is the result of integrating over the surface of the body under consideration. Such an average Nusselt number $\overline{\text{Nu}}$, would be more familiar, and is given in functional form as

$$\overline{\text{Nu}} = f_5\left(N_{ReL}, Pr\right) \tag{8.4.20}$$

Also, the Sherwood number, Sh, which is the mass transfer analog of the Nusselt number, is derivable from

$$C_A^* = f_6\left(x^*, y^*, N_{ReL}, Sc, \frac{dp^*}{dx^*}\right) \tag{8.4.21}$$

That is,

$$h_m = +\frac{D_{AB}}{L}\left(\frac{\partial C_A^*}{\partial y^*}\right)_{y^*=0} \tag{8.4.22}$$

but

$$\text{Sh} = \frac{h_m L}{D_{AB}} = +\left(\frac{\partial C_A^*}{\partial y^*}\right)_{y^*=0} = f_7\left(x^*, N_{ReL}, Sc\right) \tag{8.4.23}$$

or an average Sherwood number, $\overline{\text{Sh}}$

$$\overline{\text{Sh}} = f_8\left(N_{ReL}, Sc\right). \tag{8.4.24}$$

From the above discussion, we observe that scaling offers significant advantage in identifying important dimensionless groups. In fluid flow, the friction coefficient is important and for a given geometry, only the Reynolds number is needed (Equation 8.4.17). In the case of heat (or mass) transfer, the Nusselt (Sherwood) number can be correlated through the Reynolds and Prandtl (Schmidt) numbers, Equation 8.4.20 or Equation 8.4.24.

Application of Equations 8.4.17, 8.4.19, 8.4.20, 8.4.23, and 8.4.24 are not limited to the boundary layers. These are general results that are repeatedly used in other situations, such as for flow in conduits.

Another advantage resulting from scaling is the ability to reduce the order of a given differential equation. For example, Equation 8.3.15 could be reduced to a first order equation if the product of the Reynolds and Schmidt numbers is large enough. This reduction would not be as straightforward based on Equation 8.4.8 alone. To further emphasize this application, consider the steady-state laminar flow in a tube of radius R in which a solute A contained in the fluid undergoes a first order reaction at the wall (Krantz and Sczechowski[14]). Suppose one wants to determine the conditions justifying two different approximations:

1. Assume that the reaction causes total depletion of A at the pipe wall, and
2. Assume the "plug flow reactor" approximation, for which the radial concentration gradient is ignored while the flow is taken, to be equal to the average velocity.

The mathematical statement of the problem is as follows (see Figure 8.4).

$$V_z \frac{\partial C_A}{\partial z} = D_{AB}\left(\frac{1}{r}\frac{\partial}{\partial r}r\frac{\partial C_A}{\partial r}\right)$$

$$C_A = C_{A0} \quad \text{at} \quad z = 0$$

$$\frac{\partial C_A}{\partial r} = 0 \quad \text{at} \quad r = 0$$

$$-D_{AB}\frac{\partial C_A}{\partial r} = k_1 C_A \quad \text{at} \quad r = R \qquad 0 \le z \le L$$

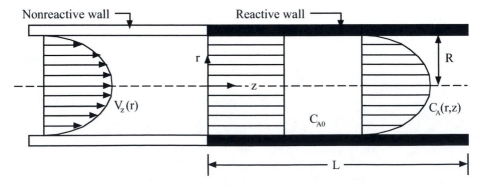

**FIGURE 8.4**
Schematic of steady-state laminar flow in a tube with reactive walls.

where $C_{A0}$ is the initial concentration of the solute, $D_{AB}$ is the binary diffusivity, and $k_1$ is the first order reaction rate coefficient.

The laminar flow velocity $v_z$ is given by

$$v_z = \frac{3}{2}\overline{v}\left[1 - \left(\frac{r}{R}\right)^2\right]$$

where $\overline{v}$ is the average velocity.

$$\text{Let } C_A^* = \frac{C_A}{C_s}, r^* = \frac{r}{r_s} \text{ and } z^* = \frac{z}{z_s}$$

Then substitute these scaled values into the differential equations and their boundary conditions. Following the substitution, divide through by the dimensional coefficient of one term. In this case it is convenient to divide by $\overline{v}\,C_s/z_s$ to obtain

$$\frac{3}{2}\left[1 - \left(\frac{r_s}{R}\right)^2 r^{*2}\right]\frac{\partial C_A^*}{\partial z^*} = \frac{D_{AB}}{\overline{v}}\frac{z_s}{r_s^2}\left(\frac{1}{r^*}\frac{\partial}{\partial r^*}r^*\frac{\partial C_A^*}{\partial r^*}\right)$$

$$C_A^* = \frac{C_{A0}}{C_s} \quad \text{at} \quad z^* = 0$$

$$\frac{\partial C_A^*}{\partial r^*} = 0 \quad \text{at} \quad r^* = 0$$

$$-\frac{\partial C_A^*}{\partial r^*} = \frac{k_1 r_s}{D_{AB}}C_A^* \quad \text{at} \quad r^* = \frac{R}{r_s} \quad 0 \le z^* \le \frac{L}{z_s}$$

In this problem, the dimensionless groups suggest the following choices for the scale factors:

$$\frac{C_{A0}}{C_s} = 1, \quad \frac{r_s}{R} = 1 \quad \text{and} \quad \frac{L}{z_s} = 1$$

Therefore, the dimensionless equations become

$$\frac{3}{2}\left[1 - r^{*2}\right]\frac{\partial C_A^*}{\partial z^*} = \frac{L}{R}\frac{1}{Pe}\left(\frac{1}{r^*}\frac{\partial}{\partial r^*}r^*\frac{\partial C_A^*}{\partial r^*}\right)$$

$$C_A^* = 1 \quad \text{at} \quad z^* = 0$$

$$\frac{\partial C_A^*}{\partial r^*} = 0 \quad \text{at} \quad r^* = 0$$

$$-\frac{\partial C_A^*}{\partial r^*} = N_{Sh} C_A^* \quad \text{at} \quad r^* = 1 \quad 0 \leq z^* \leq \frac{L}{Z_s}$$

where the quantity $Pe = \bar{v} R / D_{AB}$ is the Peclet number and is the ratio of convective mass transfer to molecular mass transfer, and

$$N_{Sh} = \frac{k_1 R}{D_{AB}}$$

is a Sherwood number or a Damkohler number.

If, $N_{Sh} \gg 1$, then $C_A^* \approx 0$ in order to assure that $\partial C_A^* / \partial r^*$ is of order one at $r^* = 1$. Therefore, for this limiting case (very fast reaction), the boundary condition can be replaced by

$$C_A^* \approx 0 \quad \text{at} \quad r^* = 1$$

If $N_{Sh} \ll 1$, then since $C_A^*$ is of order one, $\partial C_A^* / \partial r^* \ll 1$ at $r^* = 1$. However, $\partial C_A^* / \partial r^*$ is largest at $r^* = 1$. Therefore, $\partial C_A^* / \partial r^* \ll 1$ throughout the tube, and one can conclude that $C_A^* \approx C_A^*(z^*)$. Since the radial concentration gradient is negligible, the heterogeneous reaction term can be directly included into the species mass balance to obtain

$$\bar{v} \frac{dC_A}{dz} = -\frac{2k_1}{R} C_A \quad \text{for} \quad 0 \leq z \leq L$$

subject to

$$C_A = C_{A0} \text{ at } z = 0$$

which describes the classic plug flow reaction assumption.

In summary, scaling analysis can be reduced to a stepwise procedure. Following Krantz and Sczechowski[14]:

1. Write down the dimensional differential equations and their initial and boundary conditions appropriate to the transport or reactor design process being considered.

2. Form dimensionless variables by introducing unspecified scale factors for each dependent and independent variable: this also may involve introducing unspecified reference factors for some variables whose values we seek to normalize to zero.

3. Introduce these dimensionless variables into the describing differential equations and their initial and boundary conditions.

4. Divide through by the dimensional coefficient of one of the terms (preferably, one which will be retained) in each of the describing equations and their initial and boundary conditions.

5. Determine the scale and reference factors by ensuring that the principal terms in the describing equations are of order one; identifying the principal terms is dependent on the particular conditions for which the scaling is being done (e.g., a highly viscous flow, a conductive heat transfer process, etc. This step may require introducing a "region of influence," wherein the dependent variable(s) goes through a characteristic change in value).

6. The above steps result in the minimum number of dimensionless groups (parameters) for the problem. The describing equations can now be explored for very small or very large values of the dimensionless group.

Finally, a second example due to Krantz and Sczechowski[14]: consider the problem of steady-state, fully developed, laminar flow between two infinitely wide parallel plates shown in Figure 8.5. The lower plate is stationary, while the upper plate moves at a constant velocity $V_p$. The flow is also subjected to a constant axial pressure gradient such that $\Delta P > 0$. Determine the conditions for which the effect of the upper plate velocity $V_p$ can be neglected.

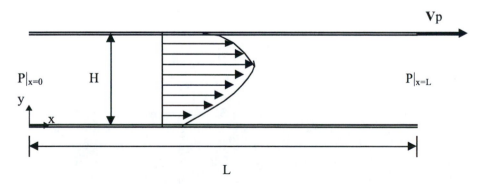

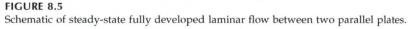

**FIGURE 8.5**
Schematic of steady-state fully developed laminar flow between two parallel plates.

*Solution*

The equations of motion and their boundary conditions are:

$$0 = -\frac{\partial P}{\partial x} + \mu \frac{d^2 v_x}{dy^2} \tag{8.4.25}$$

$$0 = -\frac{\partial P}{\partial y} + \rho g \tag{8.4.26}$$

$$v_x = 0 \quad \text{at} \quad y = 0 \tag{8.4.27}$$

$$v_x = V_P \quad \text{at} \quad y = H \tag{8.4.28}$$

Equation 8.4.26 can be integrated and combined with Equation 8.4.25 to obtain

$$0 = \frac{\Delta P}{L} + \mu \frac{d^2 v_x}{dy^2} \tag{8.4.29}$$

where $\Delta P = P_{X=0} - P_{X=L}$. Define the following dimensionless variables

$$v_x^* = \frac{v_x}{U_s} \quad \text{and} \quad y^* = \frac{y}{y_s} \tag{8.4.30}$$

Substitute Equation 8.4.30 into Equations 8.4.27 to 8.4.29 to yield

$$0 = \frac{\Delta P}{L} + \frac{\mu U_s}{y_s^2} \frac{d^2 v_x^*}{dy^{*2}} \tag{8.4.31}$$

$$U_s v_x^* = 0 \quad \text{at} \quad y_s y^* = 0 \tag{8.4.32}$$

$$U_s v_x^* = V_P \quad \text{at} \quad y_s y^* = H \tag{8.4.33}$$

Since the viscous term in Equation 8.4.31 must be retained in order to satisfy the two no-slip conditions at the solid boundaries, divide through by its dimensional coefficient. Similarly, in the boundary conditions, divide through by the dimensional coefficient of the dimensionless dependent variable. This results in

$$0 = \frac{\Delta P y_s^2}{L \mu U_s} + \frac{d^2 v_x^*}{dy^{*2}} \tag{8.4.34}$$

$$v_x^* = 0 \quad \text{at} \quad y^* = 0 \tag{8.4.35}$$

$$v_x^* = \frac{V_P}{U_s} \quad \text{at} \quad y^* = \frac{H}{y_s} \tag{8.4.36}$$

Since this problem is being scaled for conditions such that the flow is caused principally by the pressure gradient, the pressure force is to be balanced by the viscous term given in Equation 8.4.34. That is

$$\frac{\Delta P y_s^2}{L \mu U_s} = 1 \tag{8.4.37}$$

which ensures that the magnitude of the dimensionless derivative, $d^2v_x/dy^{*2}$, is of order one. Further, the dimensionless variable, $y^*$, will be bounded of order one if we impose the condition that

$$\frac{H}{y_s} = 1 \tag{8.4.38}$$

Therefore, using Equations 8.4.37 and 8.4.38 gives

$$U_s = \frac{H^2 \Delta P}{\mu L} \tag{8.4.39}$$

The velocity scale given by Equation 8.4.39 is directly proportional to the maximum velocity for flow between two flat plates driven only by a pressure gradient. This scaling ensures that dimensionless velocity goes from its minimum value of zero to its maximum value of one.

Finally, the dimensionless equations are:

$$0 = 1 + \frac{d^2v_x}{dy^{*2}} \tag{8.4.40}$$

$$v_x^* = 0 \quad \text{at} \quad y^* = 0 \tag{8.4.35}$$

$$v_x^* = \frac{V_P \mu L}{H^2 \Delta P} \quad \text{at} \quad y^* = 1 \tag{8.4.41}$$

Therefore, in order to ignore the effect of the moving upper plate on the flow relative to that of the imposed pressure gradient,

$$\frac{V_P \mu L}{H^2 \Delta P} \ll 1 \tag{8.4.42}$$

If one is interested in determining conditions such that flow is caused principally by the upper moving boundary, then the velocity scale would be

$$U_s = V_p$$

according to Equation 8.4.36. In this case, the dimensionless equations would become

$$0 = \frac{\Delta PH^2}{L\mu V_p} + \frac{d^2 v_x{}^*}{dy^{*2}}$$

Then, in order to neglect the effect of the pressure gradient on the flow, the criterion

$$\frac{H^2 \Delta P}{\mu V_p L} \ll 1$$

must be satisfied.

The analytical solution of this problem permits assessing the error incurred by neglecting the plate velocity. For example, if $(V_p\, \mu L/H^2\, \Delta P)$ $\leq 0.1$, comparison with the exact analytical solution shows that the error is 20% in the drag at the wall, while a 2% error in the drag results if $(V_p\, \mu L/H^2\, \Delta P) \leq 0.01$.

From this discussion, one can see that scaling can provide the criteria for simplifying the equations of change. Scaling is also helpful in assessing the incurred error in making these simplifying assumptions. Generally, one can observe that the error incurred in using the discussed scaling technique will be of the same order as the dimensionless group, which must be small enough to be neglected.

---

## 8.5  Problems

1.  Consider a fluid flowing in a closed conduit at some average velocity, v, with a temperature difference existing between the fluid and the tube wall. If the important variables, their symbols, and dimensional representations are tabulated below

    a.  Determine the number of pi groups.

    b.  List the contents of the recurring set.

    c.  What are the dimensionless groups?

| Variable | Symbol | Dimensions |
|---|---|---|
| Tube diameter | D | L |
| Fluid density | $\rho$ | $M/L^3$ |
| Fluid viscosity | $\mu$ | $M/L\,t$ |
| Fluid heat capacity | $c_p$ | $Q/M\,T$ |
| Fluid thermal conductivity | k | $Q/t\,L\,T$ |
| Velocity | v | $L/t$ |
| Heat transfer coefficient | h | $Q/t\,L^2 T$ |

The quantity Q represents heat, while T is temperature.

Answer:

$$Nu = f(N_{Re},\ Pr) \text{ or } N_{St} = f(N_{Re},\ Pr)$$

2. In the study of natural convection heat transfer from a vertical plane wall to an adjacent fluid, the variables are:

| Variable | Symbol | Dimensions |
|---|---|---|
| Significant length | L | L |
| Fluid density | $\rho$ | $M/L^3$ |
| Fluid viscosity | $\mu$ | $M/L\,t$ |
| Fluid heat capacity | $c_p$ | $Q/M\,T$ |
| Fluid thermal conductivity | k | $Q/t\,L\,T$ |
| Fluid thermal expansion | $\beta$ | $1/T$ |
| Gravitational acceleration | g | $L/t^2$ |
| Temperature difference | $\Delta T$ | T |
| Heat transfer coefficient | h | $Q/t\,L^2\,T$ |

a. Determine the number of pi groups.

b. List the contents of the recurring set.

c. What are the dimensionless groups?

Answer:

$$Nu = f(Gr,\ Pr)$$

3. Consider the transfer of mass from the walls of a circular conduit to a fluid flowing through the conduit. The important variables are:

| Variable | Symbol | Dimensions |
|---|---|---|
| Tube diameter | D | L |
| Fluid density | $\rho$ | $M/L^3$ |
| Fluid viscosity | $\mu$ | $M/L\,t$ |
| Fluid velocity | v | $L/t$ |
| Fluid diffusivity | $D_{AB}$ | $L^2/t$ |
| Mass transfer coefficient | $k_c$ | $L/t$ |

    a. Determine the number of pi groups.

    b. List the contents of the recurring set.

    c. What are the dimensionless groups?

Answer:

$$Nu_{AB} = f(N_{Re}, Sc)$$

# References

1. Taylor, E.S., *Dimensional Analysis for Engineers*, Oxford University Press, London, 1974.
2. Massey, B.S., *Measures in Science and Engineering*, Ellis Horwood Limited, Chichester West, UK, 1986.
3. Massey, B.S., *Units, Dimensional Analysis and Physical Similarity*, Van Nostrand, 1971.
4. Pankhurst, R.C., *Dimensional Analysis and Scale Factors*, Chapman and Hall, New York, 1964.
5. Hansen, A.G., *Similarity Analysis of Boundary Value Problems in Engineering*, Prentice-Hall, Englewood Cliffs, NJ, 1964.
6. Murphy, G., *Similitude in Engineering*, Ronald Press Co., New York, 1950.
7. Isaacson, E. de St Q. and Isaacson, M. de St Q., *Dimensional Methods in Engineering and Physics*, John Wiley, New York, 1975.
8. Welty, J.R., Wicks, C.E., and Wilson, R.E., *Fundamentals of Momentum, Heat and Mass Transfer*, 3rd ed., John Wiley, New York, 1984.
9. Bird, R.B., Stewart, W.E., and Lightfoot, E.N., *Transport Phenomena*, John Wiley, New York, 1960.
10. Incropera, F.P. and DeWitt, D.P., *Fundamentals of Heat and Mass Transfer*, 4th ed., John Wiley, New York, 1996.
11. Kays, W. M. and Crawford, M.E, *Convective Heat and Mass Transfer*, 3rd ed., McGraw-Hill, New York, 1993.
12. White, F. M., *Viscous Fluid Flow*, 2nd ed., McGraw-Hill, New York, 1991.
13. Schlichting, H., *Boundary Layer Theory*, 7th ed., McGraw-Hill, New York, 1979.
14. Krantz, W.B. and Sczechowski, J.G., Scaling initial and boundary value problems, *Chemical Engineering Education*, 28, 236, 1994.

# 9

## Selected Numerical Methods and Available Software Packages

### 9.1 Introduction and Philosophy

Numerical methods will play a more significant role in engineering science as computer technology continues to improve. Methods that were previously thought to be too effort consuming are now commercially available in any number of software packages such as Mathematica® (Wolfram Research, Inc.), Maple® (Waterloo Maple, Inc.) or Matlab® (The Math Works, Inc.). For example, finding the roots of an equation involving several confluent hypergeometric functions can be done with a single *Mathematica* command.

In this chapter, the notion will be to alert the reader to some of these commercially available packages and identify what numerical methods are being applied. There is no intent to produce an exposé on numerical analysis. However, it may be helpful to be aware of what kind of results a particular method might yield and be able to make informed decisions. To this end, the chapter will briefly review various methods and highlight their central ideas in order to allow a user to be mindful of the methods' virtues as well as their limitations.

### 9.2 Solution of Nonlinear Algebraic Equations

Nonlinear algebraic equations turn up quite frequently in chemical engineering and may appear in several different forms. For example, in thermodynamics, pressure-volume-temperature relationships of real gases are often described by equations of state, such as

$$PV^4 - RTV^3 - \beta V^2 - \gamma V - \delta = 0 \qquad (9.2.1)$$

where P, V, and T are the pressure, specific volume, and temperature, respectively. R is the gas constant and $\beta$, $\gamma$, and $\delta$ are empirical functions of temperature, specific to each gas. Equation 9.2.1 is the Beattie-Bridgeman equation of state and is a fourth-degree polynomial in the specific volume.

In multicomponent distillation, it may be necessary to estimate the minimum reflux ratio using classical methods (Underwood[1] and Treybal[2]). This estimate usually requires the solution of a polynomial in $\phi$ of degree n, such as the equation

$$\sum_{j=1}^{n} \frac{\alpha_j z_{jF} F}{\alpha_j - \phi} - F(1-q) = 0 \tag{9.2.2}$$

F   = the molar feed flow rate.
n   = the number of components in the feed.
$z_{jF}$ = the mole fraction of each component in the feed.
q   = the feed quality.
$\alpha_j$ = the relative volatility of each component at some average column conditions.
$\phi$   = the sought after root of the equation.

Fanning friction factor, f, for turbulent flow of an incompressible fluid in a smooth pipe is

$$\sqrt{\frac{2}{f}} = \frac{1}{k} \ln\left(N_{Re}\sqrt{\frac{f}{8}}\right) + B - A \tag{9.2.3}$$

The quantities A, B, and k are constants. $N_{Re}$ is the Reynolds number. Equation 9.2.3 is not in polynomial form, but can be rearranged to have all nonzero terms on one side of the equation. In fact, all three of the equations mentioned can be represented by the general form

$$f(x) = 0 \tag{9.2.4}$$

where x is the variable, which can have multiple values that satisfy the equation. It is important to remember that the roots of Equation 9.2.4 can be real and distinct or real and repeated, or have complex conjugates or some mixture of real and complex. Regardless of the type of roots, it may be helpful to have some sense as to which method one should use to attack the problem. That is, one would like to know if, by choosing a particular iterative procedure, there would be convergence.

A standard approach rearranges Equation 9.2.4 to develop a convergence criterion (Stark[3]).

Suppose $\bar{x}$ is a root of Equation 9.2.4, then

$$f(\bar{x}) = 0$$

and the equation is satisfied. However, a value x, which is not a root when substituted into Equation 9.2.4, would result in

$$f(\dot{x}) \neq 0$$

This guessing could go on indefinitely. But suppose we perform some algebraic manipulations on Equation 9.2.4 to extract an x out of it and bring it to the other side of the equation to get

$$x = g(x) \tag{9.2.5}$$

Then, if $\bar{x}$ is substituted into Equation 9.2.5 we would get the result

$$\bar{x} = g(\bar{x}) \tag{9.2.6}$$

while

$$\dot{x} \neq g(\dot{x})$$

Notice that g(x) is a new function that is different from the original f(x).

Because we usually have no idea where the roots of f(x) lie, it is typical to make a guess, and check that guess to determine which way to go next. However, it turns out that the criterion (Stark[3])

$$\left| g'(x) \right| < 1 \tag{9.2.7}$$

can be used to increase the chance of finding the roots of Equation 9.2.4 in an efficient way. That is, the absolute value of the derivative of g(x) needs to be less than one in the region containing the root $\bar{x}$ in order to guarantee convergence. For example, if

$$f(x) = e^x - 3x = 0$$

then

$$x = \frac{e^x}{3} \equiv g(x)$$

such that

$$g'(x) = \frac{e^x}{3}$$

and the range of x, for which the absolute value of the derivative is less than one, is given by

$$\left| \frac{e^x}{3} \right| < 1$$

Therefore,

$$x < \ln 3 \cong 1.1$$

Essentially, one should guess starting values of x smaller than 1.1 to expect convergence.

While this rearranging of the independent variable seems straightforward, one should exercise caution in accepting what appears to be obvious. For suppose the model

$$\frac{dC_A}{dt} = \frac{F_0}{V} C_{A0} - \frac{F_0}{V} C_A - kC_A^2$$

is used to represent the material balance of species A in an isothermal, constant volume chemical reactor system with second order reaction. If one is interested in the steady state concentration, $C_{A,s}$, then for an initial concentration, $C_{A0}$ of 1 mol/l, a feed to volume ratio, $F_0/V$ of 1 min$^{-1}$, and a rate coefficient, k = 1 l/(mol min), the model becomes

$$1 - C_{A,s} - C_{A,s}^2 = 0$$

In the form of Equation 9.2.4, substituting $x = C_{A,s}$, we get

$$1 - x - x^2 = 0$$

A quadratic whose direct solutions are x = −1.618 and 0.618. Of course, being a physical system, the negative result is ignored. For solving this problem by a substitution method, one would develop the form of Equation 9.2.5. In this example, there are two possibilities:

1. $g_1(x) \equiv x = \sqrt{1-x}$

2. $g_2(x) \equiv x = -x^2 + 1$

The first case results in

$$g'(x) = \frac{-1}{2\sqrt{1-x}}$$

such that $x < 3/4$ satisfies the requirement of Equation 9.2.7. After several iterations using direct substitution, the result $x = 0.618$ is achieved.

The second possibility results in

$$g'(x) = -2x$$

such that $x < 1/2$ satisfies the requirement of Equation 9.2.7. However, initial guess values of $x < 1/2$ would be useless, since convergence to the correct value of 0.618 could never be achieved.

If one were to solve either of these examples explicitly using an iterative procedure, the *Mathematica* routine "FindRoot" (Wolfram[4]) could perform this task with a starting value, $x_0$, a minimum value, $x_{min}$, and a maximum value, $x_{max}$ selected from the suggested range. FindRoot employs either of two approaches, depending on the number of starting values of x. If only one starting value is given, FindRoot employs Newton's method (Stark,[3] Burden, Faires, and Reynolds,[5] and Constatinides[6]). On the other hand, if one specifies two starting values, then a variant of the Secant method is employed (Burden, Faires, and Reynolds[5]).

As a demonstration of the use of FindRoot, consider the following equation, which defines the eigenvalues in a model of controlled release from a hollow fiber (Ramraj, Farrell, and Loney[7]):

$$\left[ kJ_1(x) + \beta x J_0(x) \right] \left\{ Y_1\left(x\frac{a}{b}\right) J_0\left(\frac{ax}{b\sqrt{k}}\right) - m\sqrt{k} Y_0\left(x\frac{a}{b}\right) J_1\left(\frac{ax}{b\sqrt{k}}\right) \right\} +$$

$$\left[ kY_1(x) + \beta x Y_0(x) \right] \left\{ m\sqrt{k} J_1\left(\frac{ax}{b\sqrt{k}}\right) J_0\left(x\frac{a}{b}\right) - J_0\left(\frac{ax}{b\sqrt{k}}\right) J_1\left(x\frac{a}{b}\right) \right\} = 0 \qquad (9.2.8)$$

where a, b, $\beta$, k, and m are given constants and $J_0(.)$, $Y_0(.)$, $J_1(.)$, and $Y_1(.)$ are Bessel functions. With given values of the constants, the line

$$
\text{FindRoot}\left[\begin{array}{l}
\left(\begin{array}{c}kBesselJ[1,x]\\+\beta x BesselJ[0,x]\end{array}\right)\left(\begin{array}{c}BesselY\left[1,x\dfrac{a}{b}\right]BesselJ\left[0,\dfrac{ax}{b\sqrt{k}}\right]\\-m\sqrt{k}BesselY\left[0,x\dfrac{a}{b}\right]BesselJ\left[1,\dfrac{ax}{b\sqrt{k}}\right]\end{array}\right)+\\[3em]
\left(\begin{array}{c}kBesselY[1,x]\\+\beta x BesselY[0,x]\end{array}\right)\left(\begin{array}{c}m\sqrt{k}BesselJ\left[1,\dfrac{ax}{b\sqrt{k}}\right]BesselJ\left[0,x\dfrac{a}{b}\right]-\\BesselJ\left[0,\dfrac{ax}{b\sqrt{k}}\right]BesselJ\left[1,x\dfrac{a}{b}\right]\end{array}\right)==0,\\[3em]
\left\{x,\left\{x_0,x_1\right\}\right\}
\end{array}\right]
$$

returns a numerical value for x. It is clear from this example that very complicated nonlinear functions can be dealt with in a straightforward manner by this routine. Previously, several lines of FORTRAN codes with careful attention to stability issues would be required to effect a solution. In addition to a single nonlinear equation, this routine can also be used to produce numerical solutions to systems of equations without regard to linearity.

In addition to FindRoot, there is also "NSolve" (Wolfram[4]), which is more suited to finding the roots of polynomials. Following the *Mathematica* system, polynomial root finding is based on the Jenkins-Traub algorithm (Jenkins and Traub[8]).

As mentioned previously, FindRoot can give results for systems of equations without regard to linearity. Therefore, similar to Equation 9.2.7, a criterion for systems can be established (Stark[3]). For a system of two equations, we have

$$
\left|\frac{\partial g_1}{\partial x}\right|+\left|\frac{\partial g_2}{\partial x}\right|\leq M<1
$$

(9.2.9)

$$
\left|\frac{\partial g_1}{\partial y}\right|+\left|\frac{\partial g_2}{\partial y}\right|\leq M<1
$$

for all x and y in the region containing all the values $x_i$ and $y_i$ the correct values $\bar{x}$ and $\bar{y}$. When the bound, M, on the partial derivatives are very small in the entire region, the iteration converges very quickly.

The quantities $x_i$ and $y_i$ are the iterants, while $g_1$ and $g_2$ are formed exactly the way Equation 9.2.5 was developed. Two common methods for finding roots to nonlinear systems are (1) Newton-Raphson, and (2) the modified Newton-Raphson. Both approaches are briefly discussed in the subsections below.

### 9.2.1 The Newton-Raphson Method

Suppose one wishes to solve the simultaneous equations

$$f_1(x,y) = 0$$

$$f_2(x,y) = 0$$

If both functions are continuous and differentiable, the following system can be developed (Stark[3]):

$$\frac{\partial f_1}{\partial x} h + \frac{\partial f_1}{\partial y} k = -f_1(x_i, y_i)$$

$$\frac{\partial f_2}{\partial x} h + \frac{\partial f_2}{\partial y} k = -f_2(x_i, y_i)$$

(9.2.10)

This system is the heart of the Newton-Raphson procedure. Here the partial derivatives are evaluated at $x_i$ and $y_i$. The quantities h and k are the unknowns, and are defined as

$$x_{i+1} = x_i + h$$

$$y_{i+1} = y_i + k$$

(9.2.11)

The system given by Equation 9.2.10 is very straightforward to solve, but requires the Jacobian determinant, J (see Appendix)

$$J = \begin{vmatrix} \dfrac{\partial f_1}{\partial x} & \dfrac{\partial f_1}{\partial y} \\ \dfrac{\partial f_2}{\partial x} & \dfrac{\partial f_2}{\partial y} \end{vmatrix}$$

(9.2.12)

to be nonsingular.

For example, if one were to solve the system

$$y = \cos x; \quad x = \sin y$$

using the Newton-Raphson procedure, the first step is to rewrite the system as

$$f_1(x,y) = \cos x - y = 0$$

$$f_2(x,y) = x - \sin y = 0$$

Then, differentiate to get

$$\frac{\partial f_1}{\partial x} = -\sin x; \frac{\partial f_2}{\partial x} = 1$$

$$\frac{\partial f_1}{\partial y} = -1; \frac{\partial f_2}{\partial x} = -\cos y$$

such that the Jacobian is

$$J = \begin{vmatrix} -\sin x & -1 \\ 1 & -\cos y \end{vmatrix} = \sin x \cos y + 1$$

Then, in order for the Jacobian to vanish, either sin x or cos y must be negative; that is

$$\sin x \cos y = -1$$

Therefore, as long as the iterants $x_i$ and $y_i$ are selected from the first quadrant, the Jacobian will not vanish. Then h and k can be determined for each iteration until some tolerance is met.

The need to know if the Jacobian is nonzero *a priori* cannot be met without some difficulty. In most cases, the procedure has to be carried out before such knowledge is in hand. Also, many computations are required for the use of this procedure; that is, for n simultaneous equations in n unknowns, $n^2$ partial derivatives must be determined. One must also solve for n increments h, k, ... by solving n simultaneous equations for n unknowns. These two disadvantages make it attractive to select other approaches to solving systems of nonlinear equations. One such approach involves a modification to the Newton-Raphson procedure and is highlighted in the next section.

### 9.2.2 The Modified Newton-Raphson Method

This procedure is essentially applying the single variable Newton-Raphson method n times, once for each variable. Each time the other variables are held constant. The approach is, given

$$f_1(x, y) = 0$$

$$f_2(x, y) = 0$$

then

$$x_1 = x_0 - \frac{f_1(x_0, y_0)}{\partial f_1 / \partial x} \tag{9.2.13}$$

where $\partial f_1 / \partial x$ is evaluated at $x_0$ and $y_0$. Next, $f_2$ and the most recent values of x and y, in this case $x_1$ and $y_0$, are used to calculate $y_1$:

$$y_1 = y_0 - \frac{f_2(x_1, y_0)}{\partial f_2 / \partial y} \tag{9.2.14}$$

$\partial f_2 / \partial y$ being evaluated at $x_1$ and $y_0$. With $x_i$ and $y_i$, $f_1$ is reused to calculate $x_2$. Then $f_2$ and the most recent values of x and y are used to calculate $y_2$ (Stark[3]). Notice that the choice of using $f_1$ to calculate a new x and using $f_2$ to calculate a new y appears arbitrary. Sometimes, that is, indeed, the case. However, the choice is supposed to be based on choosing the function with the steeper slope at the solution point $(\bar{x}, \bar{y})$ to determine the next x. Otherwise, if both functions have the same slope in a neighborhood of the solution, the iteration will oscillate about a true solution.

The real disadvantage in the modified Newton-Raphson method is the need to know *a priori* which function has the steeper slope at the solution point.

In practice, a graph of the functions can help to identify the curve with the steeper slope at the solution point. If a graph is not available, then one has to resort to an arbitrary choice of functions, and if divergence occurs, switch the roles of $f_1$ and $f_2$.

In summary, whether we are faced with a single nonlinear equation or a system, it is very likely that only one starting value will be available. A systematic selection of that value can improve the likelihood of a convergent solution. Equations 9.2.7 and 9.2.9 provide such a systematic approach.

At this point, it is important to point out that there are other software packages such as *Matlab*, *Maple*, and *Mathcad* that can also be employed to solve nonlinear algebraic equations. As a matter of fact, there are books providing a wide range of problems solved with *Matlab* (Constantinides and Mostoufi[9]).

*Mathematica* has packages available for calling *Matlab* from within *Mathematica* and vice versa. *Matlab* also has toolboxes that incorporate some of the symbolic capabilities of *Maple*, as pointed out in the literature (Pattee[10]).

*Mathcad*, on the other hand, is lacking in its ability to carry out symbolic computations when compared to *Mathematica* or *Matlab*. However, it is intuitive and very popular with the beginner, because there is no need to memorize commands.

*Maple* can produce program statements for a FORTRAN compiler in addition to its ability to perform symbolic algebra and numeric approximations. However, in a comparative study with *Mathematica* and *Matlab*, it was ranked behind *Mathematica* and *Matlab*, respectively (Mackenzie and Allen[11]).

---

## 9.3 Solution of Simultaneous Linear Algebraic Equations

When compared to other areas of numerical analysis, numerical linear algebra is well developed, because there is general agreement as to the best algorithms, and why they are best (Burden, Faires, and Reynolds[5] and Johnston[13]). There is a large body of work on the implementation of algorithms. This has resulted in packages of very efficient and reliable subroutines for the solution of problems in linear algebra (Burden, Faires, and Reynolds[5]). A primary goal of this section is to familiarize the reader with some of the background reasoning for the selection of algorithms.

Familiarity with matrix-vector notation and results involving matrices will be assumed, but most of what is needed is provided in the Appendix. Detailed derivations will be replaced by clarifying examples, most of which will be borrowed from the literature. These borrowed examples are chosen, because they have already been tested in the context of the algorithm or method under discussion.

Starting the discussion with the example (Riggs[14]):

$$x_1 + x_2 + 2x_3 = 9 \quad : E_1$$

$$3x_1 - 2x_2 + 3x_3 = 8 \quad : E_2 \tag{9.3.1}$$

$$4x_1 + 2x_2 - 2x_3 = 2 \quad : E_3$$

The objective is to determine the values of $x_1$, $x_2$, and $x_3$ that will simultaneously solve all three equations. In general, the system 9.3.1 is a subset of the n linear equations in n unknowns represented by:

$$Ax = b \tag{9.3.2}$$

where A is an $n \times n$ matrix and both $x$ and $b$ represent column vectors ($n \times 1$). Specifically, Equation 9.3.2 can be expanded into

$$
\begin{aligned}
a_{11}x_1 + a_{12}x_2 + \cdots a_{1n}x_n &= b_1 : E_1 \\
a_{21}x_1 + a_{22}x_2 + \cdots a_{2n}x_n &= b_2 : E_2 \\
&\vdots \\
a_{k1}x_1 + a_{k2}x_2 + \cdots a_{kn}x_n &= b_k : E_k \\
&\vdots \\
a_{n1}x_1 + a_{n2}x_2 + \cdots a_{nn}x_n &= b_n : E_n
\end{aligned}
\tag{9.3.3}
$$

or condensed using summation notation to

$$\sum_{i=1}^{n} a_{ki}x_i = b_k \quad (k = 1, 2, \cdots, n) \tag{9.3.4}$$

Regardless of the form, the objective will be the same.

Linear equations result naturally when we conduct material and energy balances, but most applications occur when we implement other numerical methods.

One of the most basic solution techniques for systems such as Equation 9.3.1 is *Gaussian elimination* (Stark,[3] Burden, Faires, and Reynolds,[5] Constantinides and Mostoufi,[9] Johnston,[13] and Riggs[14]) which is illustrated using the system (9.3.1).

We start by using the coefficient of $x_1$ in the first equation — designated as $E_1$, to eliminate the coefficients of $x_1$ in the other equations — designated as $E_2$ and $E_3$, respectively. That is (–4) times $E_1$ added to $E_3$ produces

$$0x_1 - 2x_2 - 10x_3 = -34$$

and (–3) times $E_1$ added to $E_2$ produces

$$0x_1 - 5x_2 - 3x_3 = -19$$

The *new equivalent* system of equations resulting from the above elementary operation is

$$x_1 + x_2 + 2x_3 = 9$$
$$-2x_2 - 10x_3 = -34 \tag{9.3.5}$$
$$-5x_2 - 3x_3 = -19$$

The next step is to repeat the procedure on this equivalent system, starting with the coefficient of the second row to eliminate the coefficient of $x_2$ in the third row. That is, (–5/2) times the new row two, add to the new row three, gives

$$0x_2 + 22x_3 = 66$$

The final equivalent system of equations is in *triangular form*

$$x_1 + x_2 + 2x_3 = 9$$
$$-2x_2 - 10x_3 = -34 \tag{9.3.6}$$
$$22x_3 = 66$$

It is not difficult to see how to derive the final results by solving for $x_3$ then $x_2$ and $x_1$, respectively, such that

$$\mathbf{x} = \begin{bmatrix} x_1 \\ x_2 \\ x_3 \end{bmatrix} = \begin{bmatrix} 1 \\ 2 \\ 3 \end{bmatrix}$$

or $x_1 = 1$, $x_2 = 2$, and $x_3 = 3$.

The procedure just described can be summarized into two basic steps:

1. Reduce or transform the given problem to an equivalent one, which is more easily solved.
2. Solve the reduced problem.

By *equivalent problem* we mean a problem that has the same solution as the original one.

The procedure can be generalized and is available in algorithms in many texts on numerical analysis (Burden, Faires, and Reynolds,[5] Constantinides and Mostoufi,[9] Johnston,[13] Riggs,[14] and Hanna and Sandall[15]). The first equation in 9.3.3 times $(a_{21}/a_{11})$ is subtracted from the second to eliminate the first term of the second equation; likewise, the first term of every equation thereafter, $k > 2$, is eliminated by subtracting the first equation times $(a_{k1}/a_{11})$. The result of this step should be

$$a_{11}x_1 + a_{12}x_2 + \cdots a_{1n}x_n = b_1$$
$$a'_{22}x_2 + \cdots a'_{2n}x_n = b'_2$$
$$\vdots \qquad \vdots$$
$$a'_{k2}x_2 + \cdots a'_{kn}x_n = b'_k$$
$$\vdots \qquad \vdots$$
$$a'_{n2}x_2 + \cdots a'_{nn}x_n = b'_n$$

where

$$a'_{kj} = a_{ij} - m_{k1}a_{1j}; \quad m_{kl} = (a_{k1}/a_{11})$$

Next, the second term of every equation in the third through the last equation, $k > 2$, is eliminated by subtracting the second equation times $a'_{k2} / a'_{22} = m'_{k2}$. Following this step, the third terms of the fourth through the last equation are eliminated. This is continued until the *forward elimination* process is completed and the form

$$a_{11}x_1 + a_{12}x_2 + a_{13}x_3 + \cdots + a_{1n}x_n = b_1$$
$$a'_{22}x_2 + a'_{23}x_3 + \cdots + a'_{2n}x_n = b'_2$$
$$a''_{33}x_3 + \cdots + a''_{3n}x_n = b''_3 \qquad (9.3.8)$$
$$\vdots$$
$$a_{nn}^{(n-1)}x_n = b_n'^{(n-1)}$$

results. The leading terms in each equation are called *pivots*.

The *backward substitution* procedure starts with the last equation, giving the solution

$$x_n = \frac{b_n^{(n-1)}}{a_{nn}^{(n-1)}} \qquad (9.3.8A)$$

Subsequently,

$$x_{n-1} = \frac{\left( b_{n-1}^{(n-1)} - a_{n-1,n}^{(n-1)} x_n \right)}{a_{n-1,n-1}^{(n-2)}}$$
$$\vdots$$
$$x_1 = \left( b_1 - \sum_{j=2}^{n} a_{ij}x_j \right) \Big/ a_{11}$$

completing Gaussian elimination.

In actual machine computation, the form given by Equation 9.3.2 is more useful for large systems. Here, we will rework the example using the above procedure, but with the alternate form given by Equation 9.3.2:

$$
\begin{array}{l}
x_1 + x_2 + 2x_3 = 9 \ : E_1 \\
3x_1 - 2x_2 + 3x_3 = 8: E_2 \\
4x_1 + 2x_2 - 2x_3 = 2: E_3
\end{array}
\Leftrightarrow
\begin{bmatrix} 1 & 1 & 2 \\ 4 & 2 & -2 \\ 3 & -2 & 3 \end{bmatrix}
\begin{bmatrix} x_1 \\ x_2 \\ x_3 \end{bmatrix}
=
\begin{bmatrix} 9 \\ 2 \\ 8 \end{bmatrix}
\Rightarrow
$$

$$
\begin{bmatrix}
1 & 1 & 2 & \vdots & 9 \\
4 & 2 & -2 & \vdots & 2 \\
3 & -2 & 3 & \vdots & 8
\end{bmatrix}
$$

where the first three columns are the coefficients of the equations in the system 9.3.1 and the last column is the **b**-vector as represented in Equation 9.3.2 or the nonhomogeneous term. This **b**-vector is appended unto the array of coefficients to give what is known as an *augmented array*. Notice that the second and third equations in the system 9.3.1 are switched for convenience.

To start the forward elimination process, the first row times 4 is subtracted from the second row. The first row times 3 is subtracted from the third row. The resulting augmented matrix is

$$\begin{bmatrix} 1 & 1 & 2 & \vdots & 9 \\ 0 & -2 & -10 & \vdots & -34 \\ 0 & -5 & -3 & \vdots & -19 \end{bmatrix}$$

Continuing the forward elimination, the second row times (5/2) is subtracted from the third row. The augmented matrix is now

$$\begin{bmatrix} 1 & 1 & 2 & \vdots & 9 \\ 0 & -2 & -10 & \vdots & -34 \\ 0 & 0 & 22 & \vdots & 66 \end{bmatrix}$$

which ends the forward elimination part of the procedure. Notice the form of the array; that is, each element below the main diagonal is zero. This is an example of an *upper triangular* array (see Appendix). The same triangular form resulted earlier.

The backward substitution starts with the last row. Interpreting the last row as

$$22x_3 = 66$$

results in $x_3 = 3$. Similarly, the second and first rows respectively result in

$$2x_2 - 10x_3 = -34 \quad \Rightarrow \quad x_2 = 2 \text{ and } x_1 = 1$$

Also demonstrated in this example is the fact that Gaussian elimination works when all the pivots are nonzero and in most cases are of the same order of magnitudes as the other elements. The method fails if the pivots are zero. When the pivots are relatively small, round-off error can create erroneous results unless corrective measures are introduced early. Here is an example in which the pivots are of very different orders of magnitude.

The system

$$.0030x_1 + 59.14x_2 = 59.17 : E_1$$

$$5.291x_1 - 6.130x_2 = 46.78 : E_2$$

has exact solution $x_1 = 10.00$ and $x_2 = 1.000$.

However, if we perform Gaussian elimination using four-digit arithmetic with rounding, the first pivot is $a_{11} = .0030$, and its associated multiplier, $m_{21}$, is 5.291/.0030 = 1764. Carrying out the operation $(E_2 - m_{21}E_1)$ is supposed to give the equivalent system

$$.0030x_1 + 59.14x_2 = 59.17$$

$$-104300x_2 = -104400$$

Then, backward substitution gives $x_2 = 1.001$, and $x_1 = -10.000$.

However, notice that a small error of 0.001 in $x_2$ leads to a very large error in $x_1$ (factor of 20,000).

If we revisit the system, but this time institute the corrective action by switching $E_1$ with $E_2$ to get

$$5.291x_1 - 6.130x_2 = 46.78 : E_1$$

$$.0030x_1 + 59.14x_2 = 59.17 : E_2$$

then the multiplier for this system is $m_{21} = (.0030/5.291)$ and the operation $(E_2 - m_{21}E_1)$ results in the equivalent system

$$5.291x_1 - 6.130x_2 = 46.78$$

$$59.14x_2 = 59.14$$

such that the four-digit results, using backward substitution, are $x_1 = 10.00$ and $x_2 = 1.000$.

This illustrates that Gaussian elimination can run into difficulties in cases where the pivot element $a_{kk}^{(k)}$ is small relative to the entries $a_{ij}^{(k)}$ for $k \le i \le n$ and $k \le j \le n$. Pivoting strategies are usually accomplished by selecting a new element for the pivot, $a_{pq}^{(k)}$, and then interchanging the $k$th and $p$th rows followed by the interchange of the $k$th and $q$th columns, if necessary. The strategy used in the latter example is termed *partial pivoting* or *maximal column pivoting*. Detail discussions on pivoting can be found elsewhere (Burden, Faires, and Reynolds,[5] Constantinides and Mostoufi,[9] Johnston,[13] Strang,[16] Amundson,[17] and Nakamura[18]).

A class of procedures that are not as sensitive to round-off errors is the so-called *iterative procedures*. One such procedure is the *Gauss-Seidel method*, briefly described below (Stark,[3] Burden, Faires, and Reynolds,[5] Constantinides and Mostoufi,[9] and Riggs[14]).

Consider the system 9.3.3, in which the diagonal elements $a_{11}, a_{22}, ..., a_{nn}$ are all nonzero, if necessary, the equations must be rearranged so that all diagonal elements are nonzero.

Solve the first equation for $x_1$, solve the second equation for $x_2$, and so on, to get

$$x_1 = \frac{1}{a_{11}}\left(b_1 - a_{12}x_2 - a_{13}x_3 - \cdots a_{1n}x_n\right)$$

$$x_2 = \frac{1}{a_{22}}\left(b_2 - a_{21}x_1 - a_{23}x_3 - \cdots a_{2n}x_n\right)$$

$$x_3 = \frac{1}{a_{33}}\left(b_3 - a_{31}x_1 - a_{32}x_2 - \cdots a_{3n}x_n\right) \qquad (9.3.9)$$

$$\vdots$$

$$x_n = \frac{1}{a_{nn}}\left(b_n - a_{n1}x_1 - a_{n2}x_2 - \cdots a_{n,n-1}x_{n-1}\right)$$

Then start the iteration by choosing an initial set of guesses for all the $x_i$ and inserting these into the first equation to calculate a new $x_1$. This value of $x_1$ along with the previously guessed values for $x_3, x_4, \ldots, x_n$ are inserted into the second equation to calculate $x_2$. The procedure is continued down the list until a new value of $x_n$ is determined, and then return to the top of the list to repeat the procedure if necessary.

As an illustrative example, consider the system

$$6x_1 - x_2 + x_3 = 7 : E_1$$
$$x_1 + 4x_2 - x_3 = 6 : E_2$$
$$x_1 - 2x_2 + 4x_3 = 9 : E_3$$

Then, solving for $x_1$ using the first equation, $E_1$, $x_2$ using $E_2$ and $x_3$ using $E_3$ results in

$$x_1 = \frac{7}{6} + \frac{1}{6}x_2 - \frac{1}{6}x_3$$

$$x_2 = \frac{3}{2} - \frac{1}{4}x_1 + \frac{1}{4}x_3$$

$$x_3 = \frac{9}{4} - \frac{1}{4}x_1 + \frac{1}{2}x_2$$

Next, we guess $\mathbf{x} = (0, 0, 0)^t$ as the starting point, and successive application leads to

$$x_1 = \frac{7}{6} = 1.16, \, x_2 = 3/2 - \frac{1}{4}\left(\frac{7}{6}\right) = \frac{29}{24} = 1.21, \, x_3 = 9/4 - \frac{1}{4}\left(\frac{7}{6}\right) + \frac{1}{2}\left(\frac{29}{24}\right) = 2.56$$

The vector $\mathbf{x} = (1.16, 1.21, 2.56)^t$ constitutes the values for the next iteration. This process was carried out until the result $\mathbf{x} = (1, 2, 3)^t$ was achieved. This required six iterations.

Convergence is assumed when a change in each of the answers satisfies a preset criterion. Generally, the procedure is guaranteed to converge whenever the diagonal element is larger than the sum of the row elements or column elements.

Another in the class of iterative procedure is the *Jacobi method*. This method is illustrated by solving the following system:

$$
\begin{aligned}
10x_1 - x_2 + 2x_3 \quad\;\; &= 6 : E_1 \\
-x_1 + 11x_2 - x_3 + 3x_4 &= 25 : E_2 \\
2x_1 - x_2 + 10x_3 - x_4 &= -11 : E_3 \\
3x_2 - x_3 + 8x_4 &= 15 : E_4
\end{aligned}
$$

This system has exact solution $\mathbf{x} = (1, 2, -1, 1)^t$.

The goal of the method is to convert $A\mathbf{x} = \mathbf{b}$ to the form $\mathbf{x} = T\mathbf{x} + \mathbf{c}$. This is accomplished by solving equation $E_i$ for $x_i$, for each $i = 1, 2, 3, 4$ to obtain

$$
\begin{aligned}
x_1 &= \frac{1}{10}x_2 - \frac{1}{5}x_3 + \frac{3}{5} \\
x_2 &= \frac{1}{11}x_1 + \frac{1}{11}x_3 - \frac{3}{11}x_4 + \frac{25}{11} \\
x_3 &= -\frac{1}{5}x_1 + \frac{1}{10}x_2 + \frac{1}{10}x_4 - \frac{11}{10} \\
x_4 &= -\frac{3}{8}x_2 + \frac{1}{8}x_3 + \frac{15}{8}
\end{aligned}
$$

Thus, the matrix T is given by

$$
T = \begin{bmatrix}
0 & 1/10 & -1/5 & 0 \\
1/11 & 0 & 1/11 & -3/11 \\
-1/5 & 1/10 & 0 & 1/10 \\
0 & -3/8 & 1/8 & 0
\end{bmatrix}
$$

and the vector, $\mathbf{c}$, by $\mathbf{c} = (3/5, 25/11, -11/10, 15/8)^t$.

To start this procedure, we guess $\mathbf{x}^{(0)} = (0, 0, 0, 0)^t$ and generate $x_i^{(1)}$ by

$$
\begin{aligned}
x_1^{(1)} &= \frac{1}{10}x_2^{(0)} - \frac{1}{5}x_3^{(0)} & &+ \frac{3}{5} &= 0.6000 \\
x_2^{(1)} &= \frac{1}{11}x_1^{(0)} + \frac{1}{11}x_3^{(0)} - \frac{3}{11}x_4^{(0)} &+ \frac{25}{11} &= 2.2727 \\
x_3^{(1)} &= -\frac{1}{5}x_1^{(0)} + \frac{1}{10}x_2^{(0)} + \frac{1}{10}x_4^{(0)} &- \frac{11}{10} &= -1.1000 \\
x_4^{(1)} &= -\frac{3}{8}x_2^{(0)} + \frac{1}{8}x_3^{(0)} &+ \frac{15}{8} &= 1.8750
\end{aligned}
$$

Additional iterates, k, can be generated for $\mathbf{x}^{(k)} = (x_1^{(k)}, x_2^{(k)}, x_3^{(k)}, x_4^{(k)})^t$. By continuing the procedure up to $k = 10$, we get

$$
\begin{aligned}
x_1^{(10)} &= & \frac{1}{10}x_2^{(9)} & -\frac{1}{5}x_3^{(9)} & & +\frac{3}{5} & = 1.0001 \\
x_2^{(10)} &= \frac{1}{11}x_1^{(9)} & & +\frac{1}{11}x_3^{(9)} & -\frac{3}{11}x_4^{(9)} & +\frac{25}{11} & = 1.9998 \\
x_3^{(10)} &= -\frac{1}{5}x_1^{(9)} & +\frac{1}{10}x_2^{(9)} & & +\frac{1}{10}x_4^{(9)} & -\frac{11}{10} & = -0.9998 \\
x_4^{(10)} &= & -\frac{3}{8}x_2^{(9)} & +\frac{1}{8}x_3^{(9)} & & +\frac{15}{8} & = 0.9998
\end{aligned}
$$

The decision to stop after ten iterations is due to the criterion that the relative difference between the last two iterations is to be smaller than some appropriately chosen tolerance.

Notice that one difference between the Gauss-Seidel and Jacobi methods is in their use of the newly calculated $x_i$ value. In Gauss-Seidel, the newly calculated $x_i$ value is used to determine the $x_{i+1}$ value, but this is not so in the Jacobi method. In most cases, this leads to faster convergence of the Gauss-Seidel over the Jacobi method. Generally, iteration methods are most useful for large matrices with a substantial number of zero elements.

In procedures such as the Jacobi and Gauss-Seidel methods, a residual vector, $\mathbf{r}$

$$\mathbf{r} = \mathbf{b} - A\tilde{\mathbf{x}} \tag{9.3.10}$$

where $\tilde{\mathbf{x}}$ is an approximation to the solution vector, is associated with each calculation of a component to the solution vector (Burden, Faires, and Reynolds[5]).

Then, if

$$\mathbf{r}_i^{(k)} = \left( r_{1i}^{(k)}, r_{2i}^{(k)}, \cdots, r_{ni}^{(k)} \right)^t \tag{9.3.11}$$

is the residual vector for the Gauss-Seidel method corresponding to the approximate solution vector $(x_1^{(k)}, x_2^{(k)}, \ldots, x_{i-1}^{(k)}, x_i^{(k-1)}, \ldots, x_n^{(k-1)})^t$, the $m$th component of $\mathbf{r}_i^{(k)}$ is

$$
\begin{aligned}
r_{mi}^{(k)} &= b_m - \sum_{j=1}^{i-1} a_{mj} x_j^{(k)} - \sum_{j=i}^{n} a_{mj} x_j^{(k-1)} \\[2mm]
&= b_m - \sum_{j=1}^{i-1} a_{mj} x_j^{(k)} - \sum_{j=i+1}^{n} a_{mj} x_j^{(k-1)} - a_{mj} x_i^{(k-1)}
\end{aligned}
\tag{9.3.12}
$$

for each m = 1, 2, ..., n,

Therefore, the $i$th component of $\mathbf{r}_i^{(k)}$ is

$$r_{ii}^{(k)} = b_i - \sum_{j=1}^{i-1} a_{ij} x_j^{(k)} - \sum_{j=i+1}^{n} a_{ij} x_j^{(k-1)} - a_{ii} x_i^{(k-1)} \tag{9.3.13}$$

such that

$$r_{ii}^{(k)} + a_{ii} x_i^{(k-1)} = b_i - \sum_{j=1}^{i-1} a_{ij} x_j^{(k)} - \sum_{j=i+1}^{n} a_{ij} x_j^{(k-1)} \tag{9.3.14}$$

However, in the Gauss-Seidel method,

$$x_i^{(k-1)} = \frac{1}{a_{ii}} \left( b_i - \sum_{j=1}^{i-1} a_{ij} x_j^{(k)} - \sum_{j=i+1}^{n} a_{ij} x_j^{(k-1)} \right)$$

Therefore,

$$r_{ii}^{(k)} + a_{ii} x_i^{(k-1)} = a_{ii} x_i^{(k)} \tag{9.3.15}$$

or

$$x_i^{(k)} = x_i^{(k-1)} + \frac{r_{ii}^{(k)}}{a_{ii}} \tag{9.3.16}$$

The result given by Equation 9.3.16 is obtained by requiring that the $i$th coordinate of the vector, $\mathbf{r}_{i+1}^{(k)}$ be zero at each stage. However, the reduction to zero of one coordinate of the residual vector is not the most effective way to reduce the norm of $\mathbf{r}_{i+1}^{(k)}$.

A modification that can lead to a more significant improvement in convergence speed is

$$x_i^{(k)} = x_i^{(k-1)} + \omega \frac{r_{ii}^{(k)}}{a_{ii}} \tag{9.3.17}$$

for certain choices of the positive relaxation parameter, $\omega$. Methods using Equation 9.3.17 are called *relaxation methods* (Burden, Faires, and Reynolds[5]).

For choices of $0 < \omega < 1$, we have *under-relaxation methods* which are successful for some systems that are not convergent for Gauss-Seidel. Those methods associated with $\omega > 1$ are called *over-relaxation methods* and are useful in accelerating the convergence for systems that are already convergent by

Gauss-Seidel. These over-relaxation methods are also named *Successive Over-Relaxation* (SOR), and find application in the numerical solution of certain partial differential equations.

In practice, Equation 9.3.14 and Equation 9.3.17 are appropriately combined to give

$$x_i^{(k)} = (1-\omega)x_i^{(k-1)} + \frac{\omega}{a_{ii}}\left[b_i - \sum_{j=1}^{i-1}a_{ij}x_j^{(k)} - \sum_{j=i+1}^{n}a_{ij}x_j^{(k-1)}\right] \qquad (9.3.18)$$

which is the central theme of the SOR methods.

Below is an illustration comparing the SOR, Equation 9.3.18, and the Gauss-Seidel methods. The following system

$$\begin{array}{llll} 4x_1 & +3x_2 & & = 24 & : E_1 \\ 3x_1 & +4x_2 & -x_3 & = 30 & : E_2 \\ & -x_2 & +4x_3 & = -24 & : E_3 \end{array}$$

has exact solution $x = (3, 4, -5)^t$. For both methods the initial guess is $x^{(0)} = (1, 1, -1)^t$. The equations for Gauss-Seidel are

$$x_1^{(k)} = -0.75x_2^{(k-1)} + 6$$

$$x_2^{(k)} = -0.75x_1^{(k)} + 0.25x_3^{(k-1)} + 7.5$$

$$x_3^{(k)} = 0.25x_2^{(k)} - 6$$

for each $k = 1, 2, \ldots$ and the equations for the SOR method with $\omega = 1.25$ are

$$x_1^{(k)} = -0.25x_1^{(k-1)} - 0.9375x_2^{(k-1)} + 7.5$$

$$x_2^{(k)} = -0.9375x_1^{(k)} - 0.25x_2^{(k-1)} + 0.3125x_3^{(k-1)} + 9.375$$

$$x_3^{(k)} = 1.3125x_2^{(k)} - 0.25x_3^{(k-1)} - 7.5$$

Table 9.1 summarizes the results of the comparison for up to seven iterations. From this table, we see that the SOR method is converging much faster than the Gauss-Seidel.

Not shown in the table is the fact that the Gauss-Seidel method required 34 iterations, versus 14 for the SOR, in order to obtain seven-decimal-place accuracy.

Guidelines on how one may choose $\omega$, including Kahan and Ostrowski-Reich conditions, can be found in the literature (Burden, Faires, and Reynolds[5]).

**TABLE 9.1**

Comparison of Gauss-Seidel and SOR Methods

| | Gauss-Seidel | SOR with $\omega = 1.25$ |
|---|---|---|
| $x_1^{(1)}$ | 5.2500 | 6.3125 |
| $x_2^{(1)}$ | 3.8125 | 3.5195 |
| $x_3^{(1)}$ | −5.0469 | −6.6501 |
| $x_1^{(5)}$ | 3.0343 | 3.0037 |
| $x_2^{(5)}$ | 3.9714 | 4.0029 |
| $x_3^{(5)}$ | −5.0071 | −5.0057 |
| $x_1^{(7)}$ | 3.0134 | 3.0000 |
| $x_2^{(7)}$ | 3.9888 | 4.0002 |
| $x_3^{(7)}$ | −5.0028 | −5.0003 |

## 9.3.1 Error Estimate

In addition to what has been said above about achieving a good approximation to the solution vector, $x$ in $Ax = b$, it is necessary to add a few more words regarding "very difficult to converge" systems and suggest how one may attack such problems.

It is not unreasonable to expect that if $\tilde{x}$ is an approximation to the solution $x$, then the residual $r = A\tilde{x} - b$ should have the property that the norm, $\|r\|$ is small such that the norm of the difference, $\|x - \tilde{x}\|$ would be small. Quite often, this is exactly the case. However, certain systems occurring in practice fail to display this behavior. For example, the system given by

$$\begin{bmatrix} 1 & 2 \\ 1.0001 & 2 \end{bmatrix} \begin{bmatrix} x_1 \\ x_2 \end{bmatrix} = \begin{bmatrix} 3 \\ 3.0001 \end{bmatrix}$$

has the approximate solution, $\tilde{x} = (3, 0)^t$ with the residual vector

$$r = b - A\tilde{x} \begin{bmatrix} 3 \\ 3.0001 \end{bmatrix} - \begin{bmatrix} 1 & 2 \\ 1.0001 & 2 \end{bmatrix} \begin{bmatrix} 3 \\ 0 \end{bmatrix} = \begin{bmatrix} 0 \\ .0002 \end{bmatrix}$$

Using matrix property 8, given in the Appendix (Equation A.20), we can evaluate the $\ell_\infty$ norm of the vector $r$ to be $\|r\|_\infty = .0002$. Even though the norm of this residual vector appears to be quite small, it is clear that the approximation $\tilde{x} = (3, 0)^t$ is poor; in fact, using Equation A.22, $\|x - \tilde{x}\|_\infty = 2$.

This example demonstrates what can happen when we approximate the solution to a system representing two almost parallel lines. The point (3, 0) lies on one line: $x_1 + 2x_2 = 3$ and is very close to the other line given by $1.0001x_1 + 2x_2 = 3.0001$, but is very different from the actual intersection point (1,1).

Generally, the system geometry may not be available to help us determine *a priori* when problems might occur; therefore, we must find an alternative. Such help can be obtained from the norms of the matrix A and its inverse (Burden, Faires, and Reynolds,[5] Strang,[16] Forsythe and Moler,[19] and Forsythe, Malcolm, and Moler[20]). In this section, we will use the $\ell_\infty$ norm as opposed to the $\ell_2$ norm for matrices (as opposed to vectors) where

$$\|A\|_\infty = \max_{\|x\|_\infty = 1} \|Ax\| \tag{9.3.19}$$

is the $\ell_\infty$ norm and the $\ell_2$ norm is

$$\|A\|_2 = \max_{\|x\|_2 = 1} \|Ax\|_2 \tag{9.3.20}$$

Further, if $A = [a_{ij}]$ is an $n \times n$ matrix, then

$$\|A\|_\infty = \max_{1 \le i \le n} \sum_{j=1}^{n} |a_{ij}| \tag{9.3.21}$$

To see how Equation 9.3.21 works, supposing

$$A = \begin{bmatrix} 1 & 2 & -1 \\ 0 & 3 & -1 \\ 5 & -1 & 1 \end{bmatrix}$$

then

$$\sum_{j=1}^{n} |a_{1j}| = |1| + |2| + |-1| = 4$$

$$\sum_{j=1}^{n} |a_{2j}| = |0| + |3| + |-1| = 4$$

$$\sum_{j=1}^{n} |a_{3j}| = |5| + |-1| + |1| = 7$$

therefore,

$$\|A\|_\infty = \max_{1 \le i \le n} \sum_{j=1}^{n} |a_{ij}| = \max\{4, 4, 7\} = 7$$

To continue the discussion on how one might use the norms of A and its inverse, $A^{-1}$ to anticipate accuracy based on the residual of an approximation, we need the following (Burden, Faires, and Reynolds[5]):

$$\|x - \tilde{x}\| \le \|r\| \|A^{-1}\| \tag{9.3.22}$$

and

$$\frac{\|x - \tilde{x}\|}{\|x\|} \le \|A\| \|A^{-1}\| \frac{\|r\|}{\|b\|}; \quad x \ne 0 \text{ and } b \ne 0 \tag{9.3.23}$$

A quantity called the *condition number*, K(A), is defined relative to a norm to be

$$K(A) = \|A\| \|A^{-1}\| \tag{9.3.24}$$

such that

$$\|x - \tilde{x}\| \le K(A) \frac{\|r\|}{\|A\|} \tag{9.3.25}$$

and

$$\frac{\|x - \tilde{x}\|}{\|x\|} \le K(A) \frac{\|r\|}{\|b\|} \tag{9.3.26}$$

Then, since

$$1 = \|I\| = \|A \bullet A^{-1}\| \le \|A\| \|A^{-1}\| = K(A)$$

it is expected that the matrix A will be a *well-conditioned* matrix if K(A) is close to one and an *ill-conditioned* matrix if K(A) is much greater than one.

Returning to the first example of this section

$$A = \begin{bmatrix} 1 & 2 \\ 1.0001 & 2 \end{bmatrix}$$

has $\|A\|_\infty = 3.0001$. This quantity does not appear to be too large; however,

$$A^{-1} = \begin{bmatrix} -10000 & 10000 \\ 5000.05 & -5000 \end{bmatrix}$$

such that $\|A^{-1}\|_\infty = 20000$ and $K(A) = 60002$ — indeed a very significant difference from one. We have omitted the computation details of the inverse here, and refer the reader to the Appendix for a method on how to calculate it.

In practice, the calculation of the inverse is subject to round-off error, and will depend on the accuracy with which the calculation is performed. Based on this reasoning, it is desirable to estimate the condition number by an alternate procedure. Such a procedure can be found in the literature (Burden, Faires, and Reynolds[5]).

Essentially, using t-digit arithmetic and Gaussian elimination to approximate the solution to $Ax = b$, it has been shown (Forsythe and Molder[19]) that the residual vector $r$ for the approximation $\tilde{x}$ has the property

$$\|r\| \approx 10^{-t}\|A\|\|\tilde{x}\| \tag{9.3.27}$$

From Equation 9.3.27, an estimate for the effective condition number in t-digit arithmetic can be obtained without the need to invert the matrix A. Also, Equation 9.3.27 assumes that all the arithmetic operations in the Gaussian elimination are performed using t-digit arithmetic, but that the operations that are needed to determine the residual are done in 2t-digit arithmetic.

In order to approximate the t-digit (effective) condition number, consider the system

$$Ay = r \tag{9.3.28}$$

The approximate solution to Equation 9.3.28 is readily available, since the multipliers for the Gaussian elimination method have been calculated (and retained). Now, then $\tilde{y}$, the approximate solution of $Ay = r$, satisfies

$$\tilde{y} \approx A^{-1}r = A^{-1}(b - A\tilde{x}) = A^{-1}b - A^{-1}A\tilde{x} = x - \tilde{x} \tag{9.3.29}$$

such that $\tilde{y}$ is an estimate of the error in approximating the solution to the original system. Then, using Equation 9.3.27

$$\|\tilde{y}\| \approx \|x - \tilde{x}\| = \|A^{-1}r\| \le \|A^{-1}\|\|r\| \approx \|A^{-1}\|(10^{-t}\|A\|\|\tilde{x}\|) = 10^{-t}\|\tilde{x}\|K(A)$$

resulting in

$$K(A) \approx \frac{\|\tilde{\mathbf{y}}\|}{\|\tilde{\mathbf{x}}\|} 10^t \qquad (9.3.30)$$

an estimate for the condition number involved with solving $A\mathbf{x} = \mathbf{b}$ using Gaussian elimination. Below is an illustrative example summarizing some of the main points of this section (Burden, Faires, and Reynolds[5]).
Consider the system

$$\begin{bmatrix} 3.3330 & 15920 & -10.333 \\ 2.2220 & 16.710 & 9.6120 \\ 1.5611 & 5.1791 & 1.6852 \end{bmatrix} \begin{bmatrix} x_1 \\ x_2 \\ x_3 \end{bmatrix} = \begin{bmatrix} 15913 \\ 28.544 \\ 8.4254 \end{bmatrix}$$

Using Gaussian elimination and 5-digit arithmetic, this system results in the augmented matrix

$$\begin{bmatrix} 3.3330 & 15920 & -10.333 & \vdots & 15913 \\ 0 & -10596 & 16.501 & \vdots & -10580 \\ 0 & 0 & -5.0790 & \vdots & -4.7000 \end{bmatrix}$$

whose approximate solution is $\tilde{\mathbf{x}} = (1.2001, .99991, .92538)^t$.
The residual vector corresponding to $\tilde{\mathbf{x}}$ is computed in double precision to be

$$\rho = \beta - A\tilde{\xi} = \begin{bmatrix} 15913 \\ 28.544 \\ 8.4254 \end{bmatrix} - \begin{bmatrix} 3.3330 & 15920 & -10.333 \\ 2.2220 & 16.710 & 9.6120 \\ 1.5611 & 5.1791 & 1.6852 \end{bmatrix} \begin{bmatrix} 1.2001 \\ .99991 \\ .92538 \end{bmatrix}$$

$$= \begin{bmatrix} -.00518 \\ .27413 \\ -.18616 \end{bmatrix}$$

such that $\|r\|_\infty = .27413$.
Using 5-digit arithmetic for the calculations, give the approximation for the inverse of A to be

$$A^{-1} = \begin{bmatrix} -1.1701 \times 10^{-4} & -1.4983 \times 10^{-1} & 8.5416 \times 10^{-1} \\ 6.2782 \times 10^{-5} & 1.2124 \times 10^{-4} & -3.0662 \times 10^{-4} \\ -8.6631 \times 10^{-5} & 1.3846 \times 10^{-1} & -1.9689 \times 10^{-1} \end{bmatrix}$$

and Equation 9.3.21 give $\|A^{-1}\|_\infty = 1.0041$ and $\|A\|_\infty = 15934$, such that this estimate of the condition number K(A) is 15,999 using Equation 9.3.24.

The estimate for the condition number using the t-digit approach, Equation 9.3.27 to Equation 9.3.30, involves solving the system

$$\begin{bmatrix} 3.3330 & 15920 & -10.333 \\ 2.2220 & 16.710 & 9.6120 \\ 1.5611 & 5.1791 & 1.6852 \end{bmatrix} \begin{bmatrix} y_1 \\ y_2 \\ y_3 \end{bmatrix} = \begin{bmatrix} .00518 \\ .27413 \\ -.18616 \end{bmatrix}$$

resulting in $\tilde{y} = (-.20008, 8.9987 \times 10^{-5}, .07461)^t$. Then, using Equation 9.3.30 we get

$$K(A) \approx \frac{\|\tilde{y}\|_\infty}{\|\tilde{x}\|_\infty} 10^t = 10^5 \frac{(.20008)}{1.2001} = 16672$$

which gives the same quality information without involving the computation of $A^{-1}$.

In the development of Equation 9.3.30, the estimate $\tilde{y} \approx x - \tilde{x}$ was used. A reasonable expectation is that $\tilde{x} + \tilde{y}$ is a more accurate approximation to the solution of $Ax = b$ compared to $\tilde{x}$. This notion is developed into a method called *iterative refinement* (Burden, Faires, and Reynolds[5]). To see how this improvement works, reconsider the latter example. The approximation to the given problem using 5-digit arithmetic and Gaussian elimination is $\tilde{x}^{(1)} = (1.2001, .99991, .92538)^t$ and the solution to $Ay = r^{(1)}$ is $y^{(1)} = -.20008, 8.9987 \times 10^{-5}, .07461)^t$.

Then, using the *iterative refinement* approach gives

$$\tilde{x}^{(2)} = \tilde{x}^{(1)} + y^{(1)} = (1.0000, 1.0000, .99999)^t$$

where the actual error in this approximation is $\|x - \tilde{x}^{(2)}\|_\infty = 1 \times 10^{-5}$ as compare to $\|x - \tilde{x}\|_\infty = .2001$ or

$$\frac{\|x - \tilde{x}\|_\infty}{\|x\|_\infty} = .2001$$

resulting from 5-digit arithmetic with Gaussian elimination.

One more iteration involves computing $r^{(2)} = b - A\tilde{x}^{(2)}$ and solving $Ay^{(2)} = r^{(2)}$, resulting in $y^{(2)} = (1.5002 \times 10^{-9}, 2.0951 \times 10^{-10}, 1.0000 \times 10^{-5})$. Then,

$$\tilde{x}^{(3)} = \tilde{x}^{(2)} + y^{(2)} = (1.0000, 1.0000, 1.0000)^t$$

which agrees with the actual solution of the given system.

## 9.4 Solution of Ordinary Differential Equations

In keeping with the philosophy of this chapter, we will highlight certain numerical strategies for approximating the solutions of ordinary differential equations. In doing so, we aim to point out the advantages and disadvantages of a given method or type of method, while leaving the actual algorithms and their derivations to the numerical analysis literature. Also, where appropriate, we will suggest available software packages that employ the numerical algorithm under discussion.

Following the traditional approach, this section will be discussed under two subheadings: (1) Initial value problems and (2) boundary value problems.

Supporting information, such as the algebra or calculus of finite differences, can be found in the literature (Burden, Faires, and Reynolds,[5] and Constantinides and Mostoufi[9]).

### 9.4.1   Initial Value Problems

For chemical engineers, the unsteady-state problems are of great interest, since they can help us to decide on process control strategies that can manage process upsets. As it turns out, very few of these will yield analytical results without a substantial number of simplifying assumptions. When such assumptions are unwarranted or unjustifiable, we have to turn to a numerical approach. Once this decision is made, we are then faced with choosing a method that will give a reliable approximation of the solution to the given problem. In deciding on a method, one must be mindful of the various errors that can occur and, in particular, propagation error, that can lead to instability in the method. Even though error analysis is an important aspect of numerical analysis, we will leave the in-depth discussion to the numerical analysis texts.

The discussion that follows assumes that the problem under consideration for a numerical approach will be well posed (Burden, Faires, and Reynolds,[5] and Boyce and DiPrima[21]).

Given a differential equation of the form

$$\frac{dy}{dt} = f(t, y) \tag{9.4.1}$$

and the condition that $y = b$ when $t = a$, the analytically derived solution, $F(t)$, passing through $(a, b)$ would be an equation of the form

$$y = F(t) \tag{9.4.2}$$

The issue now is, how does Equation 9.4.1 describe the solution given by Equation 9.4.2 when all we know are Equation 9.4.1 and the initial value?

Generally, the procedures will rely on the fact that given any point (t, y) on the solution curve, we can obtain the direction of the curve through that point. In principle, we have a starting point and a set of directions given by Equation 9.4.1 and without any other information, we must follow the directions to the desired final destination, the approximate solution.

The most straightforward method, the *Euler method* employs the strategy of approximating the solution by constructing a tangent through the given initial point. Then, use the given differential equation to determine the slope of this tangent. Once this information is in hand, a small step in the right direction is taken at whose end point another tangent is constructed and its slope determined, using Equation 9.4.1. By iterating this process, one can go as far as one wishes; however, round-off error usually limits the use of this technique in practice.

Formally, the procedure is as follows: we wish to generate an approximation, $\bar{y}$, corresponding to the point $\bar{t}$ such that

$$\bar{y} = F(\bar{t}) \tag{9.4.3}$$

The first step is to subdivide the interval $a \leq t \leq \bar{t}$ into n-subintervals, each of width

$$h = \frac{t - t_0}{n} \tag{9.4.4}$$

and the resulting boundaries between subintervals called *mesh points* are denoted by $t_1, t_2, \ldots, t_{n-1}, t_n$, such that $\bar{t} = t_n$ and $t_0 = a$.

Then, starting from $(t_0, b)$, draw a tangent with slope

$$\left.\frac{dy}{dt}\right|_{(t_0,b)} = f(t_0, b)$$

such that,

$$f(t_0, b) = \frac{y_1 - b}{t_1 - t_0} \tag{9.4.5}$$

Solving for $y_1$, the y-coordinate of the tangent at the location $t = t_1$, we get

$$y_1 = f(t_0, b)(t_1 - t_0) + b$$
$$= hf(t_0, b) + b \tag{9.4.6A}$$

In terms of Equation 9.4.2

$$F(t_1) \approx F(t_0) + F'(t_0)(t_1 - t_0) \tag{9.4.6B}$$

since b = F($t_0$) and f($t_0$, b) = F'($t_0$).

The approximation (as opposed to equality) comes from the fact that Equation 9.4.6B neglects second and higher-order derivatives of F(t) by comparison to the Taylor series expansion of F(t).

Whatever the size of the error created in a step, we would like to minimize it by shrinking h so that the tangent will closely approximate the actual solution, while forcing the remaining terms of the Taylor series to vanish. However, shrinking h requires increasing the number of steps to reach $\bar{t}$. This then leads to a larger round-off error and more than likely, an unacceptable result.

This brief discussion of Euler's method raises two important issues in all methods for solving the initial value problem: in order to improve the approximate solution to the initial value problem, we need to pay close attention to how the solution curve is modeled (straight line vs. polynomials, etc.), and the size of the step to use (h). How well these issues are resolved by a particular method is measured by a comparison with the Taylor series for the expected solution F(t).

Many methods are based on the direct use of the Taylor series to develop the approximate solution curve. Such methods are called *Taylor methods*. Euler's method is a Taylor method with a *local truncation error*, $\tau_i$, for the *i*th step:

$$\tau_i = \frac{y_i - y_{i-1}}{h} = f(t_{i-1}, y_{i-1}), \quad i = 1, 2, \ldots, n \qquad (9.4.7)$$

where $y_i$ is the exact value of the solution at $t_i$. The local truncation error has been defined as a measure of the amount by which the exact solution to the differential equation fails to satisfy the difference equation being used for the approximation (Stark,[3] Burden, Faires, and Reynolds,[5] Nakamura,[18] and Boyce and DiPrima[21]). Euler's method, as well as the other Taylor methods, are considered *single-step* methods. These are methods for which the approximation for the mesh point, $t_i$, involves information from only one of the previous mesh points $t_{i-1}$. In addition, Euler's method is *first order*, because the second and higher-order derivatives are neglected in the comparable Taylor series.

Taylor methods are developed for *nth order* accuracy; that is, the *(n+1)*th and higher derivatives are neglected. However, the need to determine the higher-order derivatives of f(t, y) introduces a major difficulty, and consequently, these methods are not very popular in practice.

Alternatives to the Taylor methods are the *Runge-Kutta* methods. These methods were developed to utilize the high order accuracy of the local truncation error of the Taylor methods, while avoiding computation and evaluation of the derivatives of f(t, y).

To appreciate some of the similarities between the Taylor and Runge-Kutta methods we will state the associated difference equations without the detailed derivations. Full derivations and algorithms can be found in many texts on numerical analysis (Burden, Faires, and Reynolds,[5] Constantinides and Mostoufi,[9] Nakamura,[18] Isaacson and Keller,[22] Higham,[23] and Butcher[24]).

All methods seek to obtain an approximation, w, to the well-posed initial value problem

$$\frac{dy}{dt} = f(t, y), \quad a \le t \le b, \quad y(a) = \alpha \tag{9.4.8}$$

In the case of Euler's method:

$$w_0 = \alpha$$
$$w_{i+1} = w_i + hf(t_i, w_i) \tag{9.4.9}$$

where $w_i \approx y(t_i)$ for each $i = 1, 2, \ldots, n$. For Taylor method of order n, we have

$$w_0 = \alpha$$
$$w_{i+1} = w_i + hT^{(n)}(t_i, w_i), \quad i = 0, 1, \ldots, n-1 \tag{9.4.10}$$

where

$$T^{(n)}(t_i, w_i) = f(t_i, w_i) + \frac{h}{2}f'(t_i, w_i) + \ldots + \frac{h^{n-1}}{n!}f^{(n-1)}(t_i, w_i). \tag{9.4.10A}$$

In the case of the Runge-Kutta methods, we have the so-called *midpoint method*

$$w_0 = \alpha$$
$$w_{i+1} = w_i + hf\left(t_i + \frac{h}{2}, w_i + \frac{h}{2}f(t_i, w_i)\right), \quad i = 0, 1, \ldots, n-1 \tag{9.4.11}$$

The *modified Euler method*

$$w_0 = \alpha$$
$$w_{i+1} = w_i + \frac{h}{2}\left[f(t_i, w_i) + f(t_{i+1}, w_i + hf(t_i, w_i))\right], \quad i = 0, 1, \ldots, n-1 \tag{9.4.12}$$

The *Heun method*

$$w_0 = \alpha$$
$$w_{i+1} = w_i + \frac{h}{4}\left[f(t_i, w_i) + 3f\left(t_i + \frac{2}{3}h, w_i + \frac{2}{3}hf(t_i, w_i)\right)\right] \tag{9.4.13}$$
$$i = 0, 1, \ldots, n-1$$

and the very popular Runge-Kutta order-four method, given by

$$w_0 = \alpha$$

$$k_1 = hf(t_i, w_i)$$

$$k_2 = hf\left(t_i + \frac{h}{2}, w_i + \frac{k_1}{2}\right)$$

$$k_3 = hf\left(t_i + \frac{h}{2}, w_i + \frac{k_2}{2}\right) \tag{9.4.14}$$

$$k_4 = hf(t_{i+1}, w_i + k_3)$$

$$w_{i+1} = w_i + \frac{1}{6}(k_1 + 2k_2 + 2k_3 + k_4), \quad i = 0, 1, \dots, n-1$$

Equations 9.4.11 to 9.4.13 are all second order or Runge-Kutta order-two methods.

The approach used to develop Equation 9.4.11 is typical, in that a number of parameters are evaluated by comparing their coefficients with the appropriate derivatives in the expansion of $T^{(n)}(t_i, w_i)$ given by Equation 9.4.10A. For a second order Runge-Kutta, such as the result given by Equation 9.4.11, three parameters, $a_1$, $a_2$, and $a_3$ are used. These parameters have the property that $a_1 f(t + a_2, y + a_3)$ approximates

$$T^{(2)}(t, y) = f(t, y) + \frac{h}{2} f'(t, y) \tag{9.4.15}$$

Since

$$f'(t, y) = \frac{\partial}{\partial t} f(t, y) + \frac{\partial}{\partial y} f(t, y) \frac{dy}{dt}$$

$$= \frac{\partial}{\partial t} f(t, y) + \frac{\partial}{\partial y} f(t, y) \bullet f(t, y)$$

then comparison of $T^{(2)}(t, y)$ with the truncated Taylor expansion of $f(t + a_2, y + a_3)$ about $(t, y)$ gives

$$a_1 f(t + a_2, y + a_3) = a_1 f(t, y) + a_1 a_2 \frac{\partial f(t, y)}{\partial t} + a_1 a_3 \frac{\partial f(t, y)}{\partial y}$$

$$= f(t, y) + \frac{h}{2} \frac{\partial f(t, y)}{\partial t} + \frac{h}{2} \frac{\partial f(t, y)}{\partial y} \bullet f(t, y)$$

which result in $a_1 = 1$, $a_2 = h/2$, and $a_3 = h/2$ $f(t, y)$ following the equating of coefficients. Notice that the truncated term in the Taylor expansion of $f(t + a_2, y + a_3)$ about $(t, y)$ involves the second order derivative, which implies the order of the local truncation error.

The major computational effort in applying the Runge-Kutta methods occurs in the evaluation of f. For second order methods, two functional evaluations per step are required, and for the fourth order method, four evaluations per step are required. As a consequence, the lower order methods with smaller step size may be less costly than the higher order methods using a larger step size. Relationships between evaluations per step and the order of local truncation error can be found in the literature (Butcher[24]). Also available in the numerical analysis literature (Hanna and Sandall[15]) are developments, which combine the best features of the Euler and a second order Runge-Kutta method.

All of the above methods choose the step size prior to carrying out the approximation step. Further, the step size remains fixed throughout the iterations. This feature can become a limitation when irregular regions are encountered, or when the solution to the differential equation behaves non-uniformly in parts of the interval.

An adaptive strategy that allows for changes in the step size during iterations is the *Runge-Kutta-Fehlberg*. This method also reduces the number of required evaluations of the function f per step size. Various versions of this adaptive strategy can be found in the literature (Burden, Faires, and Reynolds[5]) Constantinides and Mostoufi,[9] Johnston,[13] and Cheney and Kincaid[25]). Algorithms that suggest mechanisms for adjusting the step size are also available (Johnston[13]). The optimum step size should be such that the local errors are within a given tolerance, to ensure accuracy while not adversely affecting efficiency. To further clarify this idea, we use the following result, which is typical in the numerical analysis literature (Burden, Faires, and Reynolds,[5] Malek-Madani,[12] Nakamura,[18] Forsythe, Malcolm, and Moler,[20] Isaacson and Keller,[22] and Higham[23]):

$$|y(t) - y_n \leq k \ h^p|$$
(9.4.16)

where k is a constant that accounts for the smoothness of the solution y and the interval of approximation, p is related to the order of the difference scheme, and h is the step size. Even though Equation 9.4.16 is written for the global (entire interval) error, the essential idea remains the same. That is, reducing h by a factor of 10 reduces the error given by Equation 9.4.16 by a factor $10^p$. However, the number of algebraic operations is increased by about the same factor, which leads to an increase in the round-off error. Therefore, to get a sense of the optimal h, one that will not be too costly, a trial-and-error strategy is the usual approach taken.

Many software packages have incorporated adaptive difference schemes for ordinary differential equations as the core of their designs. "NDSolve" of *Mathematica* and "ode45" of *Matlab* are two applications that are based on an adaptive extension of the fourth-order Runge-Kutta method.

In addition to the single-step methods, there are methods that utilize the approximation at more than one previous mesh point to determine the approximation at the next point. These are the so-called *multistep* methods, and there are two types.

Generally, the multistep method for solving the initial value problem described by Equation 9.4.8 is one whose difference equation for $w_{i+1}$ at $t_{i+1}$ is given by

$$w_{i+1} = a_{m-1}w_i + a_{m-2}w_{i-1} + \dots + a_0 w_{i+1-m}$$

$$+h \begin{bmatrix} b_m f(t_{i+1}, w_{i+1}) + b_{m-1} f(t_i, w_i) \\ +\dots + b_0 f(t_{i+1-m}, w_{i+1-m}) \end{bmatrix} \qquad (9.4.17)$$

$$i = m-1, m, \dots, n-1$$

for the given starting values

$$w_0 = \alpha_0, \; w_1 = \alpha_1, \; w_2 = \alpha_2, \; \dots, \; w_{m-1} = \alpha_{m-1}$$

and an integer m > 1.

If $b_m = 0$, we have an *explicit* or *open* method, in which $w_{i+1}$ is explicitly given in terms of previously determined values. The explicit method constitutes one of the two types of multistep methods. For example,

$$w_0 = \alpha_0, w_1 = \alpha_1, w_2 = \alpha_2, w_3 = \alpha_3$$

$$w_{i+1} = w_i + \frac{h}{24} \begin{bmatrix} 55f(t_i, w_i) - 59f(t_{i-1}, w_{i-1}) \\ +37f(t_{i-2}, w_{i-2}) - 9f(t_{i-3}, w_{i-3}) \end{bmatrix}, i = 3, 4, \dots, n-1 \qquad (9.4.18)$$

defines the explicit four-step method known as the *fourth-order Adams-Bashforth technique*.

If $b_m \neq 0$, we have an *implicit* or *closed* method, since $w_{i+1}$ occurs on both sides of Equation 9.4.17. The implicit method constitutes the second of the two types of multistep methods. For example,

$$w_0 = \alpha_0, w_1 = \alpha_1, w_2 = \alpha_2$$

$$w_{i+1} = w_i + \frac{h}{24} \begin{bmatrix} 9f(t_{i+1}, w_{i+1}) + 19f(t_i, w_i) \\ -5f(t_{i-1}, w_{i-1}) - 9f(t_{i-2}, w_{i-2}) \end{bmatrix}, i = 2, 3, \dots, n-1 \qquad (9.4.19)$$

defines the implicit three-step method known as the *fourth-order Adams-Moulton technique*. Derivations of these and other multistep methods can be found in the numerical analysis literature (Burden, Faires, and Reynolds[5] and Forsythe, Malcolm, and Moler[20]).

The required starting values in either Equation 9.4.18 or Equation 9.4.19 are derived by assuming that the initial value, $\alpha$, is the same as $\alpha_0$. The remaining values, $\alpha_1$, $\alpha_2$, or $\alpha_3$, are generated by either a Runge-Kutta method or some other single-step procedure. The following example illustrates the potential use of both types of multistep methods.

Consider

$$\frac{dy}{dt} = -y + t + 1; \quad 0 \le t \le 1 \quad y(0) = 1$$

Both Equation 9.4.18 and Equation 9.4.19 will be used to approximate the solution. Also, we will use a step size of h = 0.1 together with the values from the exact solution, $y(t) = e^{-t} + t$, as starting values.

Therefore, for h = 0.1, $t_i$ = 0.1i, Equation 9.4.18 becomes

$$w_{i+1} = \frac{h}{24}\left[18.5w_i + 5.9w_{i-1} - 3.7w_{i-2} + 0.9w_{i-3} + 0.24i + 2.52\right],$$

$$i = 3, 4, \ldots, 9$$

(9.4.18A)

and Equation 9.4.19 becomes

$$w_{i+1} = \frac{h}{24}\left[-0.9w_{i+1} + 22.1w_i + 0.5w_{i-1} - 0.1w_{i-2} + 0.24i + 2.52\right],$$

$$i = 2, 3, \ldots, 9$$

(9.4.19A)

which can be rearranged to give

$$w_{i+1} = \frac{h}{24.9}\left[22.1w_i + 0.5w_{i-1} - 0.1w_{i-2} + 0.24i + 2.52\right],$$

$$i = 2, 3, \ldots, 9$$

(9.4.19B)

Results for this example are given in Table 9.2.

The table shows that the implicit method gives better results than the explicit method of the same order. Generally this is true. However, implicit methods have an inherent disadvantage in requiring that an algebraic conversion to an explicit representation be made; for example, the conversion from Equation 9.4.19A to Equation 9.4.19B. This type of algebraic conversion can be very difficult or even impossible to accomplish and is very dependent on the differential equation to be solved. For example, if we needed to solve the differential equation given by

**TABLE 9.2**

Comparison of Implicit and Explicit Methods

| $t_i$ | Adams-Bashforth $w_i$ | Error | Adams-Moulton $w_i$ | Error |
|---|---|---|---|---|
| .3 | starting value | — | 1.0408180061 | 2.146E-07 |
| .4 | 1.0703229200 | 2.874E-06 | 1.0703196614 | 3.846E-07 |
| .5 | 1.1065354755 | 4.816E-06 | 1.1065301384 | 5.213E-07 |
| .6 | 1.1488184077 | 6.772E-06 | 1.1488110076 | 6.285E-07 |
| .7 | 1.1965933934 | 8.090E-06 | 1.1965845932 | 7.106E-07 |

Adapted from Burden et al.[5]

$$\frac{dy}{dt} = e^y; \ 0 \le t \le 0.25, \ y(0) = 1$$

using Equation 9.4.19, we would get

$$w_{i+1} = w_i + \frac{h}{24}\left[9e^{w_{i+1}} + 19e^{w_i} - 5e^{w_{i-1}} + e^{w_{i-2}}\right]$$

as its difference equation. There is no clear way to solve this equation algebraically, to isolate the $w_{i+1}$ to one side.

In practice, implicit multistep methods are used to improve upon approximations obtained by explicit methods. This combination is the so-called *predictor-corrector* method. Predictor-corrector methods employ a single-step method, such as the Runge-Kutta order four, to generate the starting values to an explicit method, such as an Adams-Bashforth. Then, the approximation from the explicit method is improved upon by use of an implicit method, such as an Adams-Moulton method. Also, there are variable step size algorithms associated with the predictor-corrector strategy in the literature (Burden, Faires, and Reynolds,[5] and Cheney and Kincaid[25]).

### 9.4.2 Boundary Value Problems

Boundary value problems appear in many different areas of chemical engineering. Numerical approaches to solving them involve either a conversion to initial value problems (Case I) or, more directly, by finite difference methods (Case II).

Because there are theorems that guarantee the existence of and the uniqueness of the solutions to initial value problems, one may favor that approach. However, we must not forget that quite often, the physical situation is too complicated for us to ascertain that all the conditions of such theorems are being satisfied. A very brief discussion of each approach follows.

**Case I:**   Conversion to Initial Value Problems

The general approach of converting a given boundary value problem to a system of initial value problems is called the *Shooting method*. In the numerical analysis literature (Burden, Faires, and Reynolds,[5] Constantinides and Mostoufi,[9] Cheney and Kincaid,[25] and Na[26]), the shooting method is discussed separately for the linear and nonlinear problems. Burden et al.,[5] contains algorithms for both the linear and nonlinear problems. Following Burden et al.,[5] in order to solve the linear boundary value problem

$$y'' = p(x)y' + q(x)y + r(x), \ a \le x \le b, \ y(a) = \alpha, \ y(b) = \beta \qquad (9.4.20)$$

one would consider the two initial value problems given by

$$y'' = p(x)y' + q(x)y + r(x), \ a \le x \le b, \ y(a) = \alpha, \ y'(a) = 0 \qquad (9.4.21)$$

and

$$y'' = p(x)y' + q(x)y + r(x), \ a \le x \le b, \ y(a) = 0, \ y'(a) = 1 \qquad (9.4.22)$$

Then, if $y_1(x)$ is the solution to Equation 9.4.21 and $y_2(x)$ is the solution to Equation 9.4.22, the solution to Equation 9.4.20 is

$$y(x) = y_1(x) + \frac{\beta - y_1(b)}{y_2(b)} y_2(x), \ \ y_2(b) \ne 0 \qquad (9.4.23)$$

The solution methods discussed in the section on initial value problems can be used to approximate $y_1(x)$ and $y_2(x)$.

For the nonlinear case, the technique remains the same as that used to obtain a solution to Equation 9.4.20 except that a sequence of initial value problems of the form

$$y'' = f(x, y, y'), \ a \le x \le b, \ y(a) = \alpha, \ y'(a) = t_k \qquad (9.4.24)$$

are now required. The parameter, $t_k$, must be chosen such that

$$\lim_{k \to +\infty} y(b, t_k) = y(b) = \beta \qquad (9.4.25)$$

where $y(b, t_k)$ is the solution to Equation 9.4.24. Choosing the parameter $t_k$ to satisfy Equation 9.4.25 is not easy, and can be complicated by the fact that

$$y(b, t_k) - \beta = 0 \qquad (9.4.26)$$

is a nonlinear equation. In principle, the methodology in Section 9.2 can be employed here to solve

$$t_k = t_{k-1} - \frac{\left(y(b, t_{k-1}) - \beta\right)}{\left(dy / dt\big|_{b, t_{k-1}}\right)} \tag{9.4.27}$$

However, we cannot evaluate

$$dy/dt\big|_{b, t_{k-1}}$$

directly because we do not have an explicit representation for $y(b, t_k)$. All we know are the values $y(b, t_0)$, $y(b, t_1)$, ..., $y(b, t_{k-1})$. Burden et al. show that we can rewrite Equation 9.4.24 to reflect its dependence on $t_k$, as well as on x:

$$y''(x, t) = f(x, y(x, t), y'(x, t)), \ a \le x \le b, \ y(a, t) = \alpha, \ y'(a, t) = t_k \tag{9.4.28}$$

The subscript k is dropped inside the functional notation for convenience. Differentiating Equation 9.4.28 with respect to t and assuming that the order of differentiation of x and t is reversible gives

$$\frac{\partial}{\partial t} y''(x, t) = \frac{\partial}{\partial y} f(x, y(x, t), y'(x, t)) \frac{\partial}{\partial t} y(x, t) + \frac{\partial}{\partial y} y' f(x, y(x, t), y'(x, t)) \frac{\partial}{\partial t} y'(x, t)$$

$$\frac{\partial}{\partial t}(a, t) = 0, \frac{\partial}{\partial t} y'(a, t) = 1$$

Simplification, using $z(x, t)$ to represent $\partial y(x, t)/\partial t$, results in

$$z'' = \frac{\partial}{\partial y} f(x, y, y')z + \frac{\partial}{\partial y'}(x, y, y')z', \ a \le x \le b, \ z(a) = 0, \ z'(a) = 1$$

Then, Newton's method becomes

$$t_k = t_{k-1} - \frac{\left(y(b, t_{k-1}) - \beta\right)}{z(b, t_{k-1})} \tag{9.4.29}$$

Equation 9.4.29 requires that two initial value problems be solved at each iteration.

In practice, some version of the methods discussed under initial value problems (Section 9.4.1) is used to approximate the solution required by Newton's method.

In closing this section on the shooting methods, it is important to remember that round-off error may become very significant when we use these methods. For instance, $\beta$ may be small enough such that the term $(\beta - y_1(b))/y_2(b)$

is dominated by $-y_1(b)/y_2(b)$. When $\beta$ is too small, an alternate method must be employed.

**Case II:** Finite Difference Methods

Finite difference formulations may occur as any one of three types, namely forward, central, or backward finite difference (Burden, Faires, and Reynolds,[5] Constantinides and Mostoufi,[9] and Cheney and Kincaid[25]). Generally, these formulations lead to nonlinear systems of equations. The methods and approaches discussed in Section 9.2 can be employed. However, if the resulting system of equations is linear, then the methods of Section 9.3 apply. Next, we will briefly discuss a linear central difference and a nonlinear central difference formulation.

In the linear case, we may reconsider the boundary value problem given by

$$y'' = p(x)y' + q(x)y + r(x), \ a \le x \le b, \ y(a) = \alpha, \ y(b) = \beta \qquad (9.4.20)$$

Then the interval [a, b] is divided into $n + 1$ equal subintervals having meshpoints

$$x_i = a + i\,h, \ i = 0, 1, 2, \ldots, n + 1$$

and $h = (b - a)/(n + 1)$. At the interior mesh points, $x_i$, $i = 1, 2, \ldots, n$, the derivatives $y''$ and $y'$ are approximated by centered difference formula for $y''(x_i)$ and $y'(x_i)$ as

$$y''(x_i) \approx \frac{1}{h^2}\left[y(x_{i+1}) - 2y(x_i) + y(x_{i-1})\right] \qquad (9.4.30)$$

and

$$y'(x_i) \approx \frac{1}{2h}\left[y(x_{i+1}) - y(x_{i-1})\right] \qquad (9.4.31)$$

Substitution of Equation 9.4.30 and Equation 9.4.31 into the differential equation results in

$$\frac{1}{h^2}\left[y(x_{i+1}) - 2y(x_i) + y(x_{i-1})\right] = p(x_i)\frac{1}{2h}\left[y(x_{i+1}) - y(x_{i-1})\right] + q(x_i)y(x_i) + r(x_i)$$

which can be restated as

$$\left(\frac{2w_i - w_{i+1} - w_{i-1}}{h^2}\right) + p(x_i)\left(\frac{w_{i+1} - w_{i-1}}{2h}\right) + q(x_i)w_i = -r(x_i) \qquad (9.4.32)$$

$$w_0 = \alpha, \ w_{n+1} = \beta$$

where w is the approximation to the given boundary value problem. A further convenient rearrangement gives

$$-\left(1+\frac{h}{2}p(x_i)\right)w_{i-1} + \left(2+h^2q(x_i)\right)w_i - \left(1-\frac{h}{2}p(x_i)\right)w_{i+1} = -h^2r(x_i) \quad (9.4.33)$$

Then, if we let

$$\begin{cases} a_i = -\left(1+\dfrac{h}{2}p(x_i)\right) \\ d_i = 2+h^2q(x_i) \\ c_i = -\left(1-\dfrac{h}{2}p(x_i)\right) \\ b_i = -h^2r(x_i) \end{cases}$$

Equation 9.4.33 becomes

$$a_iw_{i-1} + d_iw_i + c_iw_{i+1} = b_i \quad (9.4.34)$$

Equation 9.4.34 can now be represented as a system of linear equations:

$$d_1w_1 + c_1w_2 = b_1 - a_1\alpha$$

$$a_iw_{i-1} + d_iw_i + c_iw_{i+1} = b_i, 2 \le i \le n-1$$

$$a_nw_{n-1} + d_nw_n = b_n - c_n\beta$$

This system may be solved by Gaussian elimination.

When the differential equation is nonlinear, an $n \times n$ nonlinear system results. For example, the boundary value problem given by

$$y''(x) = f(x, y, y'), a \le x \le b, y(a) = \alpha, y(b) = \beta \quad (9.4.35)$$

would result in the approximation

$$\frac{1}{h^2}\left[y(x_{i+1}) - 2y(x_i) + y(x_{i-1})\right] = f\left(x_i, y(x_i), \frac{y(x_{i+1}) - y(x_i)}{2h}\right) \quad (9.4.36)$$

$$i = 1, 2, \ldots, n$$

following the same steps as in the linear case.

Then, in terms of w, Equation 9.4.36 becomes

$$w_0 = \alpha, w_{n+1} = \beta$$

$$\left(\frac{2w_i - w_{i+1} - w_{i-1}}{h^2}\right) + f\left(x_i, w_i, \frac{w_{i+1} - w_{i-1}}{2h}\right) = 0, i = 1, 2, \ldots, n$$

Finally the $n \times n$ nonlinear system representation is

$$2w_1 - w_2 + h^2 f\left(x_1, w_1, \frac{w_2 - \alpha}{2h}\right) - \alpha = 0$$

$$-w_1 + 2w_2 - w_3 + h^2 f\left(x_2, w_2, \frac{w_3 - w_1}{2h}\right) = 0$$

$$\vdots \qquad\qquad \vdots$$

$$-w_{n-2} + 2w_{n-1} - w_n + h^2 f\left(x_{n-1}, w_{n-1}, \frac{w_n - w_{n-2}}{2h}\right) = 0$$

$$-w_{n-1} + 2w_n + h^2 f\left(x_n, w_n, \frac{\beta - w_{n-1}}{2h}\right) - \beta = 0$$

This system may be solved by the Newton-Raphson method.

Although finite difference formulations are usually more stable than the shooting methods, they are still approximations. The added level of approximation for the derivatives can generate errors and care must be exercised when applying these methods.

### 9.4.3   Systems of Ordinary Differential Equations

In Section 9.4.1, selected numerical methods are examined for solving the initial value problems associated with first order differential equations. Those methods are also applicable to higher order differential equations following the reduction to a system of first order equations. For example, the second order differential equation

$$\frac{d^2 y}{dt^2} = f(t, y, y') \tag{9.4.37}$$

can be reduced to a system of two first order differential equations (Burden, Faires, and Reynolds,[5] Amundson,[17] and Boyce and DiPrima[21]):

$$\frac{dy}{dt} = v \tag{9.4.37A}$$

$$\frac{dv}{dt} = f(t, y, v) \tag{9.4.37B}$$

In general, an $n$th order differential equation such as

$$y^{(n)} = f(t, y, y', \ldots, y^{(n-1)}) \tag{9.4.38}$$

can be reduced to a system of n first order differential equations of the form

$$x_1' = f_1(t, x_1, \ldots, x_n)$$
$$x_2' = f_1(t, x_1, \ldots, x_n) \tag{9.4.39}$$
$$\vdots$$
$$x_n' = f_1(t, x_1, \ldots, x_n)$$

In the particular case of the second order equation, a procedure to approximate the solution to a system of two first order equations

$$x' = f(t, x, y) \tag{9.4.40}$$

$$y' = g(t, x, y) \tag{9.4.41}$$

subject to the initial conditions

$$x(t_0) = x_0, \; y(t_0) = y_0 \tag{9.4.42}$$

is an extension of those methods discussed in Section 9.4.1.

Following Boyce and DiPrima, Euler method would be extended to

$$\begin{aligned} x_{i+1} &= x_i + hf(t_i, x_i, y_i), & y_{i+1} &= y_i hg(t_i, x_i, y_i), & i = 0, 1, \ldots, n \\ &= x_i + hx_i' & &= y_i + hy_i' \end{aligned} \tag{9.4.43}$$

for the mesh points $t_i = t_0 + i\,h$.

Similarly the fourth order Runge-Kutta method applied to Equations 9.4.40 and 9.4.41 would become

$$x_{i+1} = x_i + \frac{1}{6} h[k_{i1} + 2k_{i2} + 2k_{i3} + k_{i4}],$$

$$\tag{9.4.44}$$

$$y_{i+1} = y_i + \frac{1}{6} h[\ell_{i1} + 2\ell_{i2} + 2\ell_{i3} + \ell_{i4}],$$

where

$$k_{i1} = f(t_i, x_i, y_i), \quad \ell_{i1} = g(t_i, x_i, y_i)$$

$$k_{i2} = f\left(t_i + \frac{h}{2}, x_i + k_{i1}, y_i + \frac{1}{2}\ell_{i1}\right), \ell_{i2} = f\left(t_i + \frac{h}{2}, x_i + k_{i1}, y_i + \frac{1}{2}\ell_{i1}\right)$$

$$k_{i3} = f\left(t_i + \frac{h}{2}, x_i + k_{i2}, y_i + \frac{1}{2}\ell_{i2}\right), \ell_{i3} = f\left(t_i + \frac{h}{2}, x_i + k_{i2}, y_i + \frac{1}{2}\ell_{i2}\right)$$

$$k_{i4} = f\left(t_i + h, x_i + k_{i3}, y_i + \frac{1}{2}\ell_{i3}\right), \ell_{i4} = f\left(t_i + h, x_i + k_{i3}, y_i + \frac{1}{2}\ell_{i3}\right)$$

The predictor-corrector method described in Section 9.4.1, which involves the use of Equations 9.4.18 and 9.4.19 would become

$$x_{i+1} = x_i + \frac{1}{24}h\left[55x'_{i1} - 59x'_{i-1} + 37x'_{i-2} - 9x'_{i-3}\right],$$

$$y_{i+1} = y_i + \frac{1}{24}h\left[55y'_{i1} - 59y'_{i-1} + 37y'_{i-2} - 9y'_{i-3}\right]$$

(9.4.45)

and

$$x_{i+1} = x_i + \frac{1}{24}h\left[9x'_{i+1} + 19x'_i - 5x'_{i-1} + x'_{i-2}\right],$$

$$y_{i+1} = y_i + \frac{1}{24}h\left[9y'_{i+1} + 19y'_i - 5y'_{i-1} + y'_{i-2}\right]$$

(9.4.46)

In the above discussion, it is assumed that the functions f and g satisfy the unique solution conditions.

From Equations 9.4.44 to 9.4.46 it is evident that third and higher order equations will become cumbersome and are more manageable using matrix notations (Strang,[16] Amundson,[17] Forsythe and Moler,[19] Isaacson and Keller,[22] and Gear[27]).

Both *Mathematica* and *Matlab*, as well as other available software packages, have the capabilities to solve linear or nonlinear systems of algebraic/differential equations.

## 9.5   Solution of Partial Differential Equations

One of the most frequently occurring partial differential equation in chemical engineering is the so-called parabolic type. This equation is used to describe time-dependent diffusion processes and fluid flow. Therefore, the numerical solution methods for this type of partial differential equation are important in heat transfer, molecular diffusion, and fluid flow.

There are many numerical approaches one can use to approximate the solution to the initial and boundary value problem presented by a parabolic partial differential equation. However, our discussion will focus on just two approaches. An *explicit* finite difference method and an *implicit* finite difference method. These two approaches, as well as other numerical methods for all types of partial differential equations, can be found in the literature (Burden, Faires, and Reynolds,[5] Constantinides and Mostoufi,[9] Nakamura,[18] Isaacson and Keller,[22] Cheney and Kincaid,[25] Myint-U and Debnath,[28] Kythe, Puri, and Schaferkotter,[29] Kreyszig,[30] Oneil,[31] and Cooper[32]).

Following Burden et al.,[5] we will consider the parabolic partial differential equation in one space dimension:

$$\frac{\partial}{\partial t} u(x,t) = \alpha^2 \frac{\partial^2}{\partial x^2} u(x,t) + S(x,t), \ 0 < x < L, \ t > 0 \qquad (9.5.1)$$

subject to the initial condition

$$u(x,\,t) = f(x), \quad 0 \le x \le L \qquad (9.5.2)$$

and the boundary conditions

$$u(0,\,t) = \phi_L,\ u(L,\,t) = \phi_R,\ t > 0 \qquad (9.5.3)$$

As part of the development of a finite difference scheme to approximate the solution to Equations 9.5.1 to 9.5.3, we must select two mesh constants, h and k, such that m = L/h is an integer. The grid points are $(x_i, t_j)$, where

$$x_i = i\,h,\ i = 0,\, 1,\, \ldots,\, m \text{ and } t_j = j\,k,\ j = 0,\, 1,\, \ldots,\, T$$

Then, if we approximate the derivatives by

$$\frac{\partial}{\partial t} u(x_i, t_j) \approx \frac{u(x_i, t_j + k) - u(x_i, t_j)}{k}$$

and

$$\frac{\partial^2}{\partial x^2} u(x_i, t_j) \approx \frac{u(x_i + h, t_j) - 2u(x_i, t_j) + u(x_i - h, t_j)}{h^2}$$

and substitute into Equation 9.5.1 to get

$$\frac{w_{i,j+1} - w_{i,j}}{k} - \alpha^2 \frac{w_{i+1,j} - 2w_{i,j} + w_{i-1,j}}{h^2} = S_{i,j} \qquad (9.5.4)$$

Equation 9.5.4 can be rearranged to

$$w_{i,j+1} = \frac{\alpha^2 k}{h^2} w_{i-1,j} + \left(1 - 2\frac{\alpha^2 k}{h^2}\right) w_{i,j} + \frac{\alpha^2 k}{h^2} w_{i+1,j} + S_{i,j} \qquad (9.5.5)$$

$$i = 1, 2, \ldots, m-1, \quad j = 1, 2, \ldots, T$$

where $w_{i,j}$ approximates $u(x_i, t_j)$, the source term, $S_{i,j}$ is $S(x_i, t_j)$ and T is a convenient maximum time value. The initial and boundary conditions become respectively

$$w_{i,0} = u(x_i, 0) = f(x_i), \; i = 0, 1, 2, \ldots, m \qquad (9.5.6)$$

and

$$w_{0,j} = u(0, t_j) = \phi_L; \; w_{m,j} = u(x_m, t_j) = \phi_R \quad j = 1, 2, \ldots, T \qquad (9.5.7)$$

In the important case of no source term, and when both end conditions are zero, Equation 9.5.5 reduces to

$$w_{i,j+1} = \frac{\alpha^2 k}{h^2} w_{i-1,j} + \left(1 - 2\frac{\alpha^2 k}{h^2}\right) w_{i,j} + \frac{\alpha^2 k}{h^2} w_{i+1,j} \qquad (9.5.8)$$

$$i = 1, 2, \ldots, m-1, \quad j = 1, 2, \ldots, T$$

with

$$w_{0,j} = w_{m,j} = 0 \qquad (9.5.9)$$

The method described by Equations 9.5.5 to 9.5.9 is an explicit method and is commonly called the *forward difference* method. It is explicit because knowledge of $w_{i,j}$ for $t_j$ at all the grid points means that $w_{i,j+1}$ can be calculated for the new time $t_{j+1}$ without solving simultaneous equations.

The forward difference method is considered conditionally stable. Further, it can be shown that the method converges with a rate of convergence on the order of $(k + h^2)$ if the condition

$$\frac{\alpha^2 k}{h^2} \leq 1/2$$

is satisfied (Nakamura,[18] and Isaacson and Keller[22]). This restriction means that a small value for h requires an even smaller k-value, which can easily lead to round-off errors.

Contrasting the forward difference method is the implicit method of *Crank-Nicolson* (Burden, Faires, and Reynolds,[5] Constantinides and Mostoufi,[9] Nakamura,[18] Isaacson and Keller,[22] and Cheney and Kincaid[25]). The difference formulation for the homogeneous case of Equation 9.5.1 with zero end conditions is

$$\frac{w_{i,j+1} - w_{i,j}}{k} - \frac{\alpha^2}{2h}\left[\begin{array}{c}(w_{i+1,j} - 2w_{i,j} + w_{i-1,j}) \\ +(w_{i+1,j+1} - 2w_{i,j+1} + w_{i-1,j+1})\end{array}\right] = 0 \qquad (9.5.10)$$

Burden et al.[5] gives an algorithm involving the Crank-Nicolson method for solving Equation 9.5.10, while Constantinides et al. has *Matlab* examples involving the nonhomogeneous case of Equation 9.5.10.

The Crank-Nicolson method is unconditionally stable. However, it does require the solution of simultaneous equations. Also, this method is more accurate than the forward difference method.

---

## 9.6 Chapter Summary

Admittedly, the scope of the chapter is limited to a few of the methods that are used in chemical engineering problem solving. They certainly are not expected to produce desired results in all possible instances. However, they should provide the reader at least a cursory view of what can be expected when a problem is to be solved numerically.

A notable exclusion is a discussion on solving the linear system $Ax = 0$. However, this important case of a linear system deserves more buildup than the allowable scope of this chapter. Hopefully, the curious reader will be stimulated enough to further explore the selected references.

The chapter does highlight areas of concern to the practicing chemical engineer, namely, solution of nonlinear and linear algebraic equations, as well as solution of differential equations. Indeed, attempted numerical solution of differential equations usually leads to solution of nonlinear or linear algebraic equations. Commercially available software packages that can aid the practicing chemical engineer are also mentioned.

Principally, the chapter's thrust is to expose readers to numerical analysis methods without overwhelming them with premature details.

## 9.7 Problems

1. The equilibrium reactions $A \leftrightarrow B + D$ (1) and $A \leftrightarrow C + 2D$ (2) occur at a given temperature and pressure. At the given condition of temperature and pressure, the following relationships are observed among the mole fractions:

$$\frac{y_B y_D}{y_A} = 3.75, \quad \frac{y_C y_D^2}{y_A} = 0.135$$

a. On a basis of 100 moles of A, conduct the material balance to show that the relationships can be recasted in terms of the extent, $\xi_i$, of each reaction.

$$\frac{\xi_1(\xi_1 + 2\xi_2)}{(100 - \xi_1 - \xi_2)(100 + \xi_1 + 2\xi_2)} = 3.75$$

$$\frac{\xi_2(\xi_1 + 2\xi_2)^2}{(100 - \xi_1 - \xi_2)(100 + \xi_1 + 2\xi_2)} = 0.135$$

b. Determine $\xi_1$ and $\xi_2$, using the Newton-Raphson method.
c. Determine $\xi_1$ and $\xi_2$, using a computer algebraic system.
d. Determine $\xi_1$ and $\xi_2$, using the modified Newton-Raphson method.

2. Given the two functions

$$f_1(x) = x^2, f_2(x) = \frac{x+1}{x}$$

determine the point of intersection

a. Using the modified Newton-Raphson method with $f_1$ to find $x$ and $f_2$ to find $y$.
b. Switch the roles of $f_1$ and $f_2$.
c. Use the Newton-Raphson method.

3. Use the Gauss-Seidel method to solve the following system:

$$2x_1 + x_2 + x_3 - x_4 = -3$$

$$x_1 + 9x_2 + 8x_3 + 4x_4 = 15$$

$$-x_1 + 3x_2 + 5x_3 + 2x_4 = 10$$

$$x_2 + x_4 = 2$$

4. a.What happens if you attempt to use the Gauss-Seidel method to solve the rearranged system?

$$-x_1 + 3x_2 + 5x_3 + 2x_4 = 10$$

$$x_1 + 9x_2 + 8x_3 + 4x_4 = 15$$

$$x_2 + x_4 = 2$$

$$2x_1 + x_2 + x_3 - x_4 = -3$$

   a. Solve this system using Gaussian elimination with backward substitution.
   b. Solve by inverting the coefficient matrix.
   c. Solve using a computer algebraic system.

5. Use the Runge-Kutta order four method with step sizes of 0.02 and 0.01 to approximate the solution to

$$y' = 1 - t + 3y; \; y(0) = 1$$

   at t = 1.0. Also, compare your result to the exact answer.

6. Use the Adams-Moulton predictor-corrector method to approximate the solution to

$$\frac{dx}{dt} = x - 4y, x(0) = 1$$

$$\frac{dy}{dt} = -x + y, y(0) = 0$$

   with h = 0.1 at t = 0.4. Correct the predicted value twice.
   a. Use the exact solution values (5 digits) to start the problem.
   b. Use the Euler method to obtain the appropriate starting values $(x_1, x_2, x_3$ and $y_1, y_2, y_3)$ to the problem.
   c. Use the midpoint method to obtain the starting values.

7. Using the Runge-Kutta order four method for starting, the Adams-Bashforth method as the predictor, and the Adams-Moulton method as a corrector, approximate the solution to

$$\frac{dx}{dt} = x; x(0) = 1$$

   at t = 1.0 with a step size of 0.01.

8. Repeat problem 7 using the Runge-Kutta order four method.

9. Repeat problem 7 using the Adams-Bashforth method and the Runge-Kutta order four method to obtain the appropriate starting values.

10. Repeat problem 7 using the Adams-Moulton method and the Runge-Kutta order four method to obtain the appropriate starting values.

11. Given a slab of material 1.00m thick at a uniform temperature of 100°C. If the front surface is suddenly exposed to a constant temperature of 0°C, and the back surface is insulated, calculate the temperature profile for a time of 5000 seconds.

$$\text{Data:} \quad \frac{\partial T}{\partial t} = 2 \times 10^{-5} \, \text{m}^2 / \sec \frac{\partial^2 T}{\partial x^2} \, ; 0 \le x \le 1.00\text{m}$$

## References

1. Underwood, A.J.V., Fractional distillation of multicomponent mixtures, *Chem. Eng. Prog.*, 44, 603, 1948.
2. Treybal, R.E., *Mass-Transfer Operations*, 3rd ed., McGraw-Hill, New York, 1980.
3. Stark, P.A., *Introduction to Numerical Methods*, Macmillan, 1970.
4. Wolfram, S., *The Mathematica Book*, 3rd ed., Wolfram Media/Cambridge University Press, 1996.
5. Burden, R.L., Faires, J.D., and Reynolds, A.C., *Numerical Analysis*, 2nd ed., Prindle, Weber & Schmidt, Boston, 1978.
6. Constantinides, A., *Applied Numerical Methods with Personal Computers*, McGraw-Hill, New York, 1987.
7. Ramraj, R., Farrell, S., and Loney, N.W., Mathematical modeling of controlled release from a hollow fiber, *J. Membrane Sci.*, 162, 73, 1999.
8. Jenkins, M.A. and Traub, J.F., A three-stage algorithm for real polynomials using quadratic iteration, *SIAM J. Numerical Anal.*, 7, 545, 1970.
9. Constantinides, A. and Mostoufi, N., *Numerical Methods for Chemical Engineers with Matlab Applications*, Prentice Hall, Upper Saddle River, 1999.
10. Patee, H.A., Selecting computer mathematics, *Mech. Eng.*, September, 82, 1995.
11. Mackenzie, J. and Allen, M., Mathematical power tools, *Chem. Eng. Educ.*, Spring, 156, 1998.
12. Malek-Madani, R., *Advanced Engineering Mathematics with Mathematica and Matlab*, Addison-Wesley Longman, Reading, 1998.
13. Johnston, R.L., *Numerical Methods: A Software Approach*, John Wiley, New York, 1982.
14. Riggs, J.B., *An Introduction to Numerical Methods for Chemical Engineers*, Texas Tech University Press, Lubbock, 1988.
15. Hanna, O.T. and Sandall, O.C., *Computational Methods in Chemical Engineering*, Prentice-Hall, Upper Saddle River, 1995.
16. Strang, G., *Linear Algebra and Its Applications*, 2nd ed., Academic Press, New York, 1980.

17. Amundson, N.R., *Mathematical Methods in Chemical Engineering: Matrices and Their Applications*, Prentice Hall, Englewood Cliffs, 1966.
18. Nakamura, S., *Applied Numerical Methods*, Prentice-Hall, Englewood Cliffs, 1991.
19. Forsythe, G.E. and Moler, C.B., *Computer Solution of Linear Algebraic Systems*, Prentice-Hall, Englewood Cliffs, 1967.
20. Forsythe, G.E., Malcolm, M.A., and Moler, C.B., *Computer Methods for Mathematical Computations*, Prentice-Hall, Englewood Cliffs, 1977.
21. Boyce, W.E. and DiPrima, R.C., *Elementary Differential Equations and Boundary Value Problems*, 3rd ed., John Wiley & Sons, New York, 1977.
22. Isaacson, E. and Keller, H.B., *Analysis of Numerical Methods*, John Wiley & Sons, New York, 1966.
23. Higham, N. J., *Accuracy and Stability of Numerical Algorithms*, SIAM, Philadelphia, 1996.
24. Butcher, J.C., On the attainable order of Runge-Kutta methods, *Mathematics of Computation*, 19, 408, 1965.
25. Cheney, W. and Kincaid, D., *Numerical Math. Computing*, 2nd ed., Brooks/Cole, Monterey, 1985.
26. Na, T.Y., *Computational Methods in Engineering Boundary Value Problems*, Academic Press, New York, 1979.
27. Gear, C.W., *Numerical Initial Value Problems in Ordinary Differential Equations*, Prentice-Hall, Englewood Cliffs, 1971.
28. Myint-U, T. and Debnath, L., *Partial Differential Equations for Scientists and Engineers*, 3rd ed., Prentice-Hall, Englewood Cliffs, 1987.
29. Kythe, P.K., Puri, P., and Schaferkotter, M.R., *Partial Differential Equations and Mathematica*, CRC Press, Boca Raton, 1997.
30. Kreyszig, E., *Advanced Engineering Mathematics*, 7th ed., John Wiley & Sons, New York, 1993.
31. O'Neil, P.V., *Advanced Engineering Mathematics*, 4th ed., PWS-Kent, Boston, 1995.
32. Cooper, J.M., *Introduction to Partial Differential Equations with MATLAB*, Birkhauser, Boston, 1998.

# Appendix A

## Elementary Properties of Determinants and Matrices

### A.1 Determinants

The reader may be familiar with determinants, but a brief review is useful to the discussion on the solution of linear equations. It turns out that defining the determinant is a more difficult task than listing its properties; however, we will make an attempt to bring out somewhat of a working definition (Amundson[1]).

A determinant of the $n$th order is usually written in one of the forms

$$\det A = |A| = \begin{vmatrix} a_{11} & a_{12} & \cdots & a_{1n} \\ a_{21} & a_{22} & \cdots & a_{2n} \\ \vdots & & & \\ a_{n1} & a_{n2} & \cdots & a_{nn} \end{vmatrix} = \det \begin{bmatrix} a_{11} & a_{12} & \cdots & a_{1n} \\ a_{21} & a_{22} & \cdots & a_{2n} \\ \vdots & & & \\ a_{n1} & a_{n2} & \cdots & a_{nn} \end{bmatrix} = |a_{ij}| \quad (A.1)$$

and is an array (matrix) of $n^2$ ($n \times n$) elements, where the elements $a_{ij}$ may be real or complex numbers or functions. The determinant itself is a function, and if the $a_{ij}$ are considered variables, then $|A|$ is a function of $n^2$ variables with a particular form. Before we express the functional form of det A, a list of its useful properties are given and illustrated with a $2 \times 2$ case (Amundson[1] and Strang[2]).

1. The determinant is a linear function of the first row

$$\begin{vmatrix} a+a' & b+b' \\ c & d \end{vmatrix} = (a+a')d - (b+b')c = \begin{vmatrix} a & b \\ c & d \end{vmatrix} + \begin{vmatrix} a' & b' \\ c & d \end{vmatrix}$$

$$\begin{vmatrix} ta & tb \\ c & d \end{vmatrix} = tad - tbc = t\begin{vmatrix} a & b \\ c & d \end{vmatrix}$$

2. The determinant changes sign when two rows are exchanged.

$$\begin{vmatrix} a_{11} & a_{12} \\ a_{21} & a_{22} \end{vmatrix} = a_{11}a_{22} - a_{12}a_{21} = -\begin{vmatrix} a_{21} & a_{22} \\ a_{11} & a_{12} \end{vmatrix}$$

3. The determinant of the identity matrix (see discussion on matrices) is 1.

$$\begin{vmatrix} 1 & 0 \\ 0 & 1 \end{vmatrix} = 1$$

4. If two rows of the array A are equal, then det A = 0.

$$\begin{vmatrix} a & b \\ a & b \end{vmatrix} = ab - ab = 0$$

5. The operation of subtracting a multiple of one row from another leaves the determinant unchanged.

$$\begin{vmatrix} a_{11} - ta_{21} & a_{12} - ta_{22} \\ a_{21} & a_{22} \end{vmatrix} = \begin{vmatrix} a_{11} & a_{12} \\ a_{21} & a_{22} \end{vmatrix}$$

6. If the array A has a zero row, then det A = 0

$$\begin{vmatrix} 0 & 0 \\ a_{21} & a_{22} \end{vmatrix} = 0$$

7. If the array is triangular, then det A is the product $a_{11} a_{22} \ldots a_{nn}$ of the entries on the main diagonal.

$$\begin{vmatrix} a_{11} & a_{12} \\ 0 & a_{22} \end{vmatrix} = a_{11}a_{22} = \begin{vmatrix} a_{11} & 0 \\ a_{21} & a_{22} \end{vmatrix}$$

8. If the det A is zero, then the array A is singular. .

$$\begin{vmatrix} a_{11} & a_{12} \\ a_{21} & a_{22} \end{vmatrix} \text{ is singular if and only if } \begin{vmatrix} a_{11} & a_{12} \\ a_{21} & a_{22} \end{vmatrix} = a_{11}a_{22} - a_{12}a_{21} = 0$$

9. For any two n by n arrays (matrices), the determinant of the product is the product of the determinants, that is given two arrays A and B, then det AB = (det A)(det B).

$$\begin{vmatrix} a_{11} & a_{12} \\ a_{21} & a_{22} \end{vmatrix} \begin{vmatrix} b_{11} & b_{12} \\ b_{21} & b_{22} \end{vmatrix} = \begin{vmatrix} a_{11}b_{11} + a_{12}b_{21} & a_{11}b_{12} + a_{12}b_{22} \\ a_{21}b_{11} + a_{22}b_{12} & a_{21}b_{21} + a_{22}b_{22} \end{vmatrix}$$

10. The transpose of the array A has the same determinants as the array A

$$\det A^t = \det A$$

Now the functional form of the determinant may be given in terms of cofactor expansion

$$\det A = a_{i1}A_{i1} + a_{i2}A_{i2} + \cdots + a_{in}A_{in} \tag{A.2}$$

where $A_{ij}$ is called the *cofactor* and itself is given as

$$A_{ij} = (-1)^{i+j} \det M_{ij} \tag{A.3}$$

where $M_{ij}$ is called the *minor* and is formed by deleting row i and column j of the array A.

An example of this expansion:

$$A = \begin{bmatrix} 3 & -2 & 3 \\ 4 & 2 & -2 \\ 1 & 1 & 2 \end{bmatrix}$$

Then

$$A_{11} = (-1)^{1+1} \det \begin{bmatrix} 2 & -2 \\ 1 & 2 \end{bmatrix} = 6$$

$$A_{12} = (-1)^{1+2} \det \begin{bmatrix} 4 & -2 \\ 1 & 2 \end{bmatrix} = -10$$

$$A_{13} = (-1)^{1+3} \det \begin{bmatrix} 4 & 2 \\ 1 & 1 \end{bmatrix} = 2$$

Therefore, the det A = $a_{11}A_{11} + a_{12}A_{12} + a_{13}A_{13} = 44$

An alternative and useful formula for the determinant is

$$\det A = \det P^{-1} \det L \det D \det U = \pm(\text{product of the pivots}) \qquad \text{(A.4)}$$

where $\pm 1$ is the determinant of $P^{-1}$ or of $P$ and depends on whether the number of row exchanges is even or odd. L and U are lower and upper triangular arrays, respectively, and their determinants are 1.

Triangular arrays are those whose elements are zeros above or below the main diagonal. That is, a lower triangular array is one in which all elements above the main diagonal are zeros. The main diagonal, D, is a diagonal array whose nonzero elements appear on the main diagonal.

The array, P, is a permutation array (matrix) that reorders the rows of the array A, so that the product PA admits a factorization with nonzero pivots (Strang[2]).

While determinants themselves may have limited uses, they do facilitate the helpful applications of matrices.

## A.2  Matrices

A matrix is a rectangular array of elements arranged in rows and columns as

$$A = \begin{bmatrix} a_{11} & a_{12} & \cdots & a_{1n} \\ a_{21} & a_{22} & \cdots & a_{2n} \\ \vdots & & & \vdots \\ a_{n1} & a_{n2} & \cdots & a_{nn} \end{bmatrix} = \begin{pmatrix} a_{11} & a_{12} & \cdots & a_{1n} \\ a_{21} & a_{22} & \cdots & a_{2n} \\ \vdots & & & \vdots \\ a_{n1} & a_{n2} & \cdots & a_{nn} \end{pmatrix} = \begin{bmatrix} a_{ij} \end{bmatrix} \qquad \text{(A.5)}$$

where the square or round brackets are meant to be different from the straight lines used to represent determinants. The elements can be numbers (real or complex) or functions. The matrix given in Equation A.5 is an m by n or, more frequently, stated as m × n, which exploits the fact that there are m rows and n columns. If m = n, the matrix is said to be square or of $n$th order. Matrices will be denoted with capital letters. A special matrix containing only a single column is called a column vector, and will be denoted by a small boldface letter such as

$$\mathbf{x} = \begin{bmatrix} x_1 \\ x_2 \\ \vdots \\ x_n \end{bmatrix} = \begin{pmatrix} x_1 \\ x_2 \\ \vdots \\ x_n \end{pmatrix} \qquad \text{(A.6)}$$

The element positioned in the $i$th row and $j$th column is designated by $a_{ij}$, the first subscript identifying its row and the second its column.

Associated with each matrix A is the matrix $A^t$, known as the transpose of A. This transpose is obtained by interchanging the rows and columns of A. Therefore, if $A = [a_{ij}]$, then $A^t = [a_{ji}]$. The transpose $x^t$ of a column vector (n × 1 matrix) is a row vector (1 × n matrix). A square matrix in which all elements except those on the main diagonal are zero is called a *diagonal matrix*, and the case in which the diagonal elements are all unity is called the *identity matrix I*.

The algebraic properties of matrices can be found in the literature on linear algebra (Strang,[2] Boyce and DiPrima,[3] Schneider and Barker,[4] and Jenson and Jeffreys[5]). Here we list those that facilitate the discussions in Chapter 9. Proofs are left to the linear algebra literature.

1. *Equality*: Two m × n matrices, A and B, are said to be equal if corresponding elements are equal; that is, if $a_{ij} = b_{ij}$ for each i and j.

2. *Zero*: A matrix whose elements are all zero is called a *null matrix* and is denoted by **0**.

3. *Addition*: The sum of two m × n matrices is defined as the matrix obtained by adding corresponding elements:

$$A + B = \left[a_{ij}\right] + \left[b_{ij}\right] = \left[a_{ij} + b_{ij}\right] \tag{A.7}$$

4. *Multiplication*: The product AB of two matrices is defined whenever the number of columns of the first matrix is the same as the number of rows of the second matrix. When this condition exists, the matrices are said to be conformable. If A is n × m and B is n × r, then the product AB = C is an m × r matrix. The element in the $i$th row and $j$th column of C is obtained by multiplying each element of the $i$th row of A by the corresponding element of the $j$th column of B, and then adding the resulting products. That is,

$$c_{ij} = \sum_{k=1}^{n} a_{ik} b_{kj} \tag{A.8}$$

Also, matrix multiplication satisfies the associative law

$$(AB)C = A(BC) \tag{A.9}$$

and the distributive law

$$A(B+C) = AB + AC \tag{A.10}$$

For this reason, it is necessary to use terminology which specifies the order of multiplication. For example, if the matrices A and B are n × n, then C = AB is read as B premultiplied by A while D = BA is read as B postmultiplied by A. Further, the results AB and BA are not necessarily equal. This is a very different phenomenon from the usual symbolic algebra and may even result in the product of two matrices being the null matrix without either matrix being null. For instance,

$$A = \begin{bmatrix} 1 & 3 & 5 \\ -2 & 4 & 0 \\ 0 & 2 & 2 \end{bmatrix} \text{ and } B = \begin{bmatrix} 2 & 0 & 6 \\ 1 & 0 & 3 \\ -1 & 0 & -3 \end{bmatrix}$$

$$AB = \begin{bmatrix} 1x2+3x1+5x(-1) & 1x0+3x0+5x0 & 1x6+3x3+5x(-3) \\ -2x2+4x1+0x(-1) & -2x0+4x0+0x0 & -2x6+4x3+0x(-3) \\ 0x2+2x1+2x(-1) & 0x0+2x0+2x0 & 0x6+2x3+2x(-3) \end{bmatrix}$$

$$= \begin{bmatrix} 0 & 0 & 0 \\ 0 & 0 & 0 \\ 0 & 0 & 0 \end{bmatrix}$$

whereas,

$$BA = \begin{bmatrix} 2x1+0x(-2)+6x0 & 2x3+0x4+6x2 & 2x5+0x0+6x2 \\ 1x1+0x(-2)+3x0 & 1x3+0x4+3x2 & 1x5+0x0+3x2 \\ -1x1+0x(-2)+(-3)x0 & -1x3+0x4+-3x2 & -1x5+0x0+-3x2 \end{bmatrix}$$

$$= \begin{bmatrix} 2 & 18 & 22 \\ 1 & 9 & 11 \\ -1 & -9 & -11 \end{bmatrix}$$

This example illustrates Equation A.8, as well as the fact that the product of two conformable matrices is not necessarily commutative.

Matrix multiplication applies to the special cases of 1 × n and n × 1, row and column vectors, respectively. If we denote the 1 × n as $x^t$ and the n × 1 as **y**, then

$$x^t y = \sum_{i=1}^{n} x_i y_i \tag{A.11}$$

Also,

$$\mathbf{x}^t\mathbf{y} = \mathbf{y}^t\mathbf{x}, \quad \mathbf{x}^t(\mathbf{y}+\mathbf{z}) = \mathbf{x}^t\mathbf{y} + \mathbf{x}^t\mathbf{z}, \left(\alpha\mathbf{x}^t\right)\mathbf{y} = \alpha\mathbf{x}^t\mathbf{y} = \mathbf{x}^t(\alpha\mathbf{y}) \qquad \text{(A.12)}$$

A special product called the *scalar* or *inner product* is defined by

$$(\mathbf{x}, \mathbf{y}) = \sum_{i=1}^{n} x_i \bar{y}_i = \mathbf{x}^t\mathbf{y} \qquad \text{(A.13)}$$

where $\bar{y}$ is the conjugate of y. A particular useful form of Equation A.13 is

$$(\mathbf{x}, \mathbf{x}) = \sum_{i=1}^{n} x_i \bar{x}_i = \sum_{i=1}^{n} |x_i|^2 \qquad \text{(A.14)}$$

If $(\mathbf{x}, \mathbf{y}) = 0$ then the two vectors are said to be *orthogonal*. These vector properties can be illustrated with the following example:

$$\text{Let } \mathbf{x} = \begin{pmatrix} i \\ -2 \\ 1+i \end{pmatrix}, \quad \mathbf{y} = \begin{pmatrix} 2-i \\ i \\ 3 \end{pmatrix}$$

then

$$\mathbf{x}^t\mathbf{y} = (i)(2-i) + (-2)(i) + (1+i)(3) = 4 + 3i$$

$$(\mathbf{x}, \mathbf{y}) = (i)(2+i) + (-2)(-i) + (1+i)(3) = 2 + 7i$$

$$\mathbf{x}^t\mathbf{x} = (i)^2 + (-2)^2 + (1+i)^2 = 3 + 2i$$

$$(\mathbf{x}, \mathbf{x}) = (i)(-i) + (-2)(-2) + (1+i)(1-i) = 7$$

5. *Multiplication by a number*: The product of a matrix A by a number $\alpha$ (real or complex) is defined by

$$\alpha A = \alpha\left[a_{ij}\right] = \left[\alpha a_{ij}\right] \qquad \text{(A.15)}$$

That is, each element of the matrix is multiplied by the number $\alpha$. *This differs from the multiplication of a determinant by a number where only the first row is multiplied.*

6. *Subtraction*: The difference A − B of two conformable matrices is defined by

$$A - B = A + (-B) = \left[a_{ij}\right] - \left[b_{ij}\right] = \left[a_{ij} - b_{ij}\right] \tag{A.16}$$

7. *Inverse*: This is the operation for square matrices that is analogous to division for numbers. That is, for a given square matrix A, we need to determine another matrix B such that AB = I, the identity matrix. If the matrix B exists, it is called the inverse of A and we write B = A⁻¹. *Multiplication is commutative between any matrix and its inverse.*

$$AA^{-1} = A^{-1}A = I \tag{A.17}$$

Also, when A has an inverse, it is said to be *nonsingular*, otherwise A is said to be *singular*. Property 8 for determinants is a necessary and sufficient condition for a matrix to be singular. Direct use of the determinant to find the inverse is possible.

A procedure known as Cramer's rule may be used on small systems of equations. That is, given Ax = **b**, then **x** = A⁻¹**b** (premultiplication by A⁻¹) can be determined as illustrated below:

$$x_1 + 3x_2 = 0$$

$$2x_1 + 4x_2 = 6$$

then

$$x_1 = \frac{\begin{vmatrix} 0 & 3 \\ 6 & 4 \end{vmatrix}}{\begin{vmatrix} 1 & 3 \\ 2 & 4 \end{vmatrix}} = \frac{-18}{-2} = 9, \quad x_2 = \frac{\begin{vmatrix} 1 & 0 \\ 2 & 6 \end{vmatrix}}{\begin{vmatrix} 1 & 3 \\ 2 & 4 \end{vmatrix}} = \frac{6}{-2} = -3$$

In general, the *j*th component of **x** = A⁻¹**b** is given by

$$x_j = \frac{\det B_j}{\det A}; \quad B_j = \begin{bmatrix} a_{11} & a_{12} & b_1 & a_{1n} \\ \vdots & \vdots & \vdots & \vdots \\ a_{n1} & a_{n2} & b_n & a_{nn} \end{bmatrix} \tag{A.18}$$

The vector **b** replaces the *j*th column of A. Another way to arrive at Equation A.18 is to start with the definition given in Equation A.3. That is, if $a_{ij}$ is any element of a square matrix, and the cofactor of $a_{ij}$ in det A is $A_{ij}$, the transpose of the matrix whose elements are made up of all the $A_{ij}$s is called the *adjoint* of A and is denoted adj $A = [A_{ji}]$. It can then be shown (Strang[2] and Jenson and Jeffreys[5]) that a typical element of $A^{-1}$ is given by $A_{ji}/\det A$.

A more useful way to compute $A^{-1}$ is by means of elementary operations:

1. Interchange of two rows.
2. Multiplication of a row by a nonzero scalar.
3. Addition of any multiple of one row to another row.

Generally, any nonsingular matrix, A can be transformed into the identity I by a systematic sequence of the elementary operations. It can be shown that the same sequence of operations performed on I will yield $A^{-1}$ (Amundson,[1] Strang,[2] and Schneider and Barker[4]). An example illustrating the process is as follows:

$$A = \begin{bmatrix} 1 & -1 & -1 \\ 3 & -1 & 2 \\ 2 & 2 & 3 \end{bmatrix}$$

Step 1: Obtain zeros in the off-diagonal position in the first column by adding (–3) times the first row to the second row and adding (–2) times the first row to the third row.

$$\begin{bmatrix} 1 & -1 & -1 \\ 0 & 2 & 5 \\ 0 & 4 & 5 \end{bmatrix}$$

Step 2: Obtain a one in the diagonal position in the second column by multiplying the second row by 1/2.

$$\begin{bmatrix} 1 & -1 & -1 \\ 0 & 1 & 5/2 \\ 0 & 4 & 5 \end{bmatrix}$$

Step 3: Obtain zeros in the off-diagonal positions in the second column by adding the second row to the first row and adding (–4) times the second row to the third row.

$$\begin{bmatrix} 1 & 0 & 3/2 \\ 0 & 1 & 5/2 \\ 0 & 0 & -5 \end{bmatrix}$$

Step 4: Obtain a one in the diagonal position in the third column by multiplying the third row by (–1/5).

$$\begin{bmatrix} 1 & 0 & 3/2 \\ 0 & 1 & 5/2 \\ 0 & 0 & 1 \end{bmatrix}$$

Step 5: Obtain zeros in the off-diagonal positions in the third column by adding (–3/2) times the third row to the first row, and adding (–5/2) times the third row to the first row.

$$\begin{bmatrix} 1 & 0 & 0 \\ 0 & 1 & 0 \\ 0 & 0 & 1 \end{bmatrix}$$

If the same sequence of operations in the same order are now performed on I, the sequence of matrices:

$$\begin{bmatrix} 1 & 0 & 0 \\ 0 & 1 & 0 \\ 0 & 0 & 1 \end{bmatrix}, \begin{bmatrix} 1 & 0 & 0 \\ -3 & 1 & 0 \\ -2 & 0 & 1 \end{bmatrix}, \begin{bmatrix} 1 & 0 & 0 \\ -3/2 & 1/2 & 0 \\ -2 & 0 & 1 \end{bmatrix}, \begin{bmatrix} -1/2 & 1/2 & 0 \\ -3/2 & 1/2 & 0 \\ 4 & -2 & 1 \end{bmatrix}$$

$$\begin{bmatrix} -1/2 & 1/2 & 0 \\ -3/2 & 1/2 & 0 \\ -4/2 & 2/5 & -1/5 \end{bmatrix}, \begin{bmatrix} 7/10 & -1/10 & 3/10 \\ 1/2 & -1/2 & 1/2 \\ -4/5 & 2/5 & -1/5 \end{bmatrix}$$

The last matrix is $A^{-1}$.

8. *Distances*: To define a distance in $R^n$, we will use the idea of the *norm* of a vector. The $\ell_2$ and $\ell_\infty$ norms for the vector $\mathbf{x} = (x_1, x_2, \ldots, x_n)^t$ are defined by

$$\ell_2: \quad \|\mathbf{x}\|_2 = \left\{ \sum_{i=1}^{n} x_i^2 \right\}^{1/2} \tag{A.19}$$

which is the usual *Euclidean norm* of the vector $\mathbf{x}$ and

$$\ell_\infty: \quad \|\mathbf{x}\|_\infty = \max_{1 \le i \le n} |x_i| \tag{A.20}$$

Then if $\mathbf{x} = (x_1, x_2, \ldots, x_n)^t$ and $\mathbf{y} = (y_1, y_2, \ldots, y_n)^t$ are vectors in $\mathbb{R}^n$, the $\ell_2$ and $\ell_\infty$ distances between $\mathbf{x}$ and $\mathbf{y}$ are defined to be

$$\ell_2 : \quad \|\mathbf{x} - \mathbf{y}\|_2 = \left\{ \sum_{i=1}^{n} |x_i - y_i|^2 \right\}^{1/2} \tag{A.21}$$

and

$$\ell_\infty : \quad \|\mathbf{x} - \mathbf{y}\|_\infty = \max_{1 \le i \le n} |x_i - y_i| \tag{A.22}$$

Our major concern with matrices has to do with solving linear algebraic equations as discussed in Chapter 9. There the method of Gaussian elimination was emphasized. Below is an alternate approach to solving the system given by $A\mathbf{x} = \mathbf{b}$ in which the matrix $A^{-1}$ is exhibited. That is, the solution will be of the form $\mathbf{x} = A^{-1}\mathbf{b}$, where $\mathbf{b}$ is premultiplied by $A^{-1}$.

Consider the set of linear equations given by

$$
\begin{array}{rrrrl}
2x_1 - & 3x_2 & & -2x_4 & = 8 \\
& 3x_2 + & 2x_3 + & x_4 & = 5 \\
x_1 - & 2x_2 - & 4x_3 + & 2x_4 & = 2 \\
2x_1 + & x_2 - & 3x_3 - & x_4 & = 6
\end{array}
$$

They can be recast in the form

$$
\begin{bmatrix}
2 & -3 & 0 & -2 \\
0 & 3 & 2 & 1 \\
1 & -2 & -4 & 2 \\
2 & 1 & -3 & -1
\end{bmatrix}
\begin{bmatrix}
x_1 \\ x_2 \\ x_3 \\ x_4
\end{bmatrix}
=
\begin{bmatrix}
8 \\ 5 \\ 2 \\ 6
\end{bmatrix}
$$

where $A$ is the $4 \times 4$ matrix. We now seek the solution $\mathbf{x} = A^{-1}\mathbf{b}$ through the use of elementary operations. As a first step, we form the *augmented matrix* by appending the unit matrix of order 4 to the $4 \times 5$ matrix formed from $A$ and the vector $\mathbf{b}$:

$$
\begin{bmatrix}
2 & -3 & 0 & -2 & \vdots\,8\,\vdots & 1 & 0 & 0 & 0 \\
0 & 3 & 2 & 1 & \vdots\,5\,\vdots & 0 & 1 & 0 & 0 \\
1 & -2 & -4 & 2 & \vdots\,2\,\vdots & 0 & 0 & 1 & 0 \\
2 & 1 & -3 & -1 & \vdots\,6\,\vdots & 0 & 0 & 0 & 1
\end{bmatrix}
$$

Secondly, we carry out the three operations in the order given:
  (i) Divide the first row by 2.
  (ii) Add (–1) times the *new* first row to the third row.
  (iii) Subtract 2 times the *new* first row from the fourth row.

These elementary operations result in:

$$\begin{bmatrix} 1 & 1.5 & 0 & -1 & 4 & 0.5 & 0 & 0 & 0 \\ 0 & 3 & 2 & 1 & 5 & 0 & 1 & 0 & 0 \\ 1 & -0.5 & -4 & 3 & -2 & -0.5 & 0 & 1 & 0 \\ 2 & 1 & -3 & -1 & -2 & -1 & 0 & 0 & 1 \end{bmatrix}$$

The third step requires the following elementary operations on the newest augmented matrix:
  (i) Add the fourth row to the first row.
  (ii) Subtract the fourth row from the second row.
  (iii) Subtract 3 times the fourth row from the third row.

$$\begin{bmatrix} 1 & 2.5 & -3 & 0 & 2 & -0.5 & 0 & 0 & 1 \\ 0 & -1 & 5 & 0 & 7 & 1 & 1 & 0 & -1 \\ 0 & -12.5 & 5 & 0 & 4 & 2.5 & 0 & 1 & -3 \\ 0 & 4 & -3 & 1 & -2 & -1 & 0 & 0 & 1 \end{bmatrix}$$

The fourth step requires the operations:
  (i) Add 2 times the second row to the first row.
  (ii) Subtract 12.5 times the second row from the third row.
  (iii) Add 4 times the second row to the fourth row.

$$\begin{bmatrix} 1 & 0 & 9.5 & 0 & 19.5 & 2 & 2.5 & 0 & 1.5 \\ 0 & -1 & 5 & 0 & 7 & 1 & 1 & 0 & -1 \\ 0 & 0 & -57.5 & 0 & -83.5 & -10.0 & -12.5 & 1 & 9.5 \\ 0 & 0 & 17 & 1 & 26 & -1 & 0 & 0 & 1 \end{bmatrix}$$

The fifth step involves the following elementary operations on the newest augmented matrix:
  (i) Divide the third row by (–57.5).
  (ii) Subtract 9.5 times the *new* third row from the first row.
  (iii) Subtract 5 times the *new* third row from the second row.
  (iv) Subtract 17 times the *new* third row from the fourth row.

$$
\begin{bmatrix}
1 & 0 & 0 & 0 & \vdots 5.704\vdots & 0.348 & 0.438 & 0.165 & 0.069 \\
0 & -1 & 0 & 0 & \vdots -0.261\vdots & 0.130 & -0.087 & 0.087 & -0.174 \\
0 & 0 & 1 & 0 & \vdots 1.452\vdots & 0.174 & 0.217 & -0.017 & -0.165 \\
0 & 0 & 0 & 1 & \vdots 1.313\vdots & 0.043 & 0.304 & 0.296 & -0.191
\end{bmatrix}
$$

Finally, multiplying the second row by (–1) results in

$$
\begin{bmatrix}
1 & 0 & 0 & 0 & \vdots 5.704\vdots & 0.348 & 0.438 & 0.165 & 0.069 \\
0 & 1 & 0 & 0 & \vdots 0.261\vdots & -0.130 & 0.087 & -0.087 & 0.174 \\
0 & 0 & 1 & 0 & \vdots 1.452\vdots & 0.174 & 0.217 & -0.017 & -0.165 \\
0 & 0 & 0 & 1 & \vdots 1.313\vdots & 0.043 & 0.304 & 0.296 & -0.191
\end{bmatrix}
$$

where the 4 × 4 identity matrix has moved completely to the left, taking the place of A. This final augmented matrix represents the form $\mathbf{Ix} = \mathbf{A^{-1}b}$, where the inverse matrix is

$$
A^{-1} =
\begin{bmatrix}
0.348 & 0.438 & 0.165 & 0.069 \\
-0.130 & 0.087 & -0.087 & 0.174 \\
0.174 & 0.217 & -0.017 & -0.165 \\
0.043 & 0.304 & 0.296 & -0.191
\end{bmatrix}
$$

and the solution vector $\mathbf{Ix} = (5.704, 0.261, 1.452, 1.313)^t$.

In summary, if we use $E_i$ to denote any matrix representing the elementary operations performed above, then $\mathbf{Ax} = \mathbf{b}$ can be adjusted to a succession of equivalent forms

$$
E_n E_{n-1} \cdots E_2 E_1 \mathbf{Ax} = E_n E_{n-1} \cdots E_2 E_1 \mathbf{b} \tag{A.23}
$$

where the elementary matrices are selected to convert the square matrix A to the unit matrix I. This means that

$$
E_n E_{n-1} \cdots E_2 E_1 A = I \tag{A.24}
$$

Since the same operations are being performed on the vector $\mathbf{b}$, the column vector $\mathbf{b}$ is appended to the $n \times n$ matrix A to form an augmented matrix $\mathbf{Ab}$. Then, premultiplying $\mathbf{Ab}$ with the operators $E_i$ produces the desired result $\mathbf{x} = \mathbf{A^{-1}b}$, and, if needed, the inverse of A is

$$
E_n E_{n-1} \cdots E_2 E_1 = A^{-1} \tag{A.25}
$$

## A.3  Additional Properties of Matrices

Square matrices can appear in series (finite or infinite) and may exhibit behavior quite similar to those of scalar series. For instance, we could have

$$a_0 Y^n + a_1 Y^{n-1} + a_2 Y^{n-2} + \cdots + a_{n-1} Y + a_n I = 0 \qquad (A.26)$$

where Y is a square matrix. Matrix polynomials may be factored in ways similar to that of scalar polynomials. For example: $A^3 - 9A^2 + 26A - 24I$; can be factored into $(A - 2I)(A - 3I)(A - 4I)$.

It is also possible to use square matrices as exponents of scalar functions, such as

$$e^A = I + \frac{A}{1!} + \frac{A^2}{2!} + \frac{A^3}{3!} + \cdots \qquad (A.27)$$

Then, if B is a square matrix of the same order as A,

$$e^A e^B = e^B e^A = e^{(A+B)} \qquad (A.28)$$

and

$$e^A e^{-A} = I \qquad (A.29)$$

For a given square matrix A of order n whose elements are constants and a column vector $\mathbf{x}$, the equation

$$A\mathbf{x} = \lambda \mathbf{x} \qquad (A.30)$$

or

$$(A - \lambda I)\mathbf{x} = 0 \qquad (A.31)$$

has nontrivial solution if, and only if,

$$|(A - \lambda I)| = 0 \qquad (A.32)$$

Equation A.32 is called the *characteristic equation*.

## A.4 Calculus of Matrices

If the elements of a matrix are functions of some appropriate independent variable(s), the matrix can be differentiated or integrated with respect to the independent variable(s).

Similar to the calculus of functions, we can differentiate matrices by differentiating the elements of the matrix in each case. The usual designation for the derivative of a square matrix Y is $dY/dx$ if the matrix elements are functions of the scalar x. The differential coefficient of a product is also similar to that of a product of scalar functions; however, the order of the matrices in the product must be maintained. Therefore,

$$\frac{d}{dx}(ABC) = \frac{dA}{dx}BC + A\frac{dB}{dx}C + AB\frac{dC}{dx} \qquad (A.33)$$

where A, B, and C are conformable square matrices whose elements are functions of x.

Integration of a matrix whose elements are derivatives or functions of an independent variable is accomplished by integrating each element with respect to the independent variable. Therefore, for a typical element, $y_{ij}(x)$, of the matrix Y, the result

$$\int Y dx = \left[ \int y_{ij}(x)dx \right] \qquad (A.34)$$

## References

1. Amundson, N.R., *Mathematical Methods in Chemical Engineering*, Prentice-Hall, Englewood Cliffs, 1966.
2. Strang, G., *Linear Algebra and its Applications*, 2nd ed., Academic Press, New York, 1980.
3. Boyce, W.E. and DiPrima, R.C., *Elementary Differential Equations and Boundary Value Problems*, 3rd ed., John Wiley, New York, 1977.
4. Schneider, H. and Barker, G.P., *Matrices and Linear Algebra*, Holt, Rinehart & Winston, New York, 1968.
5. Jenson, V.G. and Jeffreys, G.V., *Mathematical Methods in Chemical Engineering*, 2nd ed., Academic Press, London, 1992.

# Index